Solution Techni for Elementary Partial Differential Equations

Second Edition

Solution Techniques
for Elementary Partial
Differential Equations

Second Edition

Christian Constanda

University of Tulsa

Oklahoma

CRC Press

Taylor & Francis Group

Boca Raton London New York

CRC Press is an imprint of the
Taylor & Francis Group an **informa** business

A CHAPMAN & HALL BOOK

Chapman & Hall/CRC
Taylor & Francis Group
6000 Broken Sound Parkway NW, Suite 300
Boca Raton, FL 33487-2742

© 2010 by Taylor and Francis Group, LLC
Chapman & Hall/CRC is an imprint of Taylor & Francis Group, an Informa business

No claim to original U.S. Government works

Printed in the United States of America on acid-free paper
10 9 8 7 6 5 4 3 2 1

International Standard Book Number: 978-1-4398-1139-9 (Paperback)

Library of Congress Cataloging-in-Publication Data

Constanda, C. (Christian)
    Solution techniques for elementary partial differential equations / Christian Constanda. -- 2nd ed.
        p. cm.
    Includes bibliographical references and index.
    ISBN 978-1-4398-1139-9 (pbk. : alk. paper)
    1. Differential equations, Partial--Numerical solutions. I. Title.

QA377.C7629 2010
518'.64--dc22   2010014608

Visit the Taylor & Francis Web site at
http://www.taylorandfrancis.com

and the CRC Press Web site at
http://www.crcpress.com

For Lia

Contents

Foreword   xi

Preface to the Second Edition   xiii

Preface to the First Edition   xv

Chapter 1. Ordinary Differential Equations: Brief Review   1

1.1. First-Order Equations   1
1.2. Homogeneous Linear Equations with Constant Coefficients   3
1.3. Nonhomogeneous Linear Equations with Constant Coefficients   5
1.4. Cauchy–Euler Equations   6
1.5. Functions and Operators   7
      Exercises   9

Chapter 2. Fourier Series   11

2.1. The Full Fourier Series   11
2.2. Fourier Sine Series   17
2.3. Fourier Cosine Series   21
2.4. Convergence and Differentiation   23
      Exercises   24

Chapter 3. Sturm–Liouville Problems   27

3.1. Regular Sturm–Liouville Problems   27
3.2. Other Problems   39
3.3. Bessel Functions   41
3.4. Legendre Polynomials   47
3.5. Spherical Harmonics   50
      Exercises   54

Chapter 4. Some Fundamental Equations of Mathematical Physics   59

4.1. The Heat Equation   59
4.2. The Laplace Equation   67
4.3. The Wave Equation   73
4.4. Other Equations   78
      Exercises   81

Chapter 5. The Method of Separation of Variables   83

5.1. The Heat Equation   83
5.2. The Wave Equation   95
5.3. The Laplace Equation   101
5.4. Other Equations   109
5.5. Equations with More than Two Variables   113
    Exercises   124

Chapter 6. Linear Nonhomogeneous Problems   131

6.1. Equilibrium Solutions   131
6.2. Nonhomogeneous Problems   136
    Exercises   140

Chapter 7. The Method of Eigenfunction Expansion   143

7.1. The Heat Equation   143
7.2. The Wave Equation   149
7.3. The Laplace Equation   152
7.4. Other Equations   155
    Exercises   159

Chapter 8. The Fourier Transformations   165

8.1. The Full Fourier Transformation   165
8.2. The Fourier Sine and Cosine Transformations   172
8.3. Other Applications   179
    Exercises   181

Chapter 9. The Laplace Transformation   187

9.1. Definition and Properties   187
9.2. Applications   192
    Exercises   202

Chapter 10. The Method of Green's Functions   205

10.1. The Heat Equation   205
10.2. The Laplace Equation   213
10.3. The Wave Equation   217
    Exercises   223

Chapter 11. General Second-Order Linear Partial Differential Equations with Two Independent Variables   227

11.1. The Canonical Form   227
11.2. Hyperbolic Equations   231
11.3. Parabolic Equations   235
11.4. Elliptic Equations   238
        Exercises   239

Chapter 12. The Method of Characteristics   241

12.1. First-Order Linear Equations   241
12.2. First-Order Quasilinear Equations   248
12.3. The One-Dimensional Wave Equation   249
12.4. Other Hyperbolic Equations   256
        Exercises   260

Chapter 13. Perturbation and Asymptotic Methods   263

13.1. Asymptotic Series   263
13.2. Regular Perturbation Problems   266
13.3. Singular Perturbation Problems   274
        Exercises   280

Chapter 14. Complex Variable Methods   285

14.1. Elliptic Equations   285
14.2. Systems of Equations   291
        Exercises   294

Answers to Odd-Numbered Exercises   297

Appendix   313

Bibliography   319

Index   321

Foreword

It is often difficult to persuade undergraduate students of the importance of mathematics. Engineering students in particular, geared towards the practical side of learning, often have little time for theoretical arguments and abstract thinking. In fact, mathematics is the language of engineering and applied science. It is the vehicle by which ideas are analyzed, developed, and communicated. It is no accident, therefore, that any undergraduate engineering curriculum requires several mathematics courses, each one designed to provide the necessary analytic tools to deal with questions raised by engineering problems of increasing complexity, for example, in the modeling of physical processes and phenomena. The most effective way to teach students how to use these mathematical tools is by example. The more worked examples and practice exercises a textbook contains, the more effective it will be in the classroom.

Such is the case with *Solution Techniques for Elementary Partial Differential Equations* by Christian Constanda. The author, a skilled classroom performer with considerable experience, understands exactly what students want and has given them just that: a textbook that explains the essence of the method briefly and then proceeds to show it in action. The book contains a wealth of worked examples and exercises (half of them with answers). An Instructor's Manual with solutions to each problem and a .pdf file for use on a computer-linked projector are also available. In my opinion, this is quite simply the best book of its kind that I have seen thus far. The book not only contains solution methods for some very important classes of PDEs, in easy-to-read format, but is also student-friendly and teacher-friendly at the same time. It is definitely a textbook that should be adopted.

Professor Peter Schiavone
Department of Mechanical Engineering
University of Alberta
Edmonton, AB, Canada

Preface to the Second Edition

In direct response to constructive suggestions received from some of the users of the book, this second edition contains a number of enhancements.

- Section 1.4 (Cauchy–Euler Equations) has been added to Chapter 1.

- Chapter 3 includes three new sections: 3.3 (Bessel Functions), 3.4 (Legendre Polynomials), and 3.5 (Spherical Harmonics).

- The new Section 4.4 in Chapter 4 lists additional mathematical models based on partial differential equations.

- Sections 5.4 and 7.4 have been added to Chapters 5 and 7, respectively, to show—by means of examples—how the methods of separation of variables and eigenfunction expansion work for equations other than heat, wave, and Laplace.

- Supplementary applications of the Fourier transformations are now shown in Section 8.3.

- The method of characteristics is applied to more general hyperbolic equations in the additional Section 12.4.

- Chapter 14 (Complex Variable Methods) is entirely new.

- The number of worked examples has increased from 110 to 143, and that of the exercises has almost quadrupled—from 165 to 604.

- The tables of Fourier and Laplace transforms in the Appendix have been considerably augmented.

- The first coefficient of the Fourier series is now $\frac{1}{2}a_0$ instead of the previous a_0. Similarly, the direct and inverse full Fourier transformations are now defined with the normalizing factor $1/\sqrt{2\pi}$ in front of the integral; the Fourier sine and cosine transformations are defined with the factor $\sqrt{2/\pi}$.

While I still believe that students should be encouraged not to use electronic computing devices in their learning of the fundamentals of partial differential equations, I have made a concession when it comes to examples and exercises involving special functions, transcendental equations, or exceedingly lengthy integration. The (new) exercises that require computational help because they are not solvable by elementary means have been given italicized numerical labels. Their answers are worked out with the *Mathematica*® software and are given in the form that package produces with full simplification. I have also included a few extra formulas in table A1 in the Appendix to assist with the evaluation of some basic integrals that occur frequently in the solution of the exercises.

The material in this edition seems to exceed what can normally be covered in a one-semester course, even when taught at a brisk pace. If a more leisurely pace is adopted, then the material might be stretched to provide work for two semesters.

I wish to thank all the readers who sent me their comments and urge them to continue to do so in the future. It is only with their help that this book may undergo further improvement.

I would also like to thank Sunil Nair, Sarah Morris, Karen Simon, and Kevin Craig at Taylor & Francis for their professional and expeditious handling of this project.

Christian Constanda
The University of Tulsa
March 2010

Preface to the First Edition

There are many textbooks on partial differential equations on the market. The great majority of them are well written and very rigorous, with full background explanations, detailed proofs, and lots of comments. But they also tend to be rather voluminous and daunting for the average student. When I ask my undergraduates what they want from a book, their most common answers are (i) to understand without excessive effort most of what is being said; (ii) to be given full yet concise explanations of the essence of the topics discussed, in simple words; (iii) to have many worked examples, preferably of the type found in test papers, so they could learn the various techniques by seeing them in action and thus improve their chances of passing examinations; and (iv) to pay as little as possible for it in the bookstore. I do not wish to comment on the validity of these answers, but I am prepared to accept that even in higher education the customer may sometimes be right.

This book is an attempt to meet all the above requirements. It is designed as a no-frills text that explains a number of major methods completely but succinctly, in everyday classroom language. It does not indulge in multi-page, multicolored spiels. It includes many practical applications with solutions, and exercises with selected answers. It has a reasonable number of pages and is produced in a format that facilitates digital reproduction, thus helping keep costs down.

Teachers have their own individual notions regarding what makes a book ideal for use in coursework. They say—with good reason—that the perfect text is the one they themselves sketched in their classroom notes but never had the time or inclination to polish up and publish. We each choose our own material, the order in which the topics are presented, and how long we spend on them. This book is no exception. It is based on my experience of the subject for many years and the feedback received from my teaching's beneficiaries. The "use in combat" of an earlier version seems to indicate that average students can work from it independently, with some occasional instructor guidance, while the high flyers get a basic and rapid grounding in the fundamentals of the subject before progressing to more advanced texts (if they are interested in further details and want to get a truly sophisticated picture of the field). A list, by no means exhaustive, of such texts can be found in the Bibliography.

This book contains no example or exercise that needs a calculating device in its solution. Computing machines are now part of everyday life and we all use them routinely and extensively. However, I believe that if you really want to learn what mathematical analysis is all about, then you should exercise your mind and hand the long way, without any electronic help. (In fact, it seems that quite a few of my students are convinced that computers are better used for surfing the Internet than for solving homework problems.) The only prerequisites for reading this book are a first course in calculus and some basic knowledge of certain types of ordinary differential equations.

The topics are arranged in the order I have found to be the most convenient. After some essential but elementary ODEs, Fourier series, and Sturm–Liouville problems are discussed briefly, the heat, Laplace, and wave equations are introduced in quick succession as mathematical models of physical phenomena, and then a number of methods (separation of variables, eigenfunction expansion, Fourier and Laplace transformations, and Green's functions) are applied in turn to specific initial/boundary value problems for each of these equations. There follows a brief discussion of the general second-order linear equation with two independent variables. Finally, the method of characteristics and perturbation (asymptotic expansion) methods are presented. A number of useful tables and formulas are listed in the Appendix.

The style of the text is terse and utilitarian. In my experience, the teacher's classroom performance does more to generate undergraduate enthusiasm and excitement for a topic than the cold words in a book, however skillfully crafted. Since the aim here is to get the students well drilled in the main solution techniques and not in the physical interpretation of the results, the latter hardly gets a mention. The examples and exercises are formal, and in many of them the chosen data may not reflect plausible real-life situations. Due to space pressure, some intermediate steps—particularly the solutions of simple ODEs—are given without full working. It is assumed that the readers know how to derive them, or that they can refer without difficulty to the summary provided in Chapter 1. Personally, in class I always go through the full solution regardless, which appears to meet with the approval of the audience. Details of a highly mathematical nature, including formal proofs, are kept to a minimum, and when they are given, an assumption is made that any conditions required by the context (for example, the smoothness and behavior of functions) are satisfied.

An Instructor's Manual containing the solutions of all the exercises is available. Also, on adoption of the book, a .pdf file of the text can be supplied to instructors for use on classroom projectors.

My own lecturing routine consists of (i) using a projector to present a skeleton of the theory, so the students do not need to take notes and can follow the live explanations, and (ii) doing a selection of examples on the board with full details, which the students take down by hand. I found that this sequence of "talking periods" and "writing periods" helps the audience maintain concentration and makes the lecture more enjoyable (if what the end-of-semester evaluations say is true).

Wanting to offer students complete, rigorous, and erudite expositions is highly laudable, but the market priorities appear to have shifted of late. With the current standards of secondary education manifestly lower than in the past, students come to us less and less equipped to tackle the learning of mathematics from a fundamental point of view. When this becomes unavoidable, they seem to prefer a concise text that shows them the method and then, without fuss and niceties of form, goes into as many worked examples as possible. Whether we like it or not, it seems that we have entered the era of the digest. It is to this uncomfortable reality that the present book seeks to offer a solution.

The last stages of preparation of this book were completed while I was a Visiting Professor in the Department of Mathematical and Computer Sciences at the University of Tulsa. I wish to thank the authorities of this institution and the faculty in the department for providing me with the atmosphere, conditions, and necessary facilities to finish the work on time. Particular thanks go to the following: Bill Coberly, the head of the department, who helped me engineer several summer visits and a couple of successful sabbatical years in Tulsa; Pete Cook, who heard my daily moans and groans from across the corridor and did not complain about it; Dale Doty, the resident *Mathematica*® wizard who drew some of the figures and showed me how to do the others; and the *sui generis* company at the lunch table in the Faculty Club for whom, in time-honored academic fashion, no discussion topic was too trivial or taboo and no explanation too implausible.

I also wish to thank Sunil Nair, Helena Redshaw, Andrea Demby, and Jasmin Naim from Chapman & Hall/CRC for their help with technical advice and flexibility over deadlines.

Finally, I would like to state for the record that this book project would not have come to fruition had I not had the full support of my wife, who, not for the first time, showed a degree of patience and understanding far beyond the most reasonable expectations.

Christian Constanda

Chapter 1
Ordinary Differential Equations: Brief Review

In the process of solving partial differential equations (PDEs) we usually reduce the problem to the solution of certain classes of ordinary differential equations (ODEs). Here we mention without proof some basic methods for integrating simple ODEs of the types encountered later in the text. We restrict our attention to *real* solutions of ODEs with *real* coefficients. In what follows, the set of real numbers is denoted by $\mathbb{R}$.

1.1. First-Order Equations

Variables separable equations. The general form of this type of ODE is

$$y' = \frac{dy}{dx} = f(x)g(y).$$

Taking standard precautions, we can rewrite the equation as

$$\frac{dy}{g(y)} = f(x)dx$$

and then integrate each side with respect to its corresponding variable.

1.1. Example. For the equation

$$y^2 y' - 2x = 0$$

the above procedure leads to

$$\int y^2 \, dy = 2 \int x \, dx,$$

which yields

$$\tfrac{1}{3} y^3 = x^2 + c, \quad c = \text{const},$$

or

$$y(x) = (3x^2 + C)^{1/3}, \quad C = \text{const.} \quad \blacksquare$$

Linear equations. Their general (normal) form is

$$y' + p(x)y = q(x),$$

where p and q are given functions. Computing an integrating factor $\mu(x)$ by means of the formula

$$\mu(x) = \exp\left\{\int p(x)\,dx\right\},$$

we obtain the general solution

$$y(x) = \frac{1}{\mu(x)}\int \mu(x)q(x)\,dx.$$

An equivalent formula for the general solution is

$$y(x) = \frac{1}{\mu(x)}\left[\int_a^x \mu(t)q(t)\,dt + C\right], \quad C = \text{const},$$

where a is any point in the domain where the ODE is satisfied.

1.2. Example. The normal form of the equation

$$xy' + 2y - x^2 = 0, \quad x \neq 0,$$

is

$$y' + \frac{2}{x}y = x.$$

Here

$$p(x) = \frac{2}{x}, \quad q(x) = x,$$

so an integrating factor is

$$\mu(x) = \exp\left\{2\int \frac{dx}{x}\right\} = e^{\ln x^2} = x^2.$$

Then the general solution of the equation is

$$y(x) = \frac{1}{x^2}\int x^3\,dx = \frac{1}{4}x^2 + \frac{C}{x^2}, \quad C = \text{const.} \quad \blacksquare$$

1.2. Homogeneous Linear Equations with Constant Coefficients

First-order equations. These are equations of the form

$$y' + ay = 0, \quad a = \text{const.}$$

Such equations can be solved by means of an integrating factor or separation of variables, or by means of the characteristic equation

$$s + a = 0,$$

whose root $s = -a$ yields the general solution

$$y(x) = Ce^{-ax}, \quad C = \text{const.}$$

1.3. Example. The characteristic equation for the ODE

$$y' - 3y = 0$$

is

$$s - 3 = 0;$$

hence, the general solution of the equation is

$$y(x) = Ce^{3x}, \quad C = \text{const.} \quad \blacksquare$$

Second-order equations. Their general form is

$$y'' + ay' + by = 0, \quad a, b = \text{const.}$$

If the characteristic equation

$$s^2 + as + b = 0$$

has two distinct real roots s_1 and s_2, then the general solution of the given ODE is

$$y(x) = C_1 e^{s_1 x} + C_2 e^{s_2 x}, \quad C_1, C_2 = \text{const.}$$

If $s_1 = s_2 = s_0$, then

$$y(x) = (C_1 + C_2 x)e^{s_0 x}, \quad C_1, C_2 = \text{const.}$$

Finally, if s_1 and s_2 are complex conjugate—that is, $s_1 = \alpha + i\beta$, $s_2 = \alpha - i\beta$, where α and β are real numbers—then the general solution is

$$y(x) = e^{\alpha x}[C_1 \cos(\beta x) + C_2 \sin(\beta x)], \quad C_1, C_2 = \text{const.}$$

1.4. Remark. When $s_1 = -s_2 = s_0$, s_0 real, the general solution of the equation can also be written as

$$y(x) = C_1 y_1(x) + C_2 y_2(x), \quad C_1, C_2 = \text{const,}$$

where $y_1(x)$ and $y_2(x)$ are any two of the functions

$$\cosh(s_0 x), \quad \sinh(s_0 x), \quad \cosh\big(s_0(x - c)\big), \quad \sinh\big(s_0(x - c)\big)$$

and c is any nonzero real number. Normally, c is chosen as the point where a boundary condition is given. ∎

1.5. Example. The characteristic equation for the ODE

$$y'' - 3y' + 2y = 0$$

is

$$s^2 - 3s + 2 = 0,$$

with roots $s_1 = 1$ and $s_2 = 2$, so the general solution of the ODE is

$$y(x) = C_1 e^x + C_2 e^{2x}, \quad C_1, C_2 = \text{const.} \quad \blacksquare$$

1.6. Example. The general solution of the equation

$$y'' - 4y = 0$$

is

$$y(x) = C_1 e^{2x} + C_2 e^{-2x}, \quad C_1, C_2 = \text{const,}$$

since the roots of its characteristic equation are $s_1 = -s_2 = 2$. According to Remark 1.4, we have alternative expressions in terms of hyperbolic functions. Thus, if $y(0)$ and $y(1)$ are prescribed, then the general solution should be written in the form

$$y(x) = C_1 \sinh(2x) + C_2 \sinh\big(2(x - 1)\big), \quad C_1, C_2 = \text{const;}$$

if $y(0)$ and $y'(3)$ are prescribed, then the preferred form is

$$y(x) = C_1 \sinh(2x) + C_2 \cosh\big(2(x-3)\big), \quad C_1, C_2 = \text{const};$$

and so on. ∎

1.7. Example. The roots of the characteristic equation for the ODE

$$y'' + 4y' + 4y = 0$$

are $s_1 = s_2 = -2$; therefore, the general solution of the ODE is

$$y = (C_1 + C_2 x)e^{-2x}, \quad C_1, C_2 = \text{const}. \ \blacksquare$$

1.8. Example. The general solution of the equation

$$y'' + 4y = 0$$

is

$$y = C_1 \cos(2x) + C_2 \sin(2x), \quad C_1, C_2 = \text{const},$$

since the roots of its characteristic equation are $s_1 = 2i$ and $s_2 = -2i$. ∎

1.9. Remark. The characteristic equation method can also be applied to find the general solution of homogeneous linear ODEs of higher order. ∎

1.3. Nonhomogeneous Linear Equations with Constant Coefficients

The first-order equations in this category are of the form

$$y' + ay = f, \quad a = \text{const};$$

the second-order equations can be written as

$$y'' + ay' + by = f, \quad a, b = \text{const}.$$

Here f is a given function. The general solution of such equations is the sum of the complementary function (the general solution of the corresponding homogeneous equation) and a particular integral (a particular solution of the nonhomogeneous equation). The latter is usually guessed from the structure of the function f or may be found by some other method, such as variation of parameters.

1.10. Example. The complementary function for the ODE

$$y' - 3y = e^{-x}$$

is

$$y_{CF} = Ce^{3x}, \quad C = \text{const.}$$

Seeking a particular integral of the form $y_{PI} = ae^{-x}$, $a = \text{const}$, we find from the equation that $a = -1/4$. Consequently, the general solution of the given ODE is

$$y = Ce^{3x} - \tfrac{1}{4}e^{-x}, \quad C = \text{const.} \ \blacksquare$$

1.11. Example. If the function on the right-hand side in Example 1.10 is replaced by e^{3x}, then we cannot find a particular integral of the form ae^{3x}, $a = \text{const}$, since this is a solution of the corresponding homogeneous equation. Instead, we try $y_{PI} = axe^{3x}$ and deduce, by replacing in the ODE, that $a = 1$; consequently, the general solution is

$$y = Ce^{3x} + xe^{3x} = (C + x)e^{3x}, \quad C = \text{const.} \ \blacksquare$$

1.12. Example. For the equation

$$y'' + 4y = 4x^2$$

we seek a particular integral of the form

$$y_{PI} = ax^2 + bx + c, \quad a, b, c = \text{const.}$$

Direct substitution into the equation yields $a = 1$, $b = 0$, and $c = -1/2$. Since the complementary function is

$$y_{CF} = C_1 \cos(2x) + C_2 \sin(2x),$$

it follows that the general solution of the given ODE is

$$y = C_1 \cos(2x) + C_2 \sin(2x) + x^2 - \tfrac{1}{2}, \quad C_1, C_2 = \text{const.} \ \blacksquare$$

1.4. Cauchy–Euler Equations

These are second-order linear equations of the form

$$x^2 y'' + \alpha x y' + \beta y = 0, \quad \alpha, \beta = \text{const},$$

where, for simplicity, we assume that $x > 0$. The solution is sought in the form

$$y = x^r, \quad r = \text{const.}$$

Substituting in the equation, we arrive at

$$r^2 + (\alpha - 1)r + \beta = 0.$$

If the roots r_1 and r_2 of this quadratic equation are real and distinct, which is the case of interest for us, then the general solution of the given ODE is

$$y = C_1 x^{r_1} + C_2 x^{r_2}, \quad C_1, C_2 = \text{const.}$$

1.13. Example. Following the above procedure, we see that the differential equation

$$2x^2 y'' + 3xy' - y = 0$$

yields

$$r_1 = 1, \quad r_2 = -\tfrac{1}{2},$$

so its general solution is

$$y = C_1 x + C_2 x^{-1/2}, \quad C_1, C_2 = \text{const.} \quad \blacksquare$$

1.5. Functions and Operators

Throughout this book we refer to a function as either f or $f(x)$, although, strictly speaking, the latter denotes the value of f at x. To avoid complicated notation, we also write $f(x) = c$ to designate a function f that takes the same value $c = \text{const}$ at all points x in its domain. When $c = 0$, we sometimes simplify this further to $f = 0$.

In the preceding sections we mentioned *linear* equations. Here we clarify the meaning of this concept.

1.14. Definition. Let $\mathcal{X}$ be a space of functions, and let L be an operator acting on the functions in $\mathcal{X}$ according to some rule. The operator L is called *linear* if

$$L(c_1 f_1 + c_2 f_2) = c_1(Lf_1) + c_2(Lf_2)$$

for any functions f_1, f_2 in $\mathcal{X}$ and any numbers c_1, c_2. Otherwise, L is called *nonlinear*. $\blacksquare$

1.15. Example. The operators of differentiation and definite integration acting on suitable functions of one independent variable are linear, since

$$L(c_1 f_1 + c_2 f_2) = (c_1 f_1 + c_2 f_2)' = c_1 f_1' + c_2 f_2' = c_1(Lf_1) + c_2(Lf_2),$$

$$L(c_1 f_1 + c_2 f_2) = \int_a^b \left[c_1 f_1(x) + c_2 f_2(x) \right] dx$$

$$= c_1 \int_a^b f_1(x)\, dx + c_2 \int_a^b f_2(x)\, dx = c_1(Lf_1) + c_2(Lf_2). \quad \blacksquare$$

1.16. Example. Let α, β, and γ be given functions. In view of the preceding example, it is easy to verify that the operator L defined by

$$Lf = \alpha f'' + \beta f' + \gamma f$$

is linear. $\blacksquare$

1.17. Example. Let L be the operator defined by

$$Lf = ff'.$$

Since

$$L(c_1 f_1 + c_2 f_2) = (c_1 f_1 + c_2 f_2)(c_1 f_1 + c_2 f_2)'$$

$$= c_1^2 f_1 f_1' + c_1 c_2 (f_1 f_2' + f_1' f_2) + c_2^2 f_2 f_2'$$

and

$$c_1(Lf_1) + c_2(Lf_2) = c_1 f_1 f_1' + c_2 f_2 f_2'$$

are not equal for all functions f_1, f_2 and all numbers c_1, c_2, the operator L is nonlinear. $\blacksquare$

1.18. Definition. A differential equation of the form

$$Lu = g,$$

where L is a linear differential operator and g is a given function, is called a *linear* equation. If the operator L is nonlinear, then the equation is also called *nonlinear*. $\blacksquare$

The following almost obvious result forms the basis of what is known as the *principle of superposition.*

1.19. Theorem. *If $Lu = g$ is a linear equation and u_1 and u_2 are solutions of this equation with $g = g_1$ and $g = g_2$, respectively, then $u_1 + u_2$ is a solution of the equation with $g = g_1 + g_2$; in other words, if*

$$Lu_1 = g_1, \quad Lu_2 = g_2,$$

then

$$L(u_1 + u_2) = g_1 + g_2.$$

Exercises

In (1)–(22) find the general solution of the given equation.

(1) $(x^2 + 1)y' = 2xy.$

(2) $y' - 3x^2(y + 1) = 0.$

(3) $(x - 1)y' + 2y = x, \; x \neq 1.$

(4) $x^2 y' - 2xy = x^5 e^x.$

(5) $2y' + 5y = 0.$

(6) $3y' - 2y = 0.$

(7) $y'' - 4y' + 3y = 0.$

(8) $2y'' - 5y' + 2y = 0.$

(9) $4y'' + 4y' + y = 0.$

(10) $y'' - 6y' + 9y = 0.$

(11) $y'' + 2y' + 5y = 0.$

(12) $y'' - 6y' + 13y = 0.$

(13) $y' + 2y = 2x + e^{4x}.$

(14) $2y' - 3y = -3x - 4 + e^x.$

(15) $2y' - y = e^{x/2}.$

(16) $y' + y = -x + 2e^{-x}.$

(17) $y'' - y = x^2 - x + 2.$

(18) $y'' - 2y' - 8y = 4 + 4x - 8x^2.$

(19) $y'' - 25y = 30e^{-5x}.$

(20) $4y'' + y = 8\cos(x/2).$

(21) $2x^2 y'' + xy' - 3y = 0.$

(22) $x^2 y'' + 2xy' - 6y = 0.$

In (23)–(26) verify whether the given ODE is linear or nonlinear.

(23) $xy'' - y' \sin x = xe^x$.

(24) $y' + 2x \sin y = 1$.

(25) $y'y'' - xy = 2x$.

(26) $y'' + \sqrt{x}\,y = \ln x$.

Chapter 2
Fourier Series

It is well known that an infinitely differentiable function $f(x)$ can be expanded in a Taylor series around a point x_0 in the interval where it is defined. This series has the form

$$f(x) \sim \sum_{n=0}^{\infty} c_n (x - x_0)^n, \tag{2.1}$$

where the coefficients c_n are given by

$$c_n = \frac{f^{(n)}(x_0)}{n!}, \quad n = 1, 2, \ldots, \quad f^{(n)} = \frac{d^n f}{dx^n}.$$

If certain conditions are satisfied, then the above series converges to f pointwise (that is, at every point x) in an open interval centered at x_0, and we can use the equality sign between the two sides in (2.1).

In this chapter we discuss a different class of expansions, which are particulary useful in the study and solution of PDEs.

2.1. The Full Fourier Series

This is an expansion of the form

$$f(x) \sim \frac{1}{2} a_0 + \sum_{n=1}^{\infty} \left(a_n \cos \frac{n\pi x}{L} + b_n \sin \frac{n\pi x}{L} \right), \tag{2.2}$$

where L is a positive number and a_0, a_n, and b_n are constant coefficients.

2.1. Definition. A function f defined on $\mathbb{R}$ is called *periodic* if there is a number $T > 0$ such that

$$f(x + T) = f(x) \quad \text{for all } x \text{ in } \mathbb{R}.$$

The smallest number T with this property is called the *fundamental period* (or, simply, the *period*) of f. It is obvious that a periodic function also satisfies

$$f(x + nT) = f(x) \quad \text{for any integer } n \text{ and all } x \text{ in } \mathbb{R}. \quad \blacksquare$$

2.2. Example. As is well known, the functions $\sin x$ and $\cos x$ are periodic with period 2π since, for all $x \in \mathbb{R}$,

$$\sin(x + 2\pi) = \sin x,$$
$$\cos(x + 2\pi) = \cos x. \quad \blacksquare$$

We see that for each positive integer $n = 1, 2, \ldots,$

$$\cos \frac{n\pi(x + 2L)}{L} = \cos\left(\frac{n\pi x}{L} + 2n\pi\right) = \cos \frac{n\pi x}{L},$$

$$\sin \frac{n\pi(x + 2L)}{L} = \sin\left(\frac{n\pi x}{L} + 2n\pi\right) = \sin \frac{n\pi x}{L},$$

so the right-hand side in (2.2) is periodic with period $2L$. This suggests the following method of construction for the full Fourier series of a given function.

Let f be defined on $[-L, L]$ (see Fig. 2.1).

Fig. 2.1. $f(x)$, $-L \le x \le L$.

We construct the periodic extension of f from $(-L, L]$ to $\mathbb{R}$, of period $2L$ (see Fig. 2.2). The value of f at $x = -L$ is left out so that the extension is correctly defined as a function.

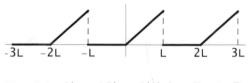

Fig. 2.2. $f(x + 2L) = f(x)$ for all x in $\mathbb{R}$.

For the extended function f it now makes sense to seek an expansion of the form (2.2). All that we need to do is compute the coefficients a_0, a_n, and b_n, and discuss the convergence of the series.

It is easy to check by direct calculation that

$$\int_{-L}^{L} \cos\frac{n\pi x}{L}\, dx = 0, \quad \int_{-L}^{L} \sin\frac{n\pi x}{L}\, dx = 0, \quad n = 1, 2, \ldots, \tag{2.3}$$

$$\int_{-L}^{L} \cos\frac{n\pi x}{L} \cos\frac{m\pi x}{L}\, dx = \begin{cases} 0, & n \neq m, \\ L, & n = m, \end{cases} \quad n, m = 1, 2, \ldots, \tag{2.4}$$

$$\int_{-L}^{L} \sin\frac{n\pi x}{L} \sin\frac{m\pi x}{L}\, dx = \begin{cases} 0, & n \neq m, \\ L, & n = m, \end{cases} \quad n, m = 1, 2, \ldots, \tag{2.5}$$

$$\int_{-L}^{L} \cos\frac{n\pi x}{L} \sin\frac{m\pi x}{L}\, dx = 0, \quad n, m = 1, 2, \ldots. \tag{2.6}$$

Regarding (2.2) formally as an equality, we integrate it term by term over $[-L, L]$ and use (2.3) to obtain

$$a_0 = \frac{1}{L} \int_{-L}^{L} f(x)\, dx. \tag{2.7}$$

If we multiply (2.2) by $\cos(m\pi x/L)$, integrate the new relation over $[-L, L]$, and take (2.3), (2.4), and (2.6) into account, we see that all the integrals on the right-hand side vanish except that for which the summation index n is equal to m, when the integral is equal to La_m. Replacing m by n, we find that

$$a_n = \frac{1}{L} \int_{-L}^{L} f(x) \cos\frac{n\pi x}{L}\, dx, \quad n = 1, 2, \ldots. \tag{2.8}$$

Clearly, (2.8) incorporates (2.7) if we allow n to take the value 0 as well; that is, $n = 0, 1, 2, \ldots.$

Finally, we multiply (2.2) by $\sin(m\pi x/L)$ and repeat the above procedure, where this time we use (2.3), (2.5), and (2.6). The result is

$$b_n = \frac{1}{L} \int_{-L}^{L} f(x) \sin\frac{n\pi x}{L}\, dx, \quad n = 1, 2, \ldots, \tag{2.9}$$

which completes the construction of the (full) Fourier series for f. The numbers a_0, a_n, and b_n given by (2.7)–(2.9) are called the *Fourier coefficients* of f.

The series on the right-hand side in (2.2) is of interest to us only on $[-L, L]$, where the original function f is defined, and may be divergent, or may have a different sum than $f(x)$.

2.3. Definition. A function f is said to be *piecewise continuous* on an interval $[a, b]$ if it is continuous at all but finitely many points in $[a, b]$, where it has jump discontinuities—that is, at any discontinuity point x the function has distinct right-hand side and left-hand side (finite) limits $f(x+)$ and $f(x-)$.

If both f and f' are continuous on $[a, b]$, then f is called *smooth* on $[a, b]$. If at least one of f, f' is piecewise continuous on $[a, b]$, then f is said to be *piecewise smooth* on $[a, b]$. ∎

2.4. Remarks. (i) The function shown on the left in Fig. 2.3 is piecewise smooth; the one on the right is not, since $f(0-)$ does not exist: the graph indicates that the function increases without bound as the variable approaches the origin from the left.

(ii) If f is piecewise continuous on $[-L, L]$, then the values of f at its points of discontinuity do not affect the construction of its Fourier series. More precisely, $\int\limits_{-L}^{L} f(x)\,dx$ exists for such a function and is independent of the values assigned to it at its (finitely many) discontinuity points. ∎

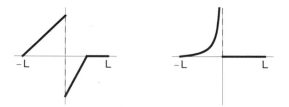

Fig. 2.3. Left: both $f(0-)$ and $f(0+)$ exist. Right: $f(0-)$ does not exist.

2.5. Theorem. *If f is piecewise smooth on $[-L, L]$, then its (full) Fourier series converges pointwise to*

(i) *the periodic extension of f to $\mathbb{R}$ at all points x where this extension is continuous;*

(ii) $\frac{1}{2}[f(x-) + f(x+)]$ *at the points x where the periodic extension of f has a discontinuity jump.*

This means that at each x in $(-L, L)$ where f is continuous, the sum of series (2.2) is equal to $f(x)$. Also, if the function f is continuous on $[-L, L]$ and such that $f(-L) = f(L)$, then the sum of (2.2) is equal to $f(x)$ at all points x in $[-L, L]$.

2.6. Example. Consider the function defined by

$$f(x) = \begin{cases} x + 2, & -2 \le x < 0, \\ 0, & 0 \le x \le 2, \end{cases}$$

whose graph is shown in Fig. 2.4. Clearly, here $L = 2$.

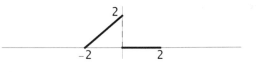

Fig. 2.4. $f(x)$, $-2 \le x \le 2$.

The periodic extension of period 4 of f, defined by $f(x + 4) = f(x)$ for all x in $\mathbb{R}$, is shown in Fig. 2.5.

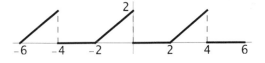

Fig. 2.5. $f(x + 4) = f(x)$ for all x in $\mathbb{R}$.

Using (2.7)–(2.9) with $L = 2$ and integration by parts, we find that

$$a_0 = \frac{1}{2} \int_{-2}^{2} f(x)\, dx = \frac{1}{2} \int_{-2}^{0} (x + 2)\, dx = 1,$$

$$a_n = \frac{1}{2} \int_{-2}^{2} f(x) \cos \frac{n\pi x}{2}\, dx = \frac{1}{2} \int_{-2}^{0} (x + 2) \cos \frac{n\pi x}{2}\, dx = \left[1 - (-1)^n\right] \frac{2}{n^2 \pi^2},$$

$$b_n = \frac{1}{2} \int_{-2}^{2} f(x) \sin \frac{n\pi x}{2}\, dx = \frac{1}{2} \int_{-2}^{0} (x + 2) \sin \frac{n\pi x}{2}\, dx = -\frac{2}{n\pi}, \quad n = 1, 2, \ldots,$$

so (2.2) yields the Fourier series

$$f(x) \sim \frac{1}{2} + \sum_{n=1}^{\infty} \left\{ [1 - (-1)^n] \frac{2}{n^2\pi^2} \cos\frac{n\pi x}{2} - \frac{2}{n\pi} \sin\frac{n\pi x}{2} \right\}.$$

By Theorem 2.5, on $-2 \le x \le 2$

$$(\text{series}) = \begin{cases} x + 2, & -2 \le x < 0, \\ 1, & x = 0, \\ 0, & 0 < x \le 2; \end{cases}$$

that is, the series converges to $f(x)$ for $-2 \le x < 0$ and $0 < x \le 2$, where the periodic extension of f (see Fig. 2.5) is continuous; at the point $x = 0$, where that extension has a jump discontinuity, the series converges to

$$\tfrac{1}{2}[f(0-) + f(0+)] = \tfrac{1}{2}(2 + 0) = 1.$$

The graph of the function defined by the sum of the above series is shown in Fig. 2.6.

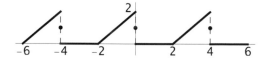

Fig. 2.6. Graphic representation of the sum of the series.

The manner in which the series approximates f is shown, with increased magnification for clarity, in Fig. 2.7, where the series has been truncated after $n = 5$. ■

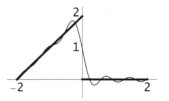

Fig. 2.7. The graph of u (heavy line) and of its 5-term approximation (light line).

2.2. Fourier Sine Series

For some classes of functions, series (2.2) has a simpler form.

2.7. Definition. A function f defined on an interval symmetric with respect to the origin is called *odd* if $f(-x) = -f(x)$ for all x in the given interval; if, on the other hand, $f(-x) = f(x)$ for all x in that interval, then f is called an *even* function. ∎

2.8. Examples. (i) The functions $\sin(n\pi x/L)$, $n = 1, 2, \ldots$, are odd. The functions $\cos(n\pi x/L)$, $n = 0, 1, 2, \ldots$, are even.

(ii) The function on the left in Fig. 2.8 is odd (its graph is symmetric with respect to the origin). The one on the right is even (its graph is symmetric with respect to the y-axis). The function graphed in Fig. 2.4 is neither odd nor even. ∎

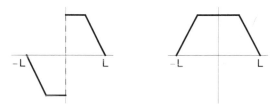

Fig. 2.8. Left: an odd function. Right: an even function.

2.9. Remarks. (i) It is easy to verify that the product of two odd functions is even, the product of two even functions is even, and the product of an odd function and an even function is odd.

(ii) If f is odd on $[-L, L]$, then

$$\int_{-L}^{0} f(x)\,dx = \int_{L}^{0} f(-x)\,d(-x) = -\int_{0}^{L} f(x)\,dx;$$

hence,

$$\int_{-L}^{L} f(x)\,dx = \int_{-L}^{0} f(x)\,dx + \int_{0}^{L} f(x)\,dx = 0.$$

If f is even on $[-L, L]$, then

$$\int_{-L}^{0} f(x)\,dx = \int_{L}^{0} f(-x)\,d(-x) = \int_{0}^{L} f(x)\,dx;$$

consequently,

$$\int_{-L}^{L} f(x)\,dx = \int_{-L}^{0} f(x)\,dx + \int_{0}^{L} f(x)\,dx = 2\int_{0}^{L} f(x)\,dx.$$

(iii) If f is odd, then so is $f(x)\cos(n\pi x/L)$ and, in view of (2.7), (2.8), and (ii) above, we have

$$a_n = 0, \quad n = 0,1,2,\ldots;$$

that is, the Fourier series of an odd function on $[-L,L]$ contains only sine terms. Similarly, if f is even, then, by Remark 2.9(i), $f(x)\sin(n\pi x/L)$ is odd, so (2.9) implies that

$$b_n = 0, \quad n = 1,2,\ldots,$$

which means that the Fourier series of an even function on $[-L,L]$ has only cosine terms, including the constant term. ∎

Remark 2.9(iii) implies that if f is defined on $[0,L]$, then it can be expanded in a *Fourier sine series*. Let f be the function whose graph is shown in Fig. 2.9.

Fig. 2.9. $f(x)$, $0 \le x \le L$.

We construct the odd extension of f from $(0,L]$ to $[-L,L]$ by setting $f(-x) = -f(x)$ for all x in $[-L,L]$, $x \neq 0$, and $f(0) = 0$ (see Fig. 2.10).

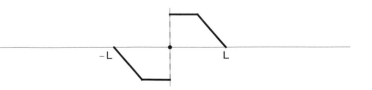

Fig. 2.10. $f(-x) = -f(x)$, $-L \le x \le L$.

Next, we construct the periodic extension with period $2L$ of this odd function from $(-L, L]$ to $\mathbb{R}$, by requiring that $f(x + 2L) = f(x)$ for all x in $\mathbb{R}$ (see Fig. 2.11).

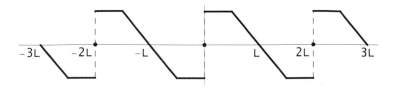

Fig. 2.11. $f(x + 2L) = f(x)$ for all x in $\mathbb{R}$.

Finally, we construct the Fourier (sine) series of this last function, which is of the form

$$f(x) \sim \sum_{n=1}^{\infty} b_n \sin \frac{n\pi x}{L},$$

where, according to Remarks 2.9(ii),(iii) and formula (2.9), the coefficients are given by

$$b_n = \frac{2}{L} \int_0^L f(x) \sin \frac{n\pi x}{L} \, dx, \quad n = 1, 2, \ldots. \tag{2.10}$$

2.10. Example. Consider the function

$$f(x) = 1 - x, \quad 0 \le x \le 1,$$

graphed in Fig. 2.12.

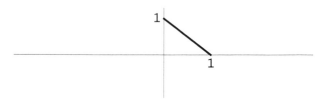

Fig. 2.12. $f(x)$, $0 \le x \le 1$.

As explained above, the odd extension of this function to $[-1, 1]$ is defined by $f(-x) = -f(x)$ for all x in $[-1, 1]$, $x \ne 0$, and $f(0) = 0$, and is shown in Fig. 2.13.

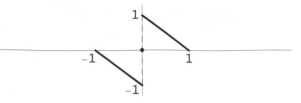

Fig. 2.13. $f(-x) = -f(x)$, $-1 \le x \le 1$.

The periodic extension to $\mathbb{R}$, of period $2L = 2$, of the function in Fig. 2.13, defined by $f(x+2) = f(x)$ for all x in $\mathbb{R}$, is shown in Fig. 2.14.

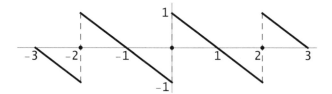

Fig. 2.14. $f(x+2) = f(x)$ for all x in $\mathbb{R}$.

Using (2.10) with $L = 1$ and integration by parts, we find that

$$b_n = 2 \int_0^1 (1 - x) \sin(n\pi x)\, dx = \frac{2}{n\pi}, \quad n = 1, 2, \ldots;$$

hence,

$$f(x) \sim \sum_{n=1}^{\infty} \frac{2}{n\pi} \sin(n\pi x).$$

By Theorem 2.5, for $0 \le x \le 1$,

$$(\text{series}) = \begin{cases} 0, & x = 0, \\ 1 - x, & 0 < x \le 1; \end{cases}$$

that is, the series converges to $f(x)$ for $0 < x \le 1$, where the periodic extension of f to $\mathbb{R}$ is continuous, and to

$$\tfrac{1}{2}\left[f(0-) + f(0+)\right] = \tfrac{1}{2}(-1 + 1) = 0$$

at $x = 0$, where the periodic extension has a jump discontinuity. ∎

2.3. Fourier Cosine Series

By Remark 2.9(iii), a function defined on $[0, L]$ can also be expanded in a *Fourier cosine series*. The construction is similar to that of a Fourier sine series.

Let f be a function defined on $[0, L]$ (see Fig. 2.15).

Fig. 2.15. $f(x)$, $0 \le x \le L$.

We extend f to an even function on $[-L, L]$ by setting $f(-x) = f(x)$ for all x in $[-L, L]$ (see Fig. 2.16).

Fig. 2.16. $f(-x) = f(x)$, $-L \le x \le L$.

Next, we construct the periodic extension of this even function from $(-L, L]$ to $\mathbb{R}$, of period $2L$, which is defined by $f(x + 2L) = f(x)$ for all x in $\mathbb{R}$ (see Fig. 2.17).

Fig. 2.17. $f(x + 2L) = f(x)$ for all x in $\mathbb{R}$.

Finally, we write the Fourier (cosine) series of this last function, which is of the form

$$f(x) \sim \frac{1}{2} a_0 + \sum_{n=1}^{\infty} a_n \cos \frac{n \pi x}{L},$$

where, according to (2.7), (2.8), and Remark 2.9(ii),

$$
a_0 = \frac{2}{L} \int_0^L f(x)\, dx,
$$

$$
a_n = \frac{2}{L} \int_0^L f(x) \cos \frac{n\pi x}{L}\, dx, \quad n = 1, 2, \ldots.
$$

(2.11)

As noticed earlier, the first equality (2.11) is covered by the second one if we let $n = 0, 1, 2, \ldots$.

2.11. Example. Consider the function

$$
f(x) = 1 - x, \quad 0 \le x \le 1,
$$

graphed in Fig. 2.18.

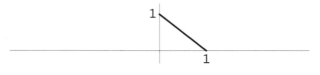

Fig. 2.18. $f(x)$, $0 \le x \le 1$.

The even extension of f to $[-1,1]$, defined by $f(-x) = f(x)$ for all x in $[-1,1]$, is shown in Fig. 2.19.

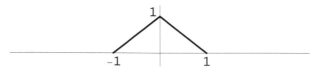

Fig. 2.19. $f(-x) = f(x)$, $-1 \le x \le 1$.

The periodic extension of period $2L = 2$ of this even function to $\mathbb{R}$, which is defined by $f(x + 2) = f(x)$ for all x in $\mathbb{R}$, is shown in Fig. 2.20.

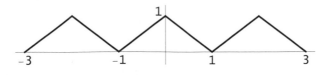

Fig. 2.20. $f(x + 2) = f(x)$ for all x in $\mathbb{R}$.

By (2.11) with $L = 1$ and integration by parts,

$$a_0 = 2 \int_0^1 f(x)\,dx = 2 \int_0^1 (1-x)\,dx = 1,$$

$$a_n = 2 \int_0^1 f(x) \cos(n\pi x)\,dx = 2 \int_0^1 (1-x) \cos(n\pi x)\,dx$$

$$= \left[1 - (-1)^n\right] \frac{2}{n^2 \pi^2}, \quad n = 1, 2, \ldots,$$

so the Fourier cosine series of f is

$$f(x) \sim \frac{1}{2} + \sum_{n=1}^{\infty} \left[1 - (-1)^n\right] \frac{2}{n^2 \pi^2} \cos(n\pi x).$$

By Theorem 2.5,

$$(\text{series}) = 1 - x \quad \text{for } 0 \le x \le 1,$$

because the periodic extension of f to $\mathbb{R}$ is continuous everywhere. ∎

2.4. Convergence and Differentiation

There are several important points that need to be made at this stage.

2.12. Remarks. (i) It is obvious that a function f defined on $[0, L]$ can be expanded in a full Fourier series, or in a sine series, or in a cosine series. While the last two are unique to the function, the first one is not: since the full Fourier series requires no symmetry, we may extend f from $[0, L]$ to $[-L, L]$ in infinitely many ways. We usually choose the type of series that seems to be the most appropriate for the problem. A function defined on $[-L, L]$ can be expanded, in general, only in a full Fourier series.

(ii) Earlier we saw that we cannot automatically use the equality sign between a function f and its Fourier series. According to Theorem 2.5, for a full series we can do so at all the points in $[-L, L]$ where f is continuous; we can also do it at $x = -L$ and $x = L$ provided that f is continuous at these points from the right and from the left, respectively, and $f(-L) = f(L)$. For a sine series, this can be done at $x = L$ if f is continuous there from the left and $f(L) = 0$. For a cosine series this is always possible if f is continuous from the left at $x = L$. Similar arguments can be used for the sine and cosine series at $x = 0$. In what follows we work exclusively with piecewise smooth functions; therefore, to avoid cumbersome notation and possible confusion, we will use the equality sign in all circumstances, assuming tacitly that equality is understood in the sense of Theorem 2.5, Remark 2.4(ii), and the above comments. ∎

In practical applications we often need to differentiate Fourier series term by term. Consequently, it is important to know when this operation may be performed.

2.13. Theorem.   (i) *If f is continuous on $[-L, L]$, $f(-L) = f(L)$, and f' is piecewise smooth on $[-L, L]$, then the full Fourier series of f can be differentiated term by term and the resulting series is the Fourier series of f', which converges to $f'(x)$ at all x where $f''(x)$ exists.*

(ii) *If f is continuous on $[0, L]$, $f(0) = f(L) = 0$, and f' is piecewise smooth on $[0, L]$, then the Fourier sine series of f can be differentiated term by term and the resulting series is the Fourier cosine series of f', which converges to $f'(x)$ at all x where $f''(x)$ exists.*

(iii) *If f is continuous on $[0, L]$ and f' is piecewise smooth on $[0, L]$, then the Fourier cosine series of f can be differentiated term by term and the resulting series is the Fourier sine series of f', which converges to $f'(x)$ at all x where $f''(x)$ exists.*

2.14. Remark. In the PDE problems that we discuss in the rest of the book, the conditions laid down in Theorem 2.13 will always be satisfied and we will formally differentiate the corresponding Fourier series term by term without explicit mention of these conditions. ∎

Exercises

In (1)–(16) construct the full Fourier series of the given function f. In each case discuss the convergence of the series on the interval $[-L, L]$ where f is defined and sketch the function to which the series converges pointwise on the interval $[-3L, 3L]$.

(1) $f(x) = \begin{cases} 1, & -1 \le x \le 0, \\ 0, & 0 < x \le 1. \end{cases}$

(2) $f(x) = \begin{cases} 0, & -\pi \leq x \leq 0, \\ -3, & 0 < x \leq \pi. \end{cases}$

(3) $f(x) = \begin{cases} -2, & -1 \leq x \leq 0, \\ 3, & 0 < x \leq 1. \end{cases}$

(4) $f(x) = \begin{cases} 1, & -\pi/2 \leq x \leq 0, \\ 2, & 0 < x \leq \pi/2. \end{cases}$

(5) $f(x) = x + 1, \quad -1 \leq x \leq 1.$

(6) $f(x) = 1 - 2x, \quad -2 \leq x \leq 2.$

(7) $f(x) = \begin{cases} -1, & -2 \leq x \leq 0, \\ 2 - x, & 0 < x \leq 2. \end{cases}$

(8) $f(x) = \begin{cases} 1 + 2x, & -1/2 \leq x \leq 0, \\ 4, & 0 < x \leq 1/2. \end{cases}$

(9) $f(x) = \begin{cases} x, & -1 \leq x \leq 0, \\ 2x - 1, & 0 < x \leq 1. \end{cases}$

(10) $f(x) = \begin{cases} x - 1, & -\pi \leq x \leq 0, \\ 2x + 1, & 0 < x \leq \pi. \end{cases}$

(11) $f(x) = x^2 - 2x + 3, \quad -1 \leq x \leq 1.$

(12) $f(x) = \begin{cases} x^2, & -1 \leq x \leq 0, \\ 1 + 2x, & 0 < x \leq 1. \end{cases}$

(13) $f(x) = e^x, \quad -2 \leq x \leq 2.$

(14) $f(x) = \begin{cases} 1, & -\pi/2 \leq x \leq 0, \\ \sin x, & 0 < x \leq \pi/2. \end{cases}$

(15) $f(x) = \begin{cases} 0, & -2 \leq x \leq -1, \\ 1 + x, & -1 < x \leq 2. \end{cases}$

(16) $f(x) = \begin{cases} 3, & -\pi \leq x \leq \pi/2, \\ 1, & \pi/2 < x \leq \pi. \end{cases}$

In (17)–(30) construct the Fourier sine series and the Fourier cosine series of the given function f. In each case discuss the convergence of the series on the interval $[0, L]$ where f is defined and sketch the functions to which the series converge pointwise on $[-3L, 3L]$.

(17) $f(x) = \begin{cases} 0, & 0 \leq x \leq 1, \\ 1, & 1 < x \leq 2. \end{cases}$

(18) $f(x) = \begin{cases} 2, & 0 \leq x \leq \pi, \\ 0, & \pi < x \leq 2\pi. \end{cases}$

(19) $f(x) = \begin{cases} 1, & 0 \le x \le 1, \\ -1, & 1 < x \le 2. \end{cases}$

(20) $f(x) = \begin{cases} -2, & 0 \le x \le \pi/2, \\ 3, & \pi/2 < x \le \pi. \end{cases}$

(21) $f(x) = 2 - x, \quad 0 \le x \le 1.$

(22) $f(x) = 3x + 1, \quad 0 \le x \le 2\pi.$

(23) $f(x) = \begin{cases} x, & 0 \le x \le 1, \\ -2, & 1 < x \le 2. \end{cases}$

(24) $f(x) = \begin{cases} 2x - 1, & 0 \le x \le 1, \\ 2x + 1, & 1 < x \le 2. \end{cases}$

(25) $f(x) = \begin{cases} 2 + x, & 0 \le x \le 1, \\ 1 - x, & 1 < x \le 2. \end{cases}$

(26) $f(x) = x^2 + x - 1, \quad 0 \le x \le 2.$

(27) $f(x) = \begin{cases} x + 1, & 0 \le x \le 1, \\ x^2 - 2x, & 1 < x \le 2. \end{cases}$

(28) $f(x) = 1 + e^x, \quad 0 \le x \le 1.$

(29) $f(x) = x + \sin x, \quad 0 \le x \le \pi.$

(30) $f(x) = \begin{cases} \cos x, & 0 \le x \le \pi/2, \\ -1, & \pi/2 < x \le \pi. \end{cases}$

Chapter 3
Sturm–Liouville Problems

There is a class of problems involving second-order ODEs, which plays an essential role in the solution of partial differential equations. Below we present the main results concerning such problems, thus laying the foundation for the methods of separation of variables and eigenfunction expansion developed in Chapters 5 and 7.

3.1. Regular Sturm–Liouville Problems

To avoid complicated notation, in what follows we denote intervals generically by I, regardless of whether they are closed, open, or half-open, finite or infinite. Specific intervals are described either in terms of their endpoints or by means of inequalities. The functions defined on such intervals are assumed to be integrable on them.

3.1. Definition. Let $\mathcal{X}$ be a space of functions defined on an interval I. A linear differential operator L acting on $\mathcal{X}$ is called *symmetric* on $\mathcal{X}$ if

$$\int_I \left[f_1(x)(Lf_2)(x) - f_2(x)(Lf_1)(x) \right] dx = 0$$

for any functions f_1 and f_2 in $\mathcal{X}$. ∎

3.2. Example. Consider the space $\mathcal{X}$ of functions that are twice continuously differentiable on $[0, 1]$ and equal to zero at $x = 0$ and $x = 1$. If L is the linear operator defined by $L \equiv d^2/dx^2$, then, using integration by parts, we find that for any f_1 and f_2 in $\mathcal{X}$,

$$\int_0^1 \left[f_1(Lf_2) - f_2(Lf_1) \right] dx = \int_0^1 (f_1 f_2'' - f_2 f_1'') \, dx$$

$$= \left[f_1 f_2' \right]_0^1 - \int_0^1 f_1' f_2' \, dx - \left[f_2 f_1' \right]_0^1 + \int_0^1 f_1' f_2' \, dx = 0.$$

Hence, L is symmetric on $\mathcal{X}$. ∎

3.3. Remark. An operator may be symmetric on a space of functions but not on another. If no restriction is imposed on the values at $x = 0$ and $x = 1$ of the functions in Example 3.2, then L is not symmetric on the new space $\mathcal{X}$. ∎

3.4. Definition. Let σ be a function defined on I, with the property that $\sigma(x) > 0$ for all x in I. Two functions f_1 and f_2, also defined on I, are called *orthogonal with weight* σ *on* I if

$$\int_I f_1(x) f_2(x) \sigma(x) \, dx = 0.$$

If $\sigma(x) = 1$, then f_1 and f_2 are simply called *orthogonal*. A set of functions that are pairwise orthogonal on I is called an *orthogonal set*. ∎

3.5. Example. The functions $f_1(x) = 1$ and $f_2(x) = 9x - 5$ are orthogonal with weight $\sigma(x) = x + 1$ on $[0,1]$ since

$$\int_0^1 (9x - 5)(x + 1) \, dx = \int_0^1 (9x^2 + 4x - 5) \, dx = 0.$$

Alternatively, we can say that the functions $f_1(x) = x + 1$ and $f_2(x) = 9x - 5$ are orthogonal on $[0,1]$. ∎

3.6. Example. The functions $\sin(3x)$ and $\cos(2x)$ are orthogonal on the interval $[-\pi, \pi]$ because

$$\int_{-\pi}^{\pi} \sin(3x) \cos(2x) \, dx = \int_{-\pi}^{\pi} \tfrac{1}{2} \big[\sin(5x) + \sin x \big] \, dx = 0. \ \blacksquare$$

3.7. Definition. Let L be a linear differential operator acting on a space $\mathcal{X}$ of functions defined on (a,b). An equation of the form

$$(Lf)(x) + \lambda \sigma(x) f(x) = 0, \quad a < x < b, \tag{3.1}$$

where λ is a (real) parameter and σ is a given function such that $\sigma(x) > 0$ for all x in (a,b), is called an *eigenvalue problem*. The numbers λ for which (3.1) has nonzero solutions in $\mathcal{X}$ are called *eigenvalues*; the corresponding nonzero solutions are called *eigenfunctions*. ∎

3.8. Theorem. *If the operator L of the eigenvalue problem (3.1) is symmetric, then*

(i) *all the eigenvalues λ are real;*

(ii) *the eigenvalues form an infinite sequence $\lambda_1 < \lambda_2 < \cdots < \lambda_n < \cdots$ such that $\lambda_n \to \infty$ as $n \to \infty$;*

(iii) *eigenfunctions associated with distinct eigenvalues are orthogonal with weight σ on (a,b).*

3.9. Definition. Let $[a,b]$ be a finite interval, let p, q, and σ be real-valued functions, and let κ_1, κ_2, κ_3, and κ_4 be real numbers such that
(i) p is continuously differentiable on $[a,b]$ and $p(x) > 0$ for all x in $[a,b]$;
(ii) q and σ are continuous on $[a,b]$ and $\sigma(x) > 0$ for all x in $[a,b]$;
(iii) κ_1, κ_2 are not both zero and κ_3, κ_4 are not both zero.
An eigenvalue problem of the form

$$\left[p(x)f'(x)\right]' + q(x)f(x) + \lambda\sigma(x)f(x) = 0, \quad a < x < b, \qquad (3.2)$$

with the boundary conditions (BCs)

$$\kappa_1 f(a) + \kappa_2 f'(a) = 0, \qquad (3.3)$$
$$\kappa_3 f(b) + \kappa_4 f'(b) = 0, \qquad (3.4)$$

is called a *regular Sturm–Liouville* (S–L) *eigenvalue problem.* ∎

3.10. Example. The choice

$$p(x) = 1, \; q(x) = 0, \; \sigma(x) = 1,$$
$$\kappa_1 = 1, \; \kappa_2 = 0, \; \kappa_3 = 1, \; \kappa_4 = 0, \; a = 0, \; b = L$$

generates the regular S–L problem

$$f''(x) + \lambda f(x) = 0, \quad 0 < x < L,$$
$$f(0) = 0, \quad f(L) = 0. \; ∎$$

3.11. Example. If p, q, σ, a, and b are as in Example 3.10 but $\kappa_1 = 0$, $\kappa_2 = 1$, $\kappa_3 = 0$, and $\kappa_4 = 1$, then we have the regular S–L problem

$$f''(x) + \lambda f(x) = 0, \quad 0 < x < L,$$
$$f'(0) = 0, \quad f'(L) = 0. \; ∎$$

3.12. Example. Taking p, q, σ, a, and b as in Example 3.10 and the coefficients

$$\kappa_1 = 1, \quad \kappa_2 = 0, \quad \kappa_3 = h, \quad \kappa_4 = 1,$$

we obtain the regular S–L problem

$$f''(x) + \lambda f(x) = 0, \quad 0 < x < L,$$
$$f(0) = 0, \quad f'(L) + hf(L) = 0. \quad \blacksquare$$

3.13. Example. The eigenvalue problem

$$f''(x) + 2f'(x) + \lambda f(x) = 0, \quad 0 < x < 1,$$
$$f(0) = 0, \quad f(1) = 0$$

is, in fact, a regular Sturm–Liouville problem. Adopting an "integrating factor"-type technique and using the coefficient of f', we choose the functions

$$p(x) = \sigma(x) = \exp\left\{ \int 2\,dx \right\} = e^{2x}, \quad q(x) = 0.$$

These functions satisfy the requirements of Definition 3.9 and, as can immediately be verified, the equality

$$\left(e^{2x} f'(x) \right)' + \lambda e^{2x} f(x) = 0, \quad 0 < x < 1,$$

is equivalent to the given equation. $\blacksquare$

3.14. Theorem. *The operator*

$$Lf = (pf')' + qf \tag{3.5}$$

defined by the left-hand side in (3.2) *and acting on a space* $\mathcal{X}$ *of functions satisfying* (3.3) *and* (3.4) *is symmetric on* $\mathcal{X}$.

Proof. Suppose, for simplicity, that $\kappa_1 \neq 0$ and $\kappa_4 \neq 0$. Then, by (3.3) and (3.4), any function f in $\mathcal{X}$ satisfies

$$f(a) = -\frac{\kappa_2}{\kappa_1} f'(a), \quad f'(b) = -\frac{\kappa_3}{\kappa_4} f(b).$$

Consequently, using (3.5) and integration by parts, we find that for any f_1 and f_2 in $\mathcal{X}$,

$$\int_a^b \left[f_1(Lf_2) - f_2(Lf_1)\right] dx$$

$$= \int_a^b \left\{ f_1\left[(pf_2')' + qf_2\right] - f_2\left[(pf_1')' + qf_1\right] \right\} dx$$

$$= \left[f_1(pf_2')\right]_a^b - \int_a^b pf_2'f_1' \, dx - \left[f_2(pf_1')\right]_a^b + \int_a^b pf_1'f_2' \, dx$$

$$= p(b)\left[f_1(b)f_2'(b) - f_2(b)f_1'(b)\right] - p(a)\left[f_1(a)f_2'(a) - f_2(a)f_1'(a)\right]$$

$$= p(b)\left[-\frac{\kappa_3}{\kappa_4} f_1(b)f_2(b) + \frac{\kappa_3}{\kappa_4} f_2(b)f_1(b)\right]$$

$$- p(a)\left[-\frac{\kappa_2}{\kappa_1} f_1'(a)f_2'(a) + \frac{\kappa_2}{\kappa_1} f_2'(a)f_1'(a)\right] = 0,$$

as required. All other combinations of nonzero constants κ_1, κ_2, κ_3, and κ_4 are treated similarly. ∎

3.15. Corollary. *The eigenvalues and eigenfunctions of a regular Sturm–Liouville problem have all the properties listed in Theorem* 3.8.

Before actually computing the eigenvalues and eigenfunctions of specific Sturm–Liouville problems, we derive a very useful formula relating these quantities. Multiplying (3.2) by f and integrating over (a,b), we easily see that

$$0 = \int_a^b \left[f(pf')' + qf^2\right] dx + \lambda \int_a^b \sigma f^2 \, dx$$

$$= \int_a^b \left[(pff')' - p(f')^2 + qf^2\right] dx + \lambda \int_a^b \sigma f^2 \, dx$$

$$= \left[pff'\right]_a^b - \int_a^b \left[p(f')^2 - qf^2\right] dx + \lambda \int_a^b \sigma f^2 \, dx.$$

Since f is an eigenfunction (nonzero solution) and $\sigma(x) > 0$ for $a < x < b$, the coefficient of λ in the last term is strictly positive. Hence, we can solve for λ and obtain

$$\lambda = \frac{\int\limits_a^b \left[p(f')^2 - qf^2\right] dx - \left[pff'\right]_a^b}{\int\limits_a^b \sigma f^2 \, dx}. \tag{3.6}$$

This expression is known as the *Rayleigh quotient*.

3.16. Example. For the regular S–L problem considered in Example 3.10 we have

$$\int\limits_a^b \left[p(x)(f')^2(x) - q(x)f^2(x)\right] dx = \int\limits_0^L (f')^2(x) \, dx \geq 0,$$

$$\left[p(x)f(x)f'(x)\right]_a^b = \left[f(x)f'(x)\right]_0^L = 0,$$

$$\int\limits_a^b \sigma(x)f^2(x) \, dx = \int\limits_0^L f^2(x) \, dx > 0,$$

so (3.6) yields

$$\lambda = \frac{\int\limits_0^L (f')^2(x) \, dx}{\int\limits_0^L f^2(x) \, dx} \geq 0. \tag{3.7}$$

It is clear that $\lambda = 0$ if and only if $f'(x) = 0$ on $[0, L]$, which means that $f(x) = \text{const}$. But this is unacceptable because, according to the BCs, the only constant solution is the zero solution. Consequently, $\lambda > 0$, and the general solution of the equation in Example 3.10 is

$$f(x) = C_1 \cos\left(\sqrt{\lambda}\,x\right) + C_2 \sin\left(\sqrt{\lambda}\,x\right), \quad C_1, C_2 = \text{const}. \tag{3.8}$$

Using the condition $f(0) = 0$, we find that $C_1 = 0$. Then the condition $f(L) = 0$ leads to

$$C_2 \sin\left(\sqrt{\lambda}\,L\right) = 0.$$

We cannot have $C_2 = 0$ because this would imply that $f = 0$ and we are seeking nonzero solutions; therefore, we must conclude that $\sin\left(\sqrt{\lambda}L\right) = 0$, from which we obtain $\sqrt{\lambda}L = n\pi$, $n = 1, 2, \ldots$, or

$$\lambda_n = \left(\frac{n\pi}{L}\right)^2, \quad n = 1, 2, \ldots.$$

These are the eigenvalues of the problem. The corresponding eigenfunctions (nonzero solutions) are

$$f_n(x) = \sin\frac{n\pi x}{L}, \quad n = 1, 2, \ldots.$$

For convenience, we have taken $C_2 = 1$; any other nonzero value of C_2 would simply produce a multiple of $\sin(n\pi x/L)$.

It is easy to see that the properties in Theorem 3.8 are satisfied. The orthogonality of the eigenfunctions on $[0, L]$ follows immediately from (2.5) since the integrand is an even function. ∎

3.17. Example. The same technique is used to compute the eigenvalues and eigenfunctions of the S–L problem in Example 3.11. Inequality (3.7) remains valid, but now we cannot reject the case $\lambda = 0$ since the functions $f(x) = c = \text{const}$ corresponding to it satisfy both BCs. Taking, say, $c = 1/2$, we conclude that the problem has the eigenvalue-eigenfunction pair

$$\lambda_0 = 0, \quad f_0(x) = 1/2.$$

As in Example 3.16, for $\lambda > 0$ the general solution of the equation is (3.8), so

$$f'(x) = -C_1\sqrt{\lambda}\sin\left(\sqrt{\lambda}x\right) + C_2\sqrt{\lambda}\cos\left(\sqrt{\lambda}x\right).$$

The condition $f'(0) = 0$ immediately yields $C_2 = 0$; therefore, $f'(L) = 0$ implies that

$$C_1\sqrt{\lambda}\sin\left(\sqrt{\lambda}L\right) = 0.$$

Since we want nonzero solutions, it follows that $\sin\left(\sqrt{\lambda}L\right) = 0$. This means that $\sqrt{\lambda}L = n\pi$, $n = 1, 2, \ldots$, so, by (3.8) with $C_2 = 0$ and $C_1 = 1$, the eigenvalue-eigenfunction pairs are

$$\lambda_n = \left(\frac{n\pi}{L}\right)^2, \quad f_n(x) = \cos\frac{n\pi x}{L}, \quad n = 1, 2, \ldots.$$

We remark that the full set of eigenvalues and eigenfunctions of the problem can be written as above, but with $n = 0, 1, 2, \ldots$.

Once again, we note that the eigenvalues and eigenfunctions satisfy the properties in Theorem 3.8. The orthogonality of the latter on $[0, L]$ follows from the first formula (2.3) and (2.4). ∎

3.18. Example. Suppose that $h > 0$ in the S–L problem considered in Example 3.12. Then, writing the second BC of that problem in the form $f'(L) = -hf(L)$, we see that

$$\left[p(x)f(x)f'(x)\right]_a^b = \left[f(x)f'(x)\right]_0^L = f(L)f'(L) - f(0)f'(0) = -hf^2(L).$$

Consequently, (3.6) yields

$$\lambda = \frac{\int\limits_0^L (f')^2(x)\,dx + hf^2(L)}{\int\limits_0^L f^2(x)\,dx} \geq 0,$$

with $\lambda = 0$ if and only if $f'(x) = 0$ on $[0, L]$ and $f(L) = 0$. But this implies that $f = 0$, which is not acceptable. Hence, the eigenvalues of this S–L problem are positive. To find them, we note that the general solution of the equation is again given by (3.8) and that the condition $f(0) = 0$ leads to $C_1 = 0$. Using the second BC, namely, $f'(L) + hf(L) = 0$, we arrive at

$$C_2\left[\sqrt{\lambda}\cos\left(\sqrt{\lambda}L\right) + h\sin\left(\sqrt{\lambda}L\right)\right] = 0.$$

For nonzero solutions we must have $C_2 \neq 0$; in other words,

$$\sqrt{\lambda}\cos\left(\sqrt{\lambda}L\right) + h\sin\left(\sqrt{\lambda}L\right) = 0.$$

Since $\cos\left(\sqrt{\lambda}L\right) \neq 0$ (otherwise $\sin\left(\sqrt{\lambda}L\right)$ would be zero as well, which is impossible), we can divide the above equality by $\cos\left(\sqrt{\lambda}L\right)$ and conclude that λ must be a solution of the transcendental equation

$$\tan\left(\sqrt{\lambda}L\right) = -\frac{\sqrt{\lambda}}{h} = -\frac{1}{hL}\left(\sqrt{\lambda}L\right).$$

Setting $\sqrt{\lambda}L = \zeta$, we see that ζ is given by the points of intersection of the graphs of the functions $y = \tan\zeta$ and $y = -\zeta/(hL)$.

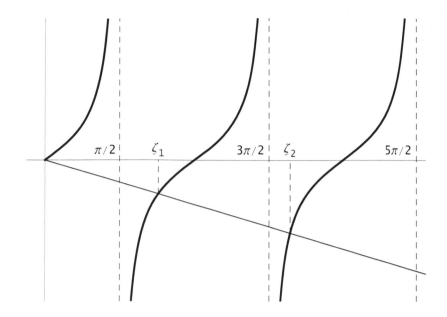

Fig. 3.1. The graphs of $y = \tan\zeta$ (heavy line) and $y = -\zeta/(hL)$ (light line).

As Fig. 3.1 shows, there are countably many such points ζ_n, so this S–L problem has an infinite sequence of positive eigenvalues

$$\lambda_n = \left(\frac{\zeta_n}{L}\right)^2, \quad n = 1, 2, \ldots, \quad \lambda_n \to \infty \text{ as } n \to \infty,$$

with corresponding eigenfunctions (given by (3.8) with $C_1 = 0$ and $C_2 = 1$)

$$f_n(x) = \sin\left(\sqrt{\lambda_n}\, x\right) = \sin\frac{\zeta_n x}{L}, \quad n = 1, 2, \ldots.$$

By Theorem 3.8(iii), the set $\{f_n\}_{n=1}^{\infty}$ is orthogonal on $[0, L]$. ∎

3.19. Example. For the S–L problem introduced in Example 3.13 we use a slightly modified approach. Here the roots of the characteristic equation $s^2 + 2s + \lambda = 0$ are

$$s_1 = -1 + \sqrt{1 - \lambda}, \quad s_2 = -1 - \sqrt{1 - \lambda}.$$

Since the nature of the roots changes according as $\lambda < 1$, $\lambda = 1$, or $\lambda > 1$, we consider these cases individually.

(i) If $\lambda < 1$, then $1 - \lambda > 0$; hence, s_1 and s_2 are real and distinct, and the general solution of the equation is

$$f(x) = C_1 e^{s_1 x} + C_2 e^{s_2 x}, \quad C_1, C_2 = \text{const.}$$

Using the conditions $f(0) = 0$ and $f(1) = 0$, we arrive at

$$C_1 + C_2 = 0, \quad C_1 e^{s_1} + C_2 e^{s_2} = 0.$$

Since $s_1 \neq s_2$, this system has only the solution $C_1 = C_2 = 0$; consequently, $f = 0$, and we conclude that there are no eigenvalues $\lambda < 1$.

(ii) If $\lambda = 1$, then

$$f(x) = (C_1 + C_2 x) e^{-x}, \quad C_1, C_2 = \text{const.}$$

The condition $f(0) = 0$ yields $C_1 = 0$, while $f(1) = 0$ gives $C_2 e^{-1} = 0$, which means that $C_2 = 0$. This generates $f = 0$, so $\lambda = 1$ is not an eigenvalue.

(iii) If $\lambda > 1$, then the general solution of the equation is

$$f(x) = e^{-x} \left[C_1 \cos\left(\sqrt{\lambda - 1}\, x\right) + C_2 \sin\left(\sqrt{\lambda - 1}\, x\right) \right], \quad C_1, C_2 = \text{const.}$$

Since $f(0) = 0$, we find that $C_1 = 0$, and from the other condition, $f(1) = 0$, it follows that

$$C_2 e^{-1} \sin\sqrt{\lambda - 1} = 0.$$

For nonzero solutions we must have

$$\sin\sqrt{\lambda - 1} = 0;$$

in other words,

$$\sqrt{\lambda - 1} = n\pi, \quad n = 1, 2, \ldots.$$

This implies that the eigenvalues of the problem are

$$\lambda_n = n^2 \pi^2 + 1, \quad n = 1, 2, \ldots,$$

with corresponding eigenfunctions

$$f_n(x) = e^{-x} \sin\left(\sqrt{\lambda_n - 1}\, x\right) = e^{-x} \sin(n\pi x), \quad n = 1, 2, \ldots. \ \blacksquare$$

Regular Sturm–Liouville problems have two important additional properties.

3.20. Theorem. (i) *Only one linearly independent eigenfunction f_n exists for each eigenvalue λ_n, $n = 1, 2, \ldots$.*

(ii) *The set $\{f_n\}_{n=1}^{\infty}$ of eigenfunctions is* complete; *that is, any piecewise smooth function u on $[a, b]$ has a* unique *generalized Fourier series (or eigenfunction) expansion of the form*

$$u(x) \sim \sum_{n=1}^{\infty} c_n f_n(x), \quad a \leq x \leq b, \tag{3.9}$$

which converges pointwise to $\frac{1}{2}\big[u(x-) + u(x+)\big]$ for $a \leq x \leq b$ (in particular, to $u(x)$ if u is continuous at x).

3.21. Remarks. (i) The coefficients c_n are computed by means of the properties mentioned in Theorem 3.20(i) and Theorem 3.8(iii). Thus, treating (3.9) formally as an equality, multiplying it by $f_m(x)\sigma(x)$, and integrating over $[a, b]$, we obtain

$$\int_a^b u(x) f_m(x) \sigma(x) \, dx = \sum_{n=1}^{\infty} c_n \int_a^b f_n(x) f_m(x) \sigma(x) \, dx$$

$$= c_m \int_a^b f_m^2(x) \sigma(x) \, dx.$$

Replacing m by n, we can now write

$$c_n = \frac{\int_a^b u(x) f_n(x) \sigma(x) \, dx}{\int_a^b f_n^2(x) \sigma(x) \, dx}, \quad n = 1, 2, \ldots. \tag{3.10}$$

The denominator in (3.10) cannot be zero for any n since the f_n are nonzero solutions of (3.2)–(3.4) and $\sigma(x) > 0$.

(ii) In the case of the S–L problems discussed in Examples 3.10 and 3.16, and in Examples 3.11 and 3.17, expansion (3.9) coincides, respectively, with the Fourier sine and cosine series of u, and (3.10) yields formulas (2.10) and (2.11). (For the problem in Examples 3.11 and 3.17, $n = 0, 1, 2, \ldots$.)

(iii) Using the method in Example 3.19, we can show that the eigenvalues and eigenfunctions of the more general regular S–L problem

$$f''(x) + af'(x) + bf(x) + \lambda cf(x) = 0, \quad 0 < x < L,$$
$$f(0) = 0, \quad f(L) = 0,$$

where a, b, and $c > 0$ are constants, are, respectively,

$$\lambda_n = \frac{1}{4c}\left(\frac{4n^2\pi^2}{L^2} + a^2 - 4b\right), \quad f_n(x) = e^{-(a/2)x}\sin\frac{n\pi x}{L}, \quad n = 1, 2, \dots. \quad \blacksquare$$

3.22. Example. Suppose that we want to expand the function

$$u(x) = x + 1, \quad 0 \le x \le 1,$$

in the eigenfunctions of the regular S–L problem discussed in Examples 3.12 and 3.18 with $L = h = 1$. The first five positive roots of the equation $\tan\zeta = -\zeta$ mentioned in the latter are, approximately,

$$\zeta_1 = 2.0288, \quad \zeta_2 = 4.9132, \quad \zeta_3 = 7.9787, \quad \zeta_4 = 11.0855, \quad \zeta_5 = 14.2074.$$

Then, using (3.10) with $\sigma(x) = 1$, we find the approximate coefficients

$$c_1 = 1.9184, \quad c_2 = 0.1572, \quad c_3 = 0.3390, \quad c_4 = 0.1307, \quad c_5 = 0.1696,$$

so (3.9) takes the form

$$u(x) \sim 1.9184\sin(2.0288x) + 0.1572\sin(4.9132x) + 0.3390\sin(7.9787x)$$
$$+ 0.1307\sin(11.0855x) + 0.1696\sin(14.2074x) + \cdots. \quad \blacksquare$$

3.23. Example. Consider the function

$$u(x) = e^{-x}, \quad 0 \le x \le 1.$$

If f_n are the eigenfunctions of the regular S–L problem discussed in Examples 3.13 and 3.19, that is,

$$f_n(x) = e^{-x}\sin(n\pi x), \quad n = 1, 2, \dots,$$

then (3.10) with $\sigma(x) = e^{2x}$ yields the coefficients

$$c_n = \left[1 - (-1)^n\right]\frac{2}{n\pi}, \quad n = 1, 2, \dots,$$

so expansion (3.9) is

$$u(x) \sim \sum_{n=1}^{\infty}\left[1 - (-1)^n\right]\frac{2}{n\pi}e^{-x}\sin(n\pi x). \quad \blacksquare$$

3.2. Other Problems

Many important Sturm–Liouville problems are not regular.

3.24. Definition. With the notation in Definition 3.9, an eigenvalue problem that consists in solving equation (3.2) with the conditions

$$f(a) = f(b), \quad f'(a) = f'(b) \tag{3.11}$$

is called a *periodic Sturm–Liouville problem.* ■

3.25. Remark. It can be shown that the differential operator L introduced in (3.5) is also symmetric on the new space of functions defined by conditions (3.11), so Theorem 3.8 is again applicable. However, in contrast to regular S–L problems, here we may have two linearly independent eigenfunctions for the same eigenvalue. Nevertheless, we can always choose a pair of eigenfunctions that are orthogonal with weight σ on $[a,b]$. For such a choice, Theorem 3.20(ii) remains valid. ■

3.26. Example. Consider the periodic Sturm–Liouville problem where $p(x) = 1$, $q(x) = 0$, $\sigma(x) = 1$, $a = -L$, and $b = L$, $L > 0$; that is,

$$f''(x) + \lambda f(x) = 0, \quad -L < x < L,$$
$$f(-L) = f(L), \quad f'(-L) = f'(L).$$

Using the Rayleigh quotient argument and the above BCs, we arrive again at inequality (3.7). We accept the case $\lambda = 0$ since, as in Example 3.17, the corresponding constant solutions satisfy both BCs; therefore, the problem has the eigenvalue-eigenfunction pair

$$\lambda_0 = 0, \quad f_0(x) = \tfrac{1}{2}.$$

For $\lambda > 0$, the general solution of the equation is

$$f(x) = C_1 \cos\left(\sqrt{\lambda}\,x\right) + C_2 \sin\left(\sqrt{\lambda}\,x\right), \quad C_1, C_2 = \text{const},$$

with derivative

$$f'(x) = -C_1\sqrt{\lambda}\sin\left(\sqrt{\lambda}\,x\right) + C_2\sqrt{\lambda}\cos\left(\sqrt{\lambda}\,x\right).$$

From the BCs and the fact that $\cos(-\alpha) = \cos\alpha$ and $\sin(-\alpha) = -\sin\alpha$ it follows that C_1 and C_2 satisfy the system

$$C_1 \cos\left(\sqrt{\lambda}L\right) - C_2 \sin\left(\sqrt{\lambda}L\right) = C_1 \cos\left(\sqrt{\lambda}L\right) + C_2 \sin\left(\sqrt{\lambda}L\right),$$
$$C_1 \sin\left(\sqrt{\lambda}L\right) + C_2 \cos\left(\sqrt{\lambda}L\right) = -C_1 \sin\left(\sqrt{\lambda}L\right) + C_2 \cos\left(\sqrt{\lambda}L\right),$$

which reduces to

$$C_1 \sin\left(\sqrt{\lambda}L\right) = 0, \quad C_2 \sin\left(\sqrt{\lambda}L\right) = 0.$$

Since we want nonzero solutions, we cannot have $\sin\left(\sqrt{\lambda}L\right) \neq 0$ because this would imply $C_1 = C_2 = 0$, that is, $f = 0$; hence, $\sin\left(\sqrt{\lambda}L\right) = 0$. As we have seen in previous examples, this yields the eigenvalues

$$\lambda_n = \left(\frac{n\pi}{L}\right)^2, \quad n = 1, 2, \dots.$$

Given that both C_1 and C_2 remain arbitrary, from the general solution we can extract (by taking $C_1 = 1$, $C_2 = 0$ and then $C_1 = 0$, $C_2 = 1$) for every λ_n the two linearly independent eigenfunctions

$$f_{1n}(x) = \cos\frac{n\pi x}{L}, \quad f_{2n}(x) = \sin\frac{n\pi x}{L}, \quad n = 1, 2, \dots,$$

which, as seen in Chapter 2, are orthogonal on $[-L, L]$. Expansion (3.9) in this case becomes

$$u(x) \sim \frac{1}{2}c_0 + \sum_{n=1}^{\infty}\left[c_{1n}f_{1n}(x) + c_{2n}f_{2n}(x)\right]$$
$$= \frac{1}{2}c_0 + \sum_{n=1}^{\infty}\left(c_{1n}\cos\frac{n\pi x}{L} + c_{2n}\sin\frac{n\pi x}{L}\right),$$

which we recognize as the full Fourier series of u. The coefficients c_0, c_{1n}, and c_{2n}, computed by means of (3.10) with $\sigma(x) = 1$, $a = -L$, $b = L$, and $n = 0, 1, 2, \dots$ (to allow for the additional eigenfunction f_0), coincide with those given by (2.7)–(2.9). ∎

3.27. Definition. An eigenvalue problem involving equation (3.2) is called a *singular Sturm–Liouville problem* if it exhibits one or more of the following features:

(i) $p(a) = 0$ or $p(b) = 0$ or both;

(ii) any of $p(x)$, $q(x)$, and $\sigma(x)$ becomes infinite as $x \to a+$ or as $x \to b-$ or both;

(iii) the interval (a, b) is infinite at a or at b or at both. ∎

3.28. Remark. In the case of a singular Sturm–Liouville problem, some new boundary conditions need to be chosen to ensure that the operator L defined by (3.5) remains symmetric. For example, if $p(a) = 0$ but $p(b) \neq 0$, then we may require $f(x)$ and $f'(x)$ to remain bounded as $x \to a+$ and to satisfy (3.4). If the interval where the equation holds is of the form (a, ∞), then we may require $f(x)$ and $f'(x)$ to be bounded as $x \to \infty$ and to satisfy (3.3). Other types of boundary conditions, dictated by analytic and/or physical necessity, may also be considered. ∎

3.29. Example. The eigenvalue problem

$$(x^2 + 1)f''(x) + 2xf'(x) + (\lambda - x)f(x) = 0, \quad x > 1,$$
$$f'(1) = 0, \quad f(x), f'(x) \text{ bounded as } x \to \infty,$$

is a singular S–L problem: the ODE can be written in the form (3.2) with $p(x) = x^2 + 1$, $q(x) = -x$, $\sigma(x) = 1$, and $(a, b) = (1, \infty)$, and the boundary conditions are of the type indicated in the latter part of Remark 3.28. ∎

A number of singular S–L problems occurring in the study of mathematical models give rise to important classes of functions. We discuss some of them explicitly.

3.3. Bessel Functions

Consider the singular S–L problem

$$x^2 f''(x) + xf'(x) + (\lambda x^2 - m^2)f(x) = 0, \quad 0 < x < a, \tag{3.12}$$
$$f(x), f'(x) \text{ bounded as } x \to 0+, \quad f(a) = 0, \tag{3.13}$$

where a is a given positive number and m is a nonnegative integer. Writing (3.12) in the form

$$\left[xf'(x) \right]' - \frac{m^2}{x} f(x) + \lambda x f(x) = 0,$$

we verify that this is (3.2) with $p(x) = x$, $q(x) = -m^2/x$, and $\sigma(x) = x$.

First, we consider the case $m = 0$, when $q = 0$ and (3.12) reduces to

$$xf''(x) + f'(x) + \lambda x f(x) = 0, \quad 0 < x < a.$$

Taking (3.13) into account, we see that

$$p(x)(f')^2(x) - q(x)f^2(x) = x(f')^2(x) \geq 0, \quad 0 < x < a,$$

$$[p(x)f(x)f'(x)]_0^a = [xf(x)f'(x)]_0^a = af(a)f'(a) - \lim_{x \to 0+} xf(x)f'(x) = 0;$$

consequently, according to the Rayleigh quotient (3.6),

$$\lambda = \frac{\int\limits_0^a x(f')^2(x)\,dx}{\int\limits_0^a xf^2(x)\,dx} \geq 0.$$

In fact, as this formula shows, we have $\lambda = 0$ if and only if $f'(x) = 0$, $0 < x \leq a$; that is, if and only if $f(x) = \text{const}$. But this is impossible, since the boundary condition $f(a) = 0$ would imply that f is the zero solution, which contradicts the definition of an eigenfunction. We must therefore conclude that $\lambda > 0$.

Now let $m \geq 1$. Multiplying (3.12) by $f(x)$ and integrating from 0 to a, we arrive at

$$\int\limits_0^a \left[x^2 f''(x)f(x) + xf'(x)f(x) + \lambda x^2 f^2(x) - m^2 f^2(x)\right] dx = 0. \quad (3.14)$$

Integration by parts, the interpretation of $f(x)$ and $f'(x)$ at $x = 0$ in the limiting sense as above, and (3.13) yield the equalities

$$\int\limits_0^a xf'(x)f(x)\,dx = \int\limits_0^a \frac{1}{2}x(f^2(x))'\,dx = \frac{1}{2}[xf^2(x)]_0^a - \frac{1}{2}\int\limits_0^a f^2(x)\,dx$$

$$= -\frac{1}{2}\int_0^a f^2(x)\,dx,$$

$$\int\limits_0^a x^2 f''(x)f(x)\,dx = [x^2 f'(x)f(x)]_0^a - \int\limits_0^a \left[x^2(f'(x))^2 + 2xf'(x)f(x)\right] dx$$

$$= -\int\limits_0^a \left[x^2(f'(x))^2 - f^2(x)\right] dx,$$

which, replaced in (3.14), lead to

$$\lambda = \frac{\int_0^a \left[x^2 (f'(x))^2 + \left(m^2 - \frac{1}{2} \right) f^2(x) \right] dx}{\int_0^a x^2 f^2(x) \, dx} \geq 0.$$

The same argument used for $m = 0$ rules out the value $\lambda = 0$; hence, $\lambda > 0$. Let $\xi = \sqrt{\lambda} x$ and $f(x) = f(\xi/\sqrt{\lambda}) = g(\xi)$. By the chain rule,

$$f'(x) = \sqrt{\lambda} g'(\xi), \quad f''(x) = \lambda g''(\xi),$$

and (3.12) becomes

$$\xi^2 g''(\xi) + \xi g'(\xi) + (\xi^2 - m^2) g(\xi) = 0, \tag{3.15}$$

which is *Bessel's equation of order m*. Two linearly independent solutions of this equation are the *Bessel functions of the first and second kind and order m* denoted by J_m and Y_m.

3.30. Remarks. (i) For $m \geq 1$, the J_m satisfy the recurrence relations

$$J_{m-1}(x) + J_{m+1}(x) = (2m/x) J_m(x), \quad J_{m-1}(x) - J_{m+1}(x) = 2 J_m'(x),$$

from which we deduce that

$$\left[x^m J_m(x) \right]' = x^m J_{m-1}(x).$$

Similar equalities are satisfied by the Y_m.

(ii) The J_m and Y_m can also be computed by means of the Bessel integrals

$$J_m(x) = \frac{1}{\pi} \int_0^\pi \cos(m\tau - x \sin \tau) d\tau = \frac{1}{2\pi} \int_{-\pi}^\pi e^{-i(m\tau - x \sin \tau)} d\tau,$$

$$Y_m(x) = \frac{1}{\pi} \int_0^\pi \sin(x \sin \tau - m\tau) d\tau - \frac{1}{\pi} \int_0^\infty \left[e^{m\tau} + (-1)^m e^{-m\tau} \right] e^{-x \sinh \tau} d\tau.$$

(iii) As $x \to 0+$,

$$J_0(x) \to 1, \quad J_m(x) \to 0, \quad m = 1, 2, \ldots,$$
$$Y_m(x) \to -\infty, \quad m = 0, 1, 2, \ldots.$$

(iv) J_m has the Taylor series expansion

$$J_m(x) = \sum_{k=0}^{\infty} \frac{(-1)^k}{k!(k+m)!} \left(\frac{x}{2}\right)^{2k+m}.$$

(v) The Bessel functions can be defined for any real, and even complex, order m. For example, when the order is a negative integer $-m$, we set

$$J_{-m}(x) = (-1)^m J_m(x).$$

(vi) When the order is an integer, the J_m are generated by the function

$$e^{(x/2)(t-1/t)} = \sum_{m=-\infty}^{\infty} J_m(x)t^m. \quad \blacksquare$$

The graphs of J_0, J_1 and Y_0, Y_1 are shown in Fig. 3.2.

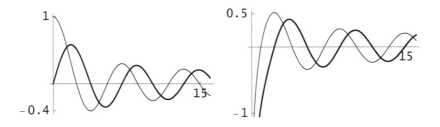

Fig. 3.2. Left: J_0 (light line) and J_1 (heavy line).
Right: Y_0 (light line) and Y_1 (heavy line).

Going back to (3.15), we can write its general solution as

$$g(\xi) = C_1 J_m(\xi) + C_2 Y_m(\xi), \quad C_1, C_2 = \text{const},$$

from which

$$f(x) = C_1 J_m\left(\sqrt{\lambda}\,x\right) + C_2 Y_m\left(\sqrt{\lambda}\,x\right).$$

In view of Remark 3.30(iii), the first condition (3.13) implies that $C_2 = 0$. The second one then yields $C_1 J_m(\sqrt{\lambda}\,a) = 0$. Since we want nonzero solutions, we must have $J_m(\sqrt{\lambda}\,a) = 0$. The function J_m has infinitely many positive zeros, which form a sequence ξ_{mn}, $n = 1, 2, \ldots$, such that $\xi_{mn} \to \infty$

as $n \to \infty$ for each $m = 0, 1, 2, \ldots$; therefore, the singular S–L problem (3.12), (3.13) has the eigenvalue-eigenfunction pairs

$$\lambda_{mn} = (\xi_{mn}/a)^2, \quad f_{mn}(x) = J_m(\xi_{mn}x/a), \quad n = 1, 2, \ldots.$$

It turns out that $\{f_{mn}\}_{n=1}^{\infty}$ is a complete system for each $m = 0, 1, 2, \ldots$, and that

$$\int_0^a J_m(\xi_{mn}x/a) J_m(\xi_{mp}x/a) x \, dx$$

$$= \begin{cases} 0, & n \neq p, \\ \frac{1}{2} a^2 J_{m+1}^2(\xi_{mn}), & n = p, \end{cases} \quad m = 0, 1, 2, \ldots. \quad (3.16)$$

These orthogonality (with weight $\sigma(x) = x$) relations allow us to expand any suitable function u in a generalized Fourier series of the form

$$u(x) \sim \sum_{n=1}^{\infty} c_{mn} J_m(\xi_{mn}x/a) \quad (3.17)$$

for each $m = 0, 1, 2, \ldots$. To find the coefficients c_{mn}, we follow the procedure described in Remark 3.21(i) with (a, b) replaced by $(0, a)$, c_n by c_{mn}, $f_n(x)$ by $J_m(\xi_{mn}x/a)$, and $\sigma(x)$ by x. Making use of (3.16), in the end we find that

$$c_{mn} = \frac{\int_0^a u(x) J_m(\xi_{mn}x/a) x \, dx}{\frac{1}{2} a^2 J_{m+1}^2(\xi_{mn})}, \quad n = 1, 2, \ldots. \quad (3.18)$$

3.31. Remark. Series (3.17) converges to $u(x)$ at all points x where u is continuous in the interval $(0, a)$. If $m = 0$, the point $x = 0$ is also included. If $m > 0$, then $x = 0$ is included if $u(0) = 0$. The point $x = a$ is included for any $m \geq 0$ if $u(a) = 0$. ∎

3.32. Example. Let

$$u(x) = 2x - 1,$$

and let $a = 1$ and $m = 0$. Keeping the computational approximation to four decimal places, we find that the first five zeros of $J_0(\xi)$ are

$$\xi_{01} = 2.4048, \quad \xi_{02} = 5.5201, \quad \xi_{03} = 8.6537,$$
$$\xi_{04} = 11.7915, \quad \xi_{05} = 14.9309,$$

which, when used in (3.18), generate the coefficients

$$c_{01} = 0.0329, \quad c_{02} = -1.1813, \quad c_{03} = 0.7452,$$
$$c_{04} = -0.7644, \quad c_{05} = 0.6146.$$

Therefore, by (3.17), we have the approximate expansion

$$u(x) \sim 0.0329\, J_0(2.4048x) - 1.1813\, J_0(5.5201x) + 0.7452\, J_0(8.6537x)$$
$$- 0.7644\, J_0(11.7915x) + 0.6146\, J_0(14.9309x) + \cdots. \quad (3.19)$$

We now construct a second expansion for u by taking $m = 1$. With the same type of approximation, the zeros of J_1 are

$$\xi_{11} = 3.8317, \quad \xi_{12} = 7.0156, \quad \xi_{13} = 10.1735,$$
$$\xi_{14} = 13.3237, \quad \xi_{15} = 16.4706,$$

so, by (3.18), we arrive at the coefficients

$$c_{11} = 0.3788, \quad c_{12} = -1.3827, \quad c_{13} = 0.4700,$$
$$c_{14} = -0.9199, \quad c_{15} = 0.4248$$

and the corresponding approximate expansion

$$u(x) \sim 0.3788\, J_1(3.8317x) - 1.3827\, J_1(7.0156x) + 0.4700\, J_1(10.1735x)$$
$$- 0.9199\, J_1(13.3237x) + 0.4248\, J_1(16.4706x) + \cdots. \quad (3.20)$$

In Fig. 3.3 the given function u (heavy line) is graphed together with the functions defined by the sum of the first five terms on the right-hand side in (3.19) and (3.20), respectively. ∎

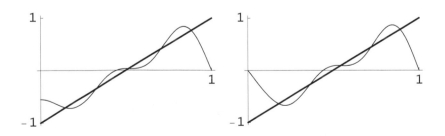

Fig. 3.3. Left: the approximation constructed with J_0.
Right: the approximation constructed with J_1.

3.4. Legendre Polynomials

Consider the singular S–L problem

$$(1 - x^2)f''(x) - 2xf'(x) + \lambda f(x) = 0, \quad -1 < x < 1, \qquad (3.21)$$

$$f(x), \ f'(x) \text{ bounded as } x \to -1+ \text{ and as } x \to 1-. \qquad (3.22)$$

Since (3.21) can be written as

$$\left[(1 - x^2)f'(x)\right]' + \lambda f(x) = 0,$$

it is clear that $p(x) = 1 - x^2$, $q(x) = 0$, and $\sigma(x) = 1$.

A detailed analysis of problem (3.21), (3.22), which is beyond the scope of this book, shows that its eigenvalues are

$$\lambda_n = n(n + 1), \quad n = 0, 1, 2, \ldots,$$

and that the general solution of (3.21) with $\lambda = \lambda_n$ is of the form

$$f_n(x) = C_1 P_n(x) + C_2 Q_n(x), \quad C_1, C_2 = \text{const}, \qquad (3.23)$$

where P_n are polynomials of degree n, called the *Legendre polynomials*. They are computed by means of the formula

$$P_n(x) = \frac{1}{2^n} \sum_{k=0}^{s} \frac{(-1)^k}{k!} \frac{(2n - 2k)!}{(n - 2k)!(n - k)!} x^{n-2k}, \quad n = 0, 1, 2, \ldots,$$

where

$$s = \begin{cases} n/2, & n \text{ even,} \\ (n - 1)/2, & n \text{ odd.} \end{cases}$$

The Q_n, called the *Legendre functions of the second kind,* have the representation

$$Q_n(x) = P_n(x) \int \frac{dx}{(1 - x^2)P_n^2(x)}, \quad n = 0, 1, 2, \ldots.$$

3.33. Remarks. (i) The P_n are given by the *Rodrigues formula*

$$P_n(x) = \frac{1}{2^n n!} \frac{d^n}{dx^n} (x^2 - 1)^n, \quad n = 0, 1, 2, \ldots;$$

thus, for $n = 0, 1, \ldots 5$,

$$P_0(x) = 1, \quad P_1(x) = x, \quad P_2(x) = \tfrac{1}{2}(-1 + 3x^2),$$
$$P_3(x) = \tfrac{1}{2}(-3x + 5x^3), \quad P_4(x) = \tfrac{1}{8}(3 - 30x^2 + 35x^4),$$
$$P_5(x) = \tfrac{1}{8}(15x - 70x^3 + 63x^5).$$

(ii) The P_n may also be computed by means of the generating function

$$(1 - 2tx + t^2)^{-1/2} = \sum_{n=0}^{\infty} P_n(x)t^n, \quad |x| < 1, \ |t| < 1.$$

(iii) The P_n satisfy the recursive relation

$$(n + 1)P_{n+1}(x) - (2n + 1)xP_n(x) + nP_{n-1}(x) = 0, \quad n = 1, 2, \ldots.$$

(iv) The set $\{P_n\}_{n=0}^{\infty}$ is orthogonal over $[-1, 1]$; specifically,

$$\int_{-1}^{1} P_n(x)P_m(x)\,dx = \begin{cases} 0, & n \neq m, \\ 2/(2n + 1), & n = m. \end{cases} \tag{3.24}$$

(v) $\{P_n\}_{n=0}^{\infty}$ is a complete system.

(vi) $Q_n(x)$ becomes unbounded as $x \to -1+$ and as $x \to 1-$. ∎

The graphs of P_4, P_5 and Q_4, Q_5 are shown in Fig. 3.4.

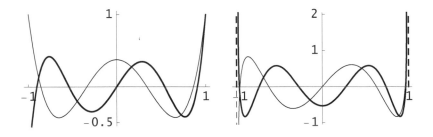

Fig. 3.4. Left: P_4 (light line) and P_5 (heavy line).
Right: Q_4 (light line) and Q_5 (heavy line).

In view of Remark 3.33(vi), f_n given by (3.23) satisfies the BCs (3.22) only if $C_2 = 0$; therefore, the eigenfunctions of the Sturm–Liouville problem

(3.21), (3.22) are

$$f_n(x) = P_n(x), \quad n = 0,1,2,\ldots.$$

As before, Remark 3.33(v) allows us to expand any suitable function u in a series of the form

$$u(x) \sim \sum_{n=0}^{\infty} c_n P_n(x), \tag{3.25}$$

where, applying the technique described in Remark 3.21(i) and taking (3.24) into account, we see that the coefficients c_n are computed by means of the formula

$$c_n = \frac{2n+1}{2} \int_{-1}^{1} f(x) P_n(x)\, dx, \quad n = 0,1,2,\ldots. \tag{3.26}$$

3.34. Remark. If u is piecewise smooth on $(-1,1)$, series (3.25) converges to $u(x)$ at all points x where u is continuous, and to $\frac{1}{2}\left[f(x-) + f(x+)\right]$ at the points x where u has a jump discontinuity. ∎

3.35. Example. Consider the function

$$u(x) = \begin{cases} 2x + 1, & -1 \le x \le 0, \\ 3, & 0 < x \le 1. \end{cases}$$

By (3.26) and the expressions of the P_n in Remark 3.33(i), we find the coefficients

$$c_0 = \tfrac{3}{2}, \quad c_1 = \tfrac{5}{2}, \quad c_2 = -\tfrac{5}{8}, \quad c_3 = -\tfrac{7}{8},$$
$$c_4 = \tfrac{3}{16}, \quad c_5 = \tfrac{11}{16},$$

so (3.25) yields the expansion

$$u(x) \sim \tfrac{3}{2} P_0(x) + \tfrac{5}{2} P_1(x) - \tfrac{5}{8} P_2(x) - \tfrac{7}{8} P_3(x)$$
$$+ \tfrac{3}{16} P_4(x) + \tfrac{11}{16} P_5(x) + \cdots. \tag{3.27}$$

The graphs of u and of the function defined by the sum of the first six terms on the right-hand side in (3.27) (the approximation of u) are shown for comparison in Fig. 3.5. ∎

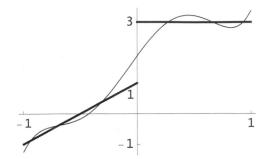

Fig. 3.5. The graphs of u (heavy line) and its approximation (light line).

3.5. Spherical Harmonics

A more general form of the singular S–L problem (3.21), (3.22) is obtained if (3.21) is replaced by the *associated Legendre equation*

$$\left[(1-x^2)f'(x)\right]' + \left(\lambda - \frac{m^2}{1-x^2}\right)f(x) = 0, \quad -1 < x < 1, \qquad (3.28)$$

where m is a nonnegative integer. Obviously, here we have $p(x) = 1 - x^2$, $q(x) = -m^2/(1-x^2)$, and $\sigma(x) = 1$.

It can be shown that the eigenvalue-eigenfunction pairs of problem (3.28), (3.22) are

$$\lambda_n = n(n+1), \quad n = m, m+1, \ldots,$$
$$f_n(x) = P_n^m(x) = (1-x^2)^{m/2} \frac{d^m}{dx^m} P_n(x) \qquad (3.29)$$
$$= \frac{(-1)^m}{2^n n!} (1-x^2)^{m/2} \frac{d^{n+m}}{dx^{n+m}} \left[(x^2-1)^n\right].$$

For each $m = 0, 1, \ldots, n$, the P_n^m, called the *associated Legendre functions*, satisfy the orthogonality formula

$$\int_{-1}^{1} P_n^m(x) P_k^m(x)\, dx = \begin{cases} 0, & n \neq k, \\ \dfrac{2(n+m)!}{(2n+1)(n-m)!}, & n = k. \end{cases} \qquad (3.30)$$

The definition of the P_n^m can be extended to negative-integer superscripts by means of the last expression on the right-hand side in (3.29). A straightforward computation shows that

$$P_n^{-m}(x) = \frac{(-1)^{-m}}{2^n n!}(1-x^2)^{-m/2}\frac{d^{n-m}}{dx^{n-m}}\left[(x^2-1)^n\right]$$

$$= (-1)^m \frac{(n-m)!}{(n+m)!}P_n^m(x). \qquad (3.31)$$

The spherical harmonics are complex-valued functions that occur in problems formulated in terms of spherical coordinates, which, as is well known, are linked to the Cartesian coordinates x, y, z by the expressions

$$x = r\cos\theta\sin\varphi, \quad y = r\sin\theta\sin\varphi, \quad z = r\cos\varphi.$$

Here, r, $r \geq 0$, is the radius, θ, $0 \leq \theta < 2\pi$, is the azimuthal angle, and φ, $0 \leq \varphi \leq \pi$, is the polar angle. The spherical harmonics are defined by

$$Y_n^m(\theta,\varphi) = P_n^m(\cos\varphi)e^{im\theta}, \quad n = 0,1,2,\ldots,$$

where $m = 0,1,\ldots,n$ and, by Euler's formula, $e^{im\theta} = \cos(m\theta) + i\sin(m\theta)$. In view of (3.30), it is not difficult to verify that

$$\int_0^{2\pi}\int_0^{\pi} Y_n^m(\theta,\varphi)\bar{Y}_k^l(\theta,\varphi)\sin\varphi\,d\varphi\,d\theta = \begin{cases} \dfrac{4\pi(n+m)!}{(2n+1)(n-m)!}, & m=l,\ n=k, \\ 0 & \text{otherwise,} \end{cases}$$

where the superposed bar denotes complex conjugation. The above equality allows us to normalize the Y_n^m by writing

$$Y_{n,m}(\theta,\varphi) = \left[\frac{(2n+1)(n-m)!}{4\pi(n+m)!}\right]^{1/2}P_n^m(\cos\varphi)e^{im\theta}. \qquad (3.32)$$

This is the form in which the spherical harmonics are usually known.

The $Y_{n,m}$ satisfy the orthonormality formula

$$\int_0^{2\pi}\int_0^{\pi} Y_{n,m}(\theta,\varphi)\bar{Y}_{k,l}(\theta,\varphi)\sin\varphi\,d\varphi\,d\theta = \begin{cases} 1, & m=l,\ n=k, \\ 0 & \text{otherwise.} \end{cases} \qquad (3.33)$$

In view of (3.31), we can also define

$$Y_{n,-m}(\theta,\varphi) = (-1)^m\bar{Y}_{n,m}(\theta,\varphi). \qquad (3.34)$$

3.36. Example. By the second formula (3.29), (3.32), and (3.34),

$$Y_{0,0}(\theta,\varphi) = \frac{1}{2}\sqrt{\frac{1}{\pi}},$$

$$Y_{1,-1}(\theta,\varphi) = \frac{1}{2}\sqrt{\frac{3}{2\pi}}\, e^{-i\theta}\sin\varphi = \frac{1}{2}\sqrt{\frac{3}{2\pi}}\,\frac{x-iy}{r},$$

$$Y_{1,0}(\theta,\varphi) = \frac{1}{2}\sqrt{\frac{3}{\pi}}\,\cos\varphi = \frac{1}{2}\sqrt{\frac{3}{\pi}}\,\frac{z}{r},$$

$$Y_{1,1}(\theta,\varphi) = -\frac{1}{2}\sqrt{\frac{3}{2\pi}}\, e^{i\theta}\sin\varphi = -\frac{1}{2}\sqrt{\frac{3}{2\pi}}\,\frac{x+iy}{r},$$

$$Y_{2,-2}(\theta,\varphi) = \frac{1}{4}\sqrt{\frac{15}{2\pi}}\, e^{-2i\theta}\sin^2\varphi = \frac{1}{4}\sqrt{\frac{15}{2\pi}}\,\frac{(x-iy)^2}{r^2},$$

$$Y_{2,-1}(\theta,\varphi) = \frac{1}{2}\sqrt{\frac{15}{2\pi}}\, e^{-i\theta}\sin\varphi\cos\varphi = \frac{1}{2}\sqrt{\frac{15}{2\pi}}\,\frac{(x-iy)z}{r^2},$$

$$Y_{2,0}(\theta,\varphi) = \frac{1}{4}\sqrt{\frac{5}{\pi}}\,(3\cos^2\varphi - 1) = \frac{1}{4}\sqrt{\frac{5}{\pi}}\,\frac{2z^2-x^2-y^2}{r^2},$$

$$Y_{2,1}(\theta,\varphi) = -\frac{1}{2}\sqrt{\frac{15}{2\pi}}\, e^{i\theta}\sin\varphi\cos\varphi = -\frac{1}{2}\sqrt{\frac{15}{2\pi}}\,\frac{(x+iy)z}{r^2},$$

$$Y_{2,2}(\theta,\varphi) = \frac{1}{4}\sqrt{\frac{15}{2\pi}}\, e^{2i\theta}\sin^2\varphi = \frac{1}{4}\sqrt{\frac{15}{2\pi}}\,\frac{(x+iy)^2}{r^2}. \quad\blacksquare$$

3.37. Remark. Any suitable function u defined on a sphere can be expanded in a spherical harmonics series of the form

$$u(\theta,\varphi) \sim \sum_{n=0}^{\infty}\sum_{m=-n}^{n} c_{n,m}Y_{n,m}(\theta,\varphi), \qquad (3.35)$$

which converges to $u(\theta,\varphi)$ at all points where u is continuous. To find the coefficients $c_{k,l}$, we multiply every term in (3.35) by $\bar{Y}_{k,l}(\theta,\varphi)\sin\varphi$, integrate over $[0,2\pi]\times[0,\pi]$, and take (3.33) into account. This (after k,l are replaced by n,m) leads to

$$c_{n,m} = \int_0^{2\pi}\int_0^{\pi} u(\theta,\varphi)\bar{Y}_{n,m}(\theta,\varphi)\sin\varphi\,d\varphi\,d\theta. \quad\blacksquare \qquad (3.36)$$

3.38. Example. Let $v(x,y,z) = x^2 + 2y$ be defined on the sphere with center at the origin and radius 1. In terms of spherical coordinates, this becomes $u(\theta,\varphi) = \cos^2\theta\sin^2\varphi + 2\sin\theta\sin\varphi$. Then, using (3.36) and the explicit expressions of the $Y_{n,m}$ given in Example 3.36, we find that

$$c_{0,0} = \frac{2\sqrt{\pi}}{3}, \quad c_{1,0} = 0, \quad c_{1,1} = c_{1,-1} = \frac{2\sqrt{2\pi}}{3}i,$$

$$c_{2,-1} = c_{2,1} = 0, \quad c_{2,0} = -\frac{2}{3}\sqrt{\frac{\pi}{5}}, \quad c_{2,2} = c_{2,-2} = \sqrt{\frac{2\pi}{15}},$$

so, by (3.35),

$$u(\theta,\varphi) = \frac{2\sqrt{\pi}}{3}Y_{0,0}(\theta,\varphi) + \frac{2\sqrt{2\pi}}{3}iY_{1,-1}(\theta,\varphi) + \frac{2\sqrt{2\pi}}{3}iY_{1,1}(\theta,\varphi)$$

$$+ \sqrt{\frac{2\pi}{15}}Y_{2,-2}(\theta,\varphi) - \frac{2}{3}\sqrt{\frac{\pi}{5}}Y_{2,0}(\theta,\varphi) + \sqrt{\frac{2\pi}{15}}Y_{2,2}(\theta,\varphi).$$

This is a terminating series because f is a linear combination of the spherical harmonics listed in Example 3.36. ∎

3.39. Example. If the function in Example 3.38 is replaced by

$$v(x,y,z) = \begin{cases} x, & z \geq 0, \\ yz, & z < 0, \end{cases}$$

then its equivalent form in spherical coordinates is

$$u(\theta,\varphi) = \begin{cases} \cos\theta\sin\varphi, & 0 \leq \varphi \leq \pi/2, \\ \sin\theta\sin\varphi\cos\varphi, & \pi/2 \leq \varphi \leq \pi. \end{cases}$$

In this case, (3.36) yields the coefficients

$$c_{0,0} = c_{1,0} = 0, \quad c_{1,1} = -\bar{c}_{1,-1} = -\sqrt{\frac{\pi}{6}}\left(2 + \frac{3}{8}i\right),$$

$$c_{2,-2} = c_{2,-2} = c_{2,2} = 0, \quad c_{2,1} = -\bar{c}_{2,-1} = -\sqrt{\frac{\pi}{30}}\left(\frac{15}{4} - i\right),$$

so expansion (3.35) becomes

$$u(\theta,\varphi) \sim \sqrt{\frac{\pi}{6}}\left(2 - \frac{3}{8}i\right)Y_{1,-1}(\theta,\varphi) - \sqrt{\frac{\pi}{6}}\left(2 + \frac{3}{8}i\right)Y_{1,1}(\theta,\varphi)$$

$$+ \sqrt{\frac{\pi}{30}}\left(\frac{15}{4} + i\right)Y_{2,-1}(\theta,\varphi) - \sqrt{\frac{\pi}{30}}\left(\frac{15}{4} - i\right)Y_{2,1}(\theta,\varphi) + \cdots. \ \blacksquare$$

3.40. Remark. A very general class of what we have called "suitable" functions in this chapter consists of square-integrable functions, namely, functions u such that

$$\int_D |u(x)|^2 \, dx < \infty,$$

where D is the domain where the functions are defined and x and dx are, respectively, a generic point in D and the element of length/area/volume, as appropriate. If series (3.9), (3.17), (3.25), and (3.35) are written collectively in the form

$$u(x) \sim \sum_{n=1}^{\infty} c_n f_n(x),$$

then they are convergent in the sense that

$$\lim_{N \to \infty} \int_D \left| u(x) - \sum_{n=1}^{N} c_n f_n(x) \right|^2 dx = 0.$$

As already mentioned, when u is sufficiently smooth, the series also converge to $u(x)$ in the classical pointwise sense at every x in D where u is continuous. To keep the notation simple, since the functions of interest to us meet the smoothness requirement, from now on we use the equality sign between them and their generalized Fourier series representations, although, strictly speaking, equality holds only at their points of continuity. ■

Exercises

In (1)–(6) verify that the given problem is a S–L eigenvalue problem and specify whether it is regular or singular.

(1) $f''(x) + \lambda f(x) = 0, \quad 0 < x < 1,$
$f(0) + 2f'(0) = 0, \quad f'(1) = 0.$

(2) $f''(x) - x f(x) + \lambda(x^2 + 1)f(x) = 0, \quad 0 < x < 1,$
$f(0) = 0, \quad f'(1) = 0.$

(3) $(x - 2)f''(x) + f'(x) + (1 + \lambda)f(x) = 0, \quad 0 < x < 2,$
$f'(0) = 0, \quad f(x), f'(x) \text{ bounded as } x \to 2-.$

(4) $f''(x) - f'(x) + \lambda f(x) = 0,$ $0 < x < 2,$
$f'(0) = 0,$ $f(2) - f'(2) = 0.$

(5) $2f''(x) + f'(x) + (\lambda + x)f(x) = 0,$ $0 < x < \infty,$
$f(0) = 0,$ $f(x), f'(x)$ bounded as $x \to \infty.$

(6) $x^2 f''(x) + x f'(x) + (2\lambda x - 1)f(x) = 0,$ $0 < x < 1,$
$f(x), f'(x)$ bounded as $x \to 0+,$ $f(1) - f'(1) = 0.$

In (7)–(16) compute the eigenvalues and eigenfunctions of the given regular S–L problem.

(7) $f''(x) + \lambda f(x) = 0,$ $0 < x < \pi,$
$f(0) = 0,$ $f'(\pi) = 0.$

(8) $f''(x) + \lambda f(x) = 0,$ $0 < x < 1,$
$f'(0) = 0,$ $f(1) = 0.$

(9) $f''(x) + \lambda f(x) = 0,$ $0 < x < 1,$
$f'(0) - f(0) = 0,$ $f(1) = 0.$

(10) $f''(x) + \lambda f(x) = 0,$ $0 < x < 1,$
$f'(0) = 0,$ $f'(1) + f(1) = 0.$

(11) $f''(x) + \lambda f(x) = 0,$ $0 < x < 1,$
$f'(0) - f(0) = 0,$ $f'(1) = 0.$

(12) $2f''(x) + (\lambda - 1)f(x) = 0,$ $0 < x < 2,$
$f(0) = 0,$ $f(2) = 0.$

(13) $f''(x) + 4f'(x) + 3\lambda f(x) = 0,$ $0 < x < 1,$
$f(0) = 0,$ $f(1) = 0.$

(14) $f''(x) - f'(x) + \lambda f(x) = 0,$ $0 < x < 1,$
$f'(0) = 0,$ $f'(1) = 0.$

(15) $2f''(x) + 3f'(x) + \lambda f(x) = 0,$ $0 < x < 1,$
$f(0) = 0,$ $f'(1) = 0.$

(16) $f''(x) - 5f'(x) + 2\lambda f(x) = 0,$ $0 < x < \pi/2,$
$f'(0) = 0,$ $f(\pi/2) = 0.$

In (17)–(24) construct the generalized Fourier series expansion for the given function u in the eigenfunctions $\left\{ \sin\left((2n - 1)x/2\right) \right\}_{n=1}^{\infty}$ of the S–L problem in Exercise 7.

(17) $u(x) = 1, \quad 0 \le x \le \pi.$

(18) $u(x) = 2x - 1, \quad 0 \le x \le \pi.$

(19) $u(x) = 3x + 2, \quad 0 \le x \le \pi.$

(20) $u(x) = \begin{cases} -1, & 0 \le x \le \pi/2, \\ 2, & \pi/2 < x \le \pi. \end{cases}$

(21) $u(x) = \begin{cases} 2, & 0 \le x \le \pi/2, \\ -3, & \pi/2 < x \le \pi. \end{cases}$

(22) $u(x) = \begin{cases} 2x - 1, & 0 \le x \le \pi/2, \\ -1, & \pi/2 < x \le \pi. \end{cases}$

(23) $u(x) = \begin{cases} 2, & 0 \le x \le \pi/2, \\ x + 1, & \pi/2 < x \le \pi. \end{cases}$

(24) $u(x) = \begin{cases} 2x + 1, & 0 \le x \le \pi/2, \\ 3 - 2x, & \pi/2 < x \le \pi. \end{cases}$

In (25)–(32) construct the generalized Fourier series expansion for the given function u in the eigenfunctions $\left\{ \cos\left((2n-1)\pi x/2\right)\right\}_{n=1}^{\infty}$ of the S–L problem in Exercise 8.

(25) $u(x) = 1/2, \quad 0 \le x \le 1.$

(26) $u(x) = x + 1, \quad 0 \le x \le 1.$

(27) $u(x) = 2 - x, \quad 0 \le x \le 1.$

(28) $u(x) = \begin{cases} 3, & 0 \le x \le 1/2, \\ 4, & 1/2 < x \le 1. \end{cases}$

(29) $u(x) = \begin{cases} -2, & 0 \le x \le 1/2, \\ 1, & 1/2 < x \le 1. \end{cases}$

(30) $u(x) = \begin{cases} 3 - 2x, & 0 \le x \le 1/2, \\ 2, & 1/2 < x \le 1. \end{cases}$

(31) $u(x) = \begin{cases} 1, & 0 \le x \le 1/2, \\ 2x - 3, & 1/2 < x \le 1. \end{cases}$

(32) $u(x) = \begin{cases} 1 - x, & 0 \le x \le 1/2, \\ 2x + 1, & 1/2 < x \le 1. \end{cases}$

In (33)–(40) compute the first five terms of the generalized Fourier series expansion for the given function u in the eigenfunctions of the S–L problem in Example 3.18 with $h = 1$ and the interval for x as indicated.

(33) $u(x) = 1, \quad 0 \le x \le 1.$

(34) $u(x) = 1 - 2x, \quad 0 \le x \le 1.$

(35) $u(x) = 2 - x, \quad 0 \le x \le 2.$

(36) $u(x) = \begin{cases} 1, & 0 \le x \le 1/2, \\ -3, & 1/2 < x \le 1. \end{cases}$

(37) $u(x) = \begin{cases} -2, & 0 \le x \le 1/2, \\ 4, & 1/2 < x \le 1. \end{cases}$

(38) $u(x) = \begin{cases} -1, & 0 \le x \le 1, \\ x + 2, & 1 < x \le 2. \end{cases}$

(39) $u(x) = \begin{cases} 2x, & 0 \le x \le 1/2, \\ 1, & 1/2 < x \le 1. \end{cases}$

(40) $u(x) = \begin{cases} 1 - x, & 0 \le x \le 1/2, \\ 1 + x, & 1/2 < x \le 1. \end{cases}$

In (41)–(48) compute the first five terms of the generalized Fourier series expansion for the given function u in the eigenfunctions $\left\{ e^{-2x} \sin(n\pi x)\right\}_{n=1}^{\infty}$ of the S–L problem in Exercise 13.

(41) $u(x) = 1, \quad 0 \le x \le 1.$

(42) $u(x) = x, \quad 0 \le x \le 1.$

(43) $u(x) = 2x + 1, \quad 0 \le x \le 1.$

(44) $u(x) = \begin{cases} -2, & 0 \le x \le 1/2, \\ -1, & 1/2 < x \le 1. \end{cases}$

(45) $u(x) = \begin{cases} 1, & 0 \le x \le 1/2, \\ -3, & 1/2 < x \le 1. \end{cases}$

(46) $u(x) = \begin{cases} 2, & 0 \le x \le 1/2, \\ 2x + 1, & 1/2 < x \le 1. \end{cases}$

(47) $u(x) = \begin{cases} x - 1, & 0 \le x \le 1/2, \\ -1, & 1/2 < x \le 1. \end{cases}$

(48) $u(x) = \begin{cases} x + 1, & 0 \le x \le 1/2, \\ 3 - 2x, & 1/2 < x \le 1. \end{cases}$

In (49)–(56) compute the first five terms of the generalized Fourier series expansion for the given function u in the eigenfunctions constructed with the Bessel functions (i) J_0 and (ii) J_1.

(49) $u(x) = 1, \quad 0 \le x \le 1.$

(50) $u(x) = 2x + 1, \quad 0 \le x \le 1.$

(51) $u(x) = 1 - 3x, \quad 0 \le x \le 1.$

(52) $u(x) = \begin{cases} 3, & 0 \le x \le 1/2, \\ 1, & 1/2 < x \le 1. \end{cases}$

(53) $u(x) = \begin{cases} -1, & 0 \le x \le 1/2, \\ 2, & 1/2 < x \le 1. \end{cases}$

(54) $u(x) = \begin{cases} x + 3, & 0 \le x \le 1/2, \\ 2, & 1/2 < x \le 1. \end{cases}$

(55) $u(x) = \begin{cases} 1, & 0 \le x \le 1/2, \\ 2 - 2x, & 1/2 < x \le 1. \end{cases}$

(56) $u(x) = \begin{cases} x - 2, & 0 \le x \le 1/2, \\ x + 1, & 1/2 < x \le 1. \end{cases}$

In (57)–(64) compute the first six terms of the generalized Fourier series expansion for the given function u in the Legendre polynomials P_n.

(57) $u(x) = x^2 - 3x + 4, \quad -1 \le x \le 1.$

(58) $u(x) = x^3 + 2x^2 - 3, \quad -1 \le x \le 1.$

(59) $u(x) = \begin{cases} -3, & -1 \le x \le 0, \\ 1, & 0 < x \le 1. \end{cases}$

(60) $u(x) = \begin{cases} 2, & -1 \le x \le 0, \\ -5, & 0 < x \le 1. \end{cases}$

(61) $u(x) = \begin{cases} -2, & -1 \le x \le 0, \\ x - 1, & 0 < x \le 1. \end{cases}$

(62) $u(x) = \begin{cases} 2x + 3, & -1 \le x \le 0, \\ 1, & 0 < x \le 1. \end{cases}$

(63) $u(x) = \begin{cases} 2x + 3, & -1 \le x \le 0, \\ 1 - 2x, & 0 < x \le 1. \end{cases}$

(64) $u(x) = \begin{cases} 3x + 1, & -1 \le x \le 0, \\ x + 4, & 0 < x \le 1. \end{cases}$

In (65)–(72) the given function v is defined on the sphere with center at the origin and radius 1. Writing it as $u(\theta, \varphi)$ in terms of spherical coordinates, compute the first nine terms of its expansion in the spherical harmonics. (Use the spherical harmonics listed in Example 3.36.)

(65) $v(x, y, z) = y^2 - 2z^2.$

(66) $v(x, y, z) = 3x^2 + xz.$

(67) $v(x,y,z) = \begin{cases} 0, & -1 \leq z < 0, \\ 1, & 0 \leq z \leq 1. \end{cases}$

(68) $v(x,y,z) = \begin{cases} 1, & -1 \leq z < 0, \\ -2, & 0 \leq z \leq 1. \end{cases}$

(69) $v(x,y,z) = \begin{cases} x - y, & -1 \leq z < 0, \\ 2, & 0 \leq z \leq 1. \end{cases}$

(70) $v(x,y,z) = \begin{cases} -1, & -1 \leq z < 0, \\ 2x + z, & 0 \leq z \leq 1. \end{cases}$

(71) $v(x,y,z) = \begin{cases} x^2 - 3y, & -1 \leq z < 0, \\ 2y + yz, & 0 \leq z \leq 1. \end{cases}$

(72) $v(x,y,z) = \begin{cases} xy, & -1 \leq z < 0, \\ xz + z^2, & 0 \leq z \leq 1. \end{cases}$

Chapter 4
Some Fundamental Equations of Mathematical Physics

The great majority of processes and phenomena in the real world are studied by means of mathematical models. An investigation of this kind generally comprises three stages.

(i) The model is set up in terms of mathematical expressions describing the quantitative relationships between the physical quantities involved.

(ii) The equations of the model are solved by means of various mathematical methods.

(iii) The mathematical results are interpreted from a physical point of view in relation to the original process.

In this chapter we show how three simple but fundamental mathematical models are derived, which involve partial differential equations. These models are important because each of them is representative of an entire class of linear second-order PDEs.

4.1. Definition. A *partial differential equation* is an equation that contains an unknown function of several variables and one or more of its partial derivatives. ∎

To simplify the notation, we will denote the partial derivatives of functions almost exclusively by means of subscripts; thus, for $u = u(x,t)$,

$$u_t \equiv \frac{\partial u}{\partial t}, \quad u_{tt} \equiv \frac{\partial^2 u}{\partial t^2}, \quad u_x \equiv \frac{\partial u}{\partial x}, \quad u_{xx} \equiv \frac{\partial^2 u}{\partial x^2}, \quad \text{etc.}$$

4.1. The Heat Equation

Heat conduction in a one-dimensional rod. Consider a heat-conducting rod in the shape of a thin cylinder described by the following physical parameters:

L : length;

A : cross-sectional area, assumed constant;

$E(x,t)$:   heat energy density (heat energy per unit volume);

$\varphi(x,t)$:   heat flux (heat energy per unit area, flowing to the right per unit time);

$q(x,t)$:   heat sources or sinks (heat energy per unit volume, generated or lost inside the rod per unit time).

Throughout what follows we assume that the lateral (cylindrical) surface of the rod is insulated; that is, no heat exchange takes place across it. We also use the generic term "sources" to describe both sources and sinks.

The physical law we are using to set up the mathematical model in this case is the law of conservation of heat energy, which states that

rate of change of heat energy in body

= heat flow across boundary per unit time

+ heat generated by sources per unit time.

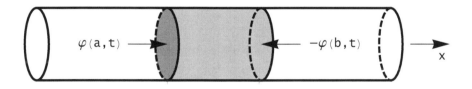

Fig. 4.1. An arbitrary segment of rod.

For an arbitrary length of rod between $x = a$ and $x = b$ on the x-axis (see Fig. 4.1), this law translates as

$$\frac{d}{dt}\int_a^b E(x,t)A\,dx = [\varphi(a,t) - \varphi(b,t)]A + \int_a^b q(x,t)A\,dx$$

$$= -\int_a^b \varphi_x(x,t)A\,dx + \int_a^b q(x,t)A\,dx, \quad t > 0.$$

basic theorem being used

the FTC (

The integral

undoes

the derivative

$$D_x = \frac{\partial x}{\partial x} = \frac{\partial}{\partial t}$$

Dividing through by A and moving all the terms to the left-hand side, we find that for $t > 0$ and any a, b, $0 < a < b < L$,

$$\int_a^b \left[E_t(x,t) + \varphi_x(x,t) - q(x,t) \right] dx = 0.$$

function of x & t

In view of the arbitrariness of a and b, this means that

Energy to temperature

$$E_t(x,t) = -\varphi_x(x,t) + q(x,t), \quad 0 < x < L, \, t > 0. \tag{4.1}$$

We need to introduce further physical parameters for the rod; thus,

$u(x,t)$:   temperature;

$c(x)$:   specific heat (the heat energy that raises the temperature of one unit of mass by one unit), assumed, for simplicity, to be independent of temperature;

$K_0(x)$:   thermal conductivity, also assumed to be independent of temperature;

$\rho(x)$:   mass density.

Then the total heat energy in the section of rod between $x = a$ and $x = b$ is

$$\int_a^b E(x,t)A\,dx = \int_a^b c(x)\rho(x)u(x,t)A\,dx, \quad t > 0;$$

as above, this means that

specific × mass × temperature
heat   density

relation with energy

$$E(x,t) = c(x)\rho(x)u(x,t), \quad 0 < x < L, \, t > 0. \tag{4.2}$$

At the same time, according to Fourier's law of heat conduction,

Flux

$$\varphi(x,t) = -K_0(x)u_x(x,t), \quad 0 < x < L, \, t > 0. \tag{4.3}$$

Using (4.2) and (4.3), we can now write the conservation of heat energy expressed by (4.1) as

constant   *source*

$$c(x)\rho(x)u_t(x,t) = \left(K_0(x)u_x(x,t) \right)_x + q(x,t), \quad 0 < x < L, \, t > 0. \tag{4.4}$$

specific heat is a constant   *if homogeneous it'll be zero*

Assuming that the rod is uniform (c, ρ, and K_0 are constants) and that there are no internal heat sources ($q = 0$), we see that (4.4) reduces to

$$u_t(x,t) = k u_{xx}(x,t), \quad 0 < x < L, \ t > 0, \tag{4.5}$$

where $k = K_0/(c\rho) = \text{const}$ is the *thermal diffusivity* of the rod. Equation (4.5) is called the *heat (diffusion) equation*.

If sources are present in the rod, then this equation is replaced by its nonhomogeneous version

$$u_t(x,t) = k u_{xx}(x,t) + q(x,t), \quad 0 < x < L, \ t > 0,$$

where q now incorporates the constant $1/(c\rho)$.

Initial condition. Since (4.5) is a first-order equation with respect to time, it needs only one initial condition, which is normally taken to be

$$u(x,0) = f(x), \quad 0 < x < L.$$

This means prescribing the initial distribution of temperature in the rod.

Boundary conditions. The equation is of second order with respect to the space variable, so we need two boundary conditions. There are three main types of physically meaningful conditions that are usually prescribed at the "near" endpoint $x = 0$ and "far" endpoint $x = L$.

(i) The temperature may be given at one endpoint; for example,

$$u(0,t) = \alpha(t), \quad t > 0.$$

(ii) If the rod is insulated at an endpoint, then the condition must mean that the heat flux there is zero. In view of (4.3), this is equivalent to the derivative u_x being equal to zero; for example,

$$u_x(L,t) = 0, \quad t > 0.$$

More generally, if the heat flux through the endpoint $x = L$ is prescribed, the above condition becomes

$$u_x(L,t) = \beta(t), \quad t > 0.$$

physical
law of heating/cooling

(iii) When one of the endpoints is in contact with another medium, we use Newton's law of cooling, which states that the heat flux at that endpoint is proportional to the difference between the temperature of the rod and the temperature of the external medium; for example,

$$K_0 u_x(0,t) = H[u(0,t) - U(t)], \quad t > 0,$$

where $U(t)$ is the (known) temperature of the external medium and $H > 0$ is the *heat transfer coefficient*. Owing to the convention concerning the direction of the heat flux, at the far endpoint this type of condition becomes

$$-K_0 u_x(L,t) = H[u(L,t) - U(t)], \quad t > 0.$$

4.2. Remarks. (i) Only one boundary condition is prescribed at each endpoint.

(ii) The boundary condition at $x = 0$ may differ from that at $x = L$.

(iii) It is easily verified that the heat equation is linear.

(iv) The one-dimensional heat equation is the simplest example of a so-called *parabolic equation*. ∎

The model. As we have seen, the mathematical model for heat conduction in a rod consists of several elements.

4.3. Definition. A partial differential equation (PDE) and the initial conditions (ICs) and boundary conditions (BCs) associated with it form an *initial boundary value problem* (IBVP). If only initial conditions or boundary conditions are present, then we have an *initial value problem* (IVP) or a *boundary value problem* (BVP), respectively. ∎ *Dragons*

4.4. Example. The IBVP modeling heat conduction in a one-dimensional uniform rod with sources, insulated lateral surface, and temperature prescribed at both endpoints is of the form

$$u_t(x,t) = k u_{xx}(x,t) + q(x,t), \quad 0 < x < L, \, t > 0, \qquad \text{(PDE)}$$

$$u(0,t) = \alpha(t), \quad u(L,t) = \beta(t), \quad t > 0, \qquad \text{(BCs)}$$

$$u(x,0) = f(x), \quad 0 < x < L, \qquad \text{(IC)}$$

where q, α, β, and f are given functions. ∎

The solution is some surface describing the functions action
(heat gain or loss)

4.5. Example. If the near endpoint is insulated and the far one is kept in a medium of constant zero temperature, and if the rod contains no sources, then the corresponding IBVP is

$$u_t(x,t) = ku_{xx}(x,t), \quad 0 < x < L,\ t > 0, \qquad \text{(PDE)}$$
$$u_x(0,t) = 0, \quad u_x(L,t) + hu(L,t) = 0, \quad t > 0, \qquad \text{(BCs)}$$
$$u(x,0) = f(x), \quad 0 < x < L, \qquad \text{(IC)}$$

where f is a given function and h is a known positive constant. ∎

4.6. Definition. By a *classical solution* of an IBVP we understand a function $u(x,t)$ that satisfies the PDE, BCs, and ICs pointwise everywhere in the region where the problem is formulated. For brevity, in what follows we will refer to such functions simply as "solutions". It is obvious that an IBVP will have solutions in this sense only if the data functions have a certain degree of smoothness. ∎

4.7. Example. Suppose that q, α, β, and f in Example 4.4 are continuous functions. Then a solution of that IBVP has the following properties:

(i) it is continuously differentiable with respect to t and twice continuously differentiable with respect to x at all points in the domain (semi-infinite strip) in the (x,t)-plane defined by

$$G = \{(x,t) : 0 < x < L,\ t > 0\},$$

and satisfies the PDE at all points in G;

(ii) its continuity extends to the "spatial" boundary lines of G, that is, to the two half-lines

$$\partial G_x = \{(x,t) : x = 0,\ t > 0 \text{ or } x = L,\ t > 0\},$$

and it satisfies the appropriate BC at every point on ∂G_x;

(iii) its continuity also extends to the "temporal" boundary line of G

$$\partial G_t = \{(x,t) : 0 < x < L,\ t = 0\},$$

and it satisfies the IC at all points on ∂G_t. ∎

4.8. Remarks. (i) We note that nothing is said in Example 4.7 about the behavior of u at the "corner" points $(0,0)$ and $(L,0)$ of G. If, for example,

$$\lim_{t \to 0} u(0,t) = \lim_{x \to 0} u(x,0),$$

that is, if

$$\lim_{t \to 0} \alpha(t) = \lim_{x \to 0} f(x),$$

then u can be defined at $(0,0)$ by the common value of the above limits. If, on the other hand, the two limits are distinct, then u cannot be defined at $(0,0)$ in the classical sense. In the applications that follow we will not concern ourselves with these two "corner" points.

(ii) Suppose that f has a jump discontinuity at a point $(x_0,0)$ on ∂G_t. In this case it is impossible to find a solution u that satisfies the IC at this point in the classical sense; however, we might find a type of solution that does so in some "average" way.

(iii) The definition of a solution can be generalized in the obvious manner to IBVPs with more than two independent variables. ∎

4.9. Theorem.   *If the functions α, β, and f are sufficiently smooth to ensure that u, u_t, u_x, and u_{xx} are continuous in G and up to the boundary of G, including the two "corner" points, then the IBVP in Example 4.4 has at most one solution.*

Proof. Suppose that there are two solutions u_1 and u_2 of the problem in question. Then, owing to linearity, it is obvious that $u = u_1 - u_2$ is a solution of the fully homogeneous IBVP; that is, for any arbitrarily fixed number $T > 0$ the function u satisfies

$$u_t(x,t) = k u_{xx}(x,t), \quad 0 < x < L, \, 0 < t < T,$$
$$u(0,t) = 0, \quad u(L,t) = 0, \quad 0 < t < T,$$
$$u(x,0) = 0, \quad 0 < x < L.$$

Multiplying the PDE by u, integrating over $[0,L]$, and using the BCs, the smoothness properties of the solutions for $0 \le x \le L$ and $0 \le t \le T$, and integration by parts, we find that

$$0 = \int_0^L \left(u u_t - k u u_{xx} \right) dx$$

$$= \int_0^L \left[\tfrac{1}{2}(u^2)_t + k u_x^2 \right] dx - k \left[u u_x \right]_{x=0}^{x=L} = \int_0^L \left[\tfrac{1}{2}(u^2)_t + k u_x^2 \right] dx;$$

therefore,

$$\frac{1}{2} \frac{d}{dt} \int_0^L u^2 \, dx = -k \int_0^L u_x^2 \, dx \le 0, \quad 0 \le t \le T,$$

which means that the function

$$W(t) = \int_0^L u^2(x,t) \, dx, \quad 0 \le t \le T,$$

is nonincreasing. Since $W(t) \ge 0$ and $W(0) = 0$ (because of the IC satisfied by u), this is possible only if $W(t) = 0$, $0 \le t \le T$. In view of the nonnegative integrand in the definition of W above and the arbitrariness of $T > 0$, we then conclude that $u = 0$; in other words, the solutions u_1 and u_2 coincide. ∎

4.10. Remarks. (i) Theorem 4.9 states that if the IBVP in Example 4.4 has a solution, then that solution is unique. The existence issue is resolved in Chapter 5, where we actually construct the solution.

(ii) The properties required of the solution in Theorem 4.9 are stronger than those mentioned in Example 4.7.

(iii) Uniqueness can also be proved for the solutions of IBVPs with other types of BCs.

(iv) It is easily seen that the change of variables

$$\xi = \frac{x}{L}, \quad \tau = \frac{k}{L^2} t, \quad u(x,t) = u(L\xi, L^2\tau/k) = v(\xi,\tau)$$

reduces the heat equation to the form

$$v_\tau(\xi,\tau) = v_{\xi\xi}(\xi,\tau), \quad 0 < \xi < 1, \ \tau > 0.$$

Hence, without loss of generality, in most of our applications we consider this simpler version, with x, t, and u in place of ξ, τ, and v, respectively. ∎

4.2. The Laplace Equation

The higher-dimensional heat equation. For simplicity, we study the case of two space dimensions. Consider a heat-conducting body occupying a finite region D in the (x,y)-plane, bounded by a smooth, simple, closed curve ∂D, and let R be an arbitrary element of D with a smooth boundary ∂R (see Fig. 4.2). We adopt the same notation for the physical parameters of this body as in Section 4.1.

Heat flow depends on direction, so here the heat flux is a vector $\vec{\varphi}$, which we can write in the form

$$\vec{\varphi} = \text{ normal component } + \text{ tangential component}$$
$$= (\vec{\varphi} \cdot \vec{n})\vec{n} + (\vec{\varphi} \cdot \vec{\tau})\vec{\tau}, \quad |\vec{n}| = |\vec{\tau}| = 1,$$

where $\vec{n}$ and $\vec{\tau}$ are the unit normal and tangent vectors to ∂R, directed as shown in Fig. 4.2. Clearly, the tangential component makes no contribution to the heat exchange between R and the rest of the body.

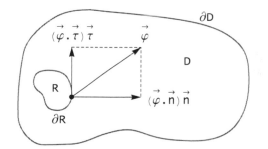

Fig. 4.2. A two-dimensional body configuration.

As in the case of a rod, the conservation of heat energy states that

rate of change of heat energy
    = heat flow across boundary per unit time
        + heat generated by sources per unit time.

This translates mathematically as

$$\frac{d}{dt} \int_R E(x,y,t)\, dA = -\int_{\partial R} \vec{\varphi}(x,y,t) \cdot \vec{n}(x,y)\, ds + \int_R q(x,y,t)\, dA,$$

where dA and ds are, respectively, the elements of area and arc length. Taking (4.2) into account and applying the divergence theorem, that is,

$$\int_R (\operatorname{div}\vec{\varphi})(x,y,t)\, dA = \int_{\partial R} \vec{\varphi}(x,y,t)\cdot \vec{n}(x,y)\, ds,$$

we can rewrite the above equality as

$$\int_R \left[c(x,y)\rho(x,y)u_t(x,y,t) + (\operatorname{div}\vec{\varphi})(x,y,t) - q(x,y,t) \right] dA = 0.$$

In view of the arbitrariness of R, we conclude that for (x,y) in D and $t > 0$,

$$c(x,y)\rho(x,y)u_t(x,y,t) = -(\operatorname{div}\vec{\varphi})(x,y,t) + q(x,y,t). \qquad (4.6)$$

Once again, we now use Fourier's law of heat conduction, which in two dimensions is

$$\vec{\varphi}(x,y,t) = -K_0(x,y)(\operatorname{grad} u)(x,y,t);$$

consequently, (4.6) becomes

$$c(x,y)\rho(x,y)u_t(x,y,t) = \operatorname{div}\big(K_0(x,y)(\operatorname{grad} u)(x,y,t)\big) + q(x,y,t).$$

For a uniform body (c, ρ, and K_0 are constants) with no internal sources ($q = 0$), the last formula reduces to

$$u_t(x,y,t) = k(\operatorname{div}\operatorname{grad} u)(x,y,t), \quad k = \frac{K_0}{c\rho} = \text{const},$$

or

$$u_t(x,y,t) = k(\Delta u)(x,y,t), \quad (x,y)\text{ in } D,\ t > 0, \qquad (4.7)$$

where the differential operator Δ, defined for smooth functions $v(x,y)$ by

$$(\Delta v)(x,y) = v_{xx}(x,y) + v_{yy}(x,y),$$

is called the *Laplacian*. Equality (4.7) is the two-dimensional heat (diffusion) equation.

If the body contains sources, then (4.7) is replaced by its nonhomogeneous counterpart

$$u_t(x,y,t) = k(\Delta u)(x,y,t) + q(x,y,t), \quad (x,y)\text{ in } D,\ t > 0,$$

where, as before, q now incorporates the factor $1/(c\rho)$.

Initial condition. Here this takes the form

$$u(x,y,0) = f(x,y), \quad (x,y) \text{ in } D,$$

and represents the initial distribution of temperature in the body.

Boundary conditions. The main types of BCs are similar in nature to those for a rod.

(i) When the temperature is prescribed on the boundary,

$$u(x,y,t) = \alpha(x,y,t), \quad (x,y) \text{ on } \partial D, \ t > 0.$$

This is called a *Dirichlet boundary condition.*

(ii) If the flux through the boundary is prescribed, then

$$(\operatorname{grad} u)(x,y,t) \cdot \vec{n}(x,y) = \beta(x,y,t), \quad (x,y) \text{ on } \partial D, \ t > 0,$$

or, equivalently,

$$u_n(x,y,t) = \beta(x,y,t), \quad (x,y) \text{ on } \partial D, \ t > 0,$$

where $\vec{n}$ is the unit outward normal to ∂D and $u_n = \partial u/\partial n$. This is called a *Neumann boundary condition.* In particular, when the boundary is insulated we have

$$u_n(x,y,t) = 0, \quad (x,y) \text{ on } \partial D, \ t > 0.$$

(iii) Newton's law of cooling takes the form

$$-K_0 u_n(x,y,t) = H\big[u(x,y,t) - U(x,y,t)\big], \quad (x,y) \text{ on } \partial D, \ t > 0.$$

This is referred to as a *Robin boundary condition.*

Sometimes we may have one type of condition prescribed on some part of the boundary, and another type on the remaining part.

Equilibrium temperature. An equilibrium (steady-state) temperature is a time-independent solution $u = u(x,y)$ of (4.7); in other words, it is a function satisfying the *Laplace equation*

$$(\Delta u)(x,y) = 0, \quad (x,y) \text{ in } D,$$

and a time-independent boundary condition, for example,

$$u(x,y) = \alpha(x,y), \quad (x,y) \text{ on } \partial D.$$

If steady-state sources $q(x,y)$ are present in the body, then the equilibrium temperature satisfies the *Poisson equation*

$$(\Delta u)(x,y) = -\frac{1}{k} q(x,y), \quad (x,y) \text{ in } D.$$

constant = usually absorbed into equation

In what follows, when we discuss the Poisson equation—that is, the nonhomogeneous Laplace equation—we omit the factor $-1/k$ on the right-hand side, regarding it as incorporated in the source term q.

An equilibrium temperature may, or may not, exist.

4.11. Remarks. (i) Problems in three space variables are formulated similarly, with $u = u(x,y,z,t)$.

(ii) In polar coordinates $x = r\cos\theta$, $y = r\sin\theta$, for $u = u(r,\theta)$ we have

$$\Delta u = r^{-1}(ru_r)_r + r^{-2}u_{\theta\theta} = u_{rr} + r^{-1}u_r + r^{-2}u_{\theta\theta}.$$

(iii) In cylindrical coordinates $x = r\cos\theta$, $y = r\sin\theta$, z, the Laplacian of $u = u(r,\theta,z)$ takes the form

$$\Delta u = r^{-1}(ru_r)_r + r^{-2}u_{\theta\theta} + u_{zz}.$$

(iv) In circularly (axially) symmetric problems the function u is independent of θ, so

$$\Delta u = r^{-1}(ru_r)_r + u_{zz} = u_{rr} + r^{-1}u_r + u_{zz}.$$

(v) In spherical coordinates $x = r\cos\theta\sin\varphi$, $y = r\sin\theta\sin\varphi$, $z = r\cos\varphi$, for $u = u(r,\theta,\varphi)$ we have

$$\Delta u = r^{-2}(r^2 u_r)_r + (r^2\sin^2\varphi)^{-1}u_{\theta\theta} + (r^2\sin\varphi)^{-1}((\sin\varphi)u_\varphi)_\varphi.$$

(vi) It is obvious that the Laplace and Poisson equations are linear.

(vii) The Laplace equation is the simplest example of a so-called *elliptic equation*. ∎

4.12. Example. The equilibrium temperature distribution in a thin, uniform finite plate with sources, insulated upper and lower faces, and prescribed temperature on the boundary is modeled by the BVP

$$(\Delta u)(x,y) = q(x,y), \quad (x,y) \text{ in } D, \tag{PDE}$$

$$u(x,y) = \alpha(x,y), \quad (x,y) \text{ on } \partial D, \tag{BC}$$

where ∂D is a simple closed curve. The Laplacian is written either in Cartesian or in polar coordinates, depending on the geometry of the plate. When polar coordinates are used, the nature of the ensuing PDE may require us to consider additional "boundary" conditions, suggested by the physics of the process. ■

4.13. Remark. A (classical) solution of the BVP in Example 4.12 has the following properties:

(i) it is twice continuously differentiable in D and satisfies the PDE at every point in D;

(ii) it is continuous up to the boundary ∂D of D and satisfies the BC at every point of ∂D.

As mentioned in Remark 4.8(ii), a boundary data function with discontinuities may generate a less smooth solution. ■

4.14. Theorem. *If u is a solution of the BVP in Example 4.12, then u attains its maximum and minimum values on the boundary ∂D of D.*

Proof. We anticipate a result established in Section 5.3, according to which (see Remark 5.16) the temperature at the center of a circular disk is equal to the average of the temperature on its boundary circle.

Suppose that the maximum of the solution u occurs at a point P inside D. Regarding P as the center of a small disk lying within D, we deduce that the value of u at P is the average of all the values of u on the boundary circle of that disk, and so it cannot be greater than all of those values. We have thus arrived at a contradiction, which implies that u must attain its maximum on the boundary ∂D. Considering $v = -u$, we also conclude that u attains its minimum on ∂D.

This assertion is known as the *maximum principle* for the Laplace equation. ■

4.15. Corollary. *If a solution u of the BVP in Example* 4.12 *is identically zero on ∂D, then u is also identically zero in D.*

Proof. By Theorem 4.14, the maximum and minimum of u are zero, since they occur at points on ∂D. Hence, u is zero in D. ■

4.16. Theorem. *The BVP in Example* 4.12 *has at most one solution.*

Proof. Suppose that there are two solutions u_1 and u_2. Because of the linearity of the PDE and BC, the difference $u = u_1 - u_2$ is a solution of the fully homogenous BVP

$$(\Delta u)(x,y) = 0, \quad (x,y) \text{ in } D,$$
$$u(x,y) = 0, \quad (x,y) \text{ on } \partial D;$$

therefore, by Corollary 4.15, $u = 0$, which means that u_1 and u_2 are the same function. ■

4.17. Definition. A solution of a BVP (IVP, IBVP) is said to *depend continuously on the data* (or to be *stable*) if a small variation in the data (BCs, ICs, nonhomogeneous term in the PDE) induces only a small variation in the solution. ■

4.18. Theorem. *The solution of the BVP in Example* 4.12 *depends continuously on the boundary data.*

Proof. Consider the BVPs

$$(\Delta u)(x,y) = q(x,y), \quad (x,y) \text{ in } D,$$
$$u(x,y) = \alpha(x,y), \quad (x,y) \text{ on } \partial D,$$

and

$$(\Delta v)(x,y) = q(x,y), \quad (x,y) \text{ in } D,$$
$$v(x,y) = \alpha(x,y) + \varepsilon(x,y), \quad (x,y) \text{ on } \partial D,$$

where ε is a small perturbation of the boundary function α. Then, taking the linearity of the PDE and BC into account, we see that $w = u - v$ satisfies

$$(\Delta w)(x,y) = 0, \quad (x,y) \text{ in } D,$$
$$w(x,y) = \varepsilon(x,y), \quad (x,y) \text{ on } \partial D.$$

By Theorem 4.14,

$$\min \varepsilon(x,y) \le w(x,y) = u(x,y) - v(x,y) \le \max \varepsilon(x,y), \quad (x,y) \text{ in } D;$$

that is, the perturbation of the solution u is small. ∎

4.19. Remark. A BVP (IVP, IBVP) is said to be *well posed* if it has
a unique solution that depends continuously on the data. In the following
chapters we construct a solution for the BVP in Example 4.12. According to
Theorem 4.16, this will be the only solution of that problem. Furthermore,
by Theorem 4.18, this solution depends continuously on the boundary data.
Thus, we conclude that the Dirichlet problem for the Laplace equation is
well posed. ∎

4.3. The Wave Equation

Vibrating strings. Consider the motion of a tightly stretched elastic
string, in which the horizontal displacement of the points of the string is
negligible (see Fig. 4.3).

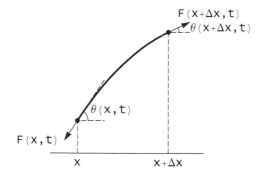

Fig. 4.3. Arbitrary small segment of an elastic string.

We define the following physical parameters for the string:

L : length;

$u(x,t)$: vertical displacement;

$\rho_0(x)$: mass density (mass per unit length);

$F(x,t)$: tension;

$q(x,t)$: vertical component of the body force per unit mass.

weight

Assuming that the vertical displacement from the equilibrium position (segment $[0, L]$ along the x-axis) is small and writing Newton's second law

$$\text{force} = \text{mass} \times \text{acceleration} \qquad F = ma$$

in vertical projection for the string segment between two close points corresponding to x and $x + \Delta x$, we obtain the approximate equality

$$\rho_0(x)(\Delta x)u_{tt}(x,t) \cong F(x + \Delta x, t)\sin\big(\theta(x + \Delta x, t)\big)$$
$$- F(x,t)\sin\big(\theta(x,t)\big) + \rho_0(x)(\Delta x)q(x,t).$$

If we divide by Δx and then let $\Delta x \to 0$, we arrive at the equality

$$\rho_0(x)u_{tt}(x,t) = \big[F(x,t)\sin\big(\theta(x,t)\big)\big]_x + \rho_0(x)q(x,t). \qquad (4.8)$$

For a small angle θ, the slope of the string satisfies

$$u_x = \tan\theta = \frac{\sin\theta}{\cos\theta} = \frac{\sin\theta}{1 - \frac{1}{2}\theta^2 + \cdots} \cong \sin\theta,$$

so we can rewrite (4.8) in the form

$$\rho_0(x)u_{tt}(x,t) = \big(F(x,t)u_x(x,t)\big)_x + \rho_0(x)q(x,t). \qquad (4.9)$$

Assuming that the string is perfectly elastic (that is, when θ is small we may take $F(x,t) = F_0 = \text{const}$) and homogeneous ($\rho_0 = \text{const}$), and that the body force is negligible compared to tension ($q = 0$), we see that (4.9) becomes

$$u_{tt}(x,t) = c^2 u_{xx}(x,t), \quad 0 < x < L, \ t > 0, \qquad (4.10)$$

where $c^2 = F_0/\rho_0$. This is the one-dimensional *wave equation*, in which c has the dimensions of velocity. If the body force is not negligible, then (4.10) is replaced by

$$u_{tt}(x,t) = c^2 u_{xx}(x,t) + q(x,t), \quad 0 < x < L, \ t > 0.$$

Initial conditions. Since (4.10) is of second order with respect to time, we need two initial conditions. These usually are

$$u(x,0) = f(x), \quad u_t(x,0) = g(x), \quad 0 < x < L;$$

that is, we are prescribing the initial position and velocity of the points in the string.

Boundary conditions. As in the case of the heat equation, there are three main types of BCs for (4.10) that are physically meaningful.

(i) The displacement may be prescribed at an endpoint; for example,

$$u(L,t) = \beta(t).$$

(ii) An endpoint may be free (no vertical tension):

$$F_0 \sin\theta \cong F_0 \tan\theta = F_0 u_x = 0,$$

$$F_0 \neq 0$$

so we may have a condition of the form

$$u_x(0,t) = 0, \quad t > 0.$$

(iii) An endpoint may have an elastic attachment, described by

$$F_0 u_x(0,t) = ku(0,t) \quad \text{or} \quad F_0 u_x(L,t) = -ku(L,t), \quad t > 0,$$

where $k = \text{const}$.

4.20. Remarks. (i) There is an obvious analogy between the BCs associated with the wave equation and those associated with the heat equation.

(ii) The condition prescribed at one endpoint may be different from that prescribed at the other.

(iii) It is clear that the wave equation is linear.

(iv) The one-dimensional wave equation is the simplest example of a linear second-order *hyperbolic equation.* ∎

The model. As we have seen, the vibrations of an elastic string are governed by the wave equation and appropriate BCs and ICs.

4.21. Example. The IBVP for a vibrating string with fixed endpoints when the effect of the body force is taken into account is of the form

* A clamped String F_0/ρ_0

Typical wave equation

$$u_{tt}(x,t) = c^2 u_{xx}(x,t) + q(x,t), \quad 0 < x < L, \, t > 0, \qquad \text{(PDE)}$$

$$u(0,t) = 0, \quad u(L,t) = 0, \quad t > 0, \qquad \text{(BCs)}$$

$$u(x,0) = f(x), \quad u_t(x,0) = g(x), \quad 0 < x < L, \qquad \text{(ICs)}$$

where q, f, and g are given functions. ∎

4.22. Example. If the string in Example 4.21 has an elastic attachment
at its near endpoint, a free far endpoint, and a negligible body force, then
the corresponding IBVP is

$$u_{tt}(x,t) = c^2 u_{xx}(x,t), \quad 0 < x < L, \, t > 0, \quad \text{(PDE)}$$
$$u_x(0,t) - ku(0,t) = 0, \quad u_x(L,t) = 0, \quad t > 0, \quad \text{(BCs)}$$
$$u(x,0) = f(x), \quad u_t(x,0) = g(x), \quad 0 < x < L, \quad \text{(ICs)}$$

where k is a known positive constant. ∎

4.23. Remark. Let the domain G and its boundary lines ∂G_x and ∂G_t be
as defined in Example 4.7. A (classical) solution u of the IBVP in Example
4.21 has the following properties:

(i) it is twice continuously differentiable with respect to x and t in G and
satisfies the PDE at every point in G;

(ii) it is continuous up to ∂G_x and satisfies the BCs at every point on
∂G_x;

(iii) it is continuous together with its first-order time derivative up to ∂G_t
and satisfies the ICs at every point on ∂G_t.

As in the case of the heat equation and the Laplace equation, the presence
of jump discontinuities in the data functions leads to a solution with reduced
smoothness. ∎

4.24. Theorem. *If the functions q, f, and g are such that u and all its
first-order and second-order derivatives are continuous up to the boundary of
G, including the "corner" points $(0,0)$ and $(L,0)$, then the IBVP in Example
4.21 has at most one solution.*

Proof. Suppose that the given IBVP has two solutions u_1 and u_2. Then
their difference $u = u_1 - u_2$ satisfies the fully homogeneous IBVP

$$u_{tt}(x,t) = c^2 u_{xx}(x,t), \quad 0 < x < L, \, t > 0,$$
$$u(0,t) = 0, \quad u(L,t) = 0, \quad t > 0,$$
$$u(x,0) = 0, \quad u_t(x,0) = 0, \quad 0 < x < L.$$

Multiplying the PDE by u_t, integrating the result with respect to x over
$[0, L]$ and with respect to t over $[0, T]$, where $T > 0$ is an arbitrarily fixed

number, using integration by parts, and taking the BCs and the smoothness properties of u into account, we see that

$$0 = \int\limits_0^T \int\limits_0^L (u_{tt}u_t - c^2 u_{xx}u_t)\, dx\, dt$$

$$= \int\limits_0^T \int\limits_0^L (u_{tt}u_t + c^2 u_x u_{xt})\, dx\, dt - c^2 \int\limits_0^T \left[u_x u_t\right]_{x=0}^{x=L}\, dt$$

$$= \frac{1}{2}\int\limits_0^T \int\limits_0^L (u_t^2 + c^2 u_x^2)_t\, dx\, dt = \frac{1}{2}\int\limits_0^L \left[u_t^2 + c^2 u_x^2\right]_{t=0}^{t=T}\, dx;$$

hence, for any $T > 0$,

$$\int\limits_0^L \left[u_t^2(x,T) + c^2 u_x^2(x,T)\right] dx = \int\limits_0^L \left[u_t^2(x,0) + c^2 u_x^2(x,0)\right] dx,$$

which implies that

$$V(t) = \int\limits_0^L \left[u_t^2(x,t) + c^2 u_x^2(x,t)\right] dx = \kappa = \text{const} \geq 0, \quad t \geq 0.$$

From the ICs and the smoothness properties of u it follows that $V(0) = 0$, so $\kappa = 0$; in other words, $V = 0$. Since the integrand in V is nonnegative, we see that this is possible if and only if

$$u_t(x,t) = 0, \quad u_x(x,t) = 0, \quad 0 \leq x \leq L,\ t \geq 0.$$

Consequently, u is constant in G and on its boundary lines. Since u is zero on these lines, we conclude that $u = 0$; thus, u_1 and u_2 coincide. ∎

4.25. Remark. (i) As for the heat and Laplace equations, in what follows we construct a solution for the IBVP mentioned in Example 4.21. By Theorem 4.24, this will be the only solution of that problem.

(ii) Uniqueness can also be proved for the solutions of IBVPs with other BCs.

(iii) The substitution

$$\xi = \frac{1}{L}x, \quad \tau = \frac{c}{L}t, \quad u(x,t) = u(L\xi, L\tau/c) = v(\xi,\tau)$$

reduces the wave equation to

$$v_{\tau\tau}(\xi,\tau) = v_{\xi\xi}(\xi,\tau), \quad 0 < \xi < 1, \ \tau > 0.$$

It is mostly this simpler form that we will use in applications. ∎

4.4. Other Equations

Below is a list of other linear partial differential equations that occur in important mathematical models. Their classification as parabolic, hyperbolic, or elliptic is explained in Chapter 11.

Brownian motion. The function $u(x,t)$ that specifies the probability that a particle undergoing one-dimensional motion in a fluid is located at point x at time t satisfies the second-order parabolic PDE

$$u_t(x,t) = au_{xx}(x,t) - bu_x(x,t),$$

where the coefficients $a, b > 0$ are related to the average displacement of the particle per unit time and the variance of the observed displacement around the average.

Diffusion–convection problems. The one-dimensional case is described by the second-order parabolic PDE

$$u_t(x,t) = ku_{xx}(x,t) - au_x(x,t) + bu(x,t),$$

where $u(x,t)$ is the temperature and the coefficients $k, a > 0$ and b are expressed in terms of the physical properties of the medium, the heat flow rate, and the strength of the source.

When $a = 0$ and $b > 0$, the above equation describes a diffusion process with a chain reaction.

Stock market prices. The price $V(S,t)$ of a derivative on the stock market is the solution of the second-order parabolic PDE

$$V_t(S,t) + \tfrac{1}{2}\sigma^2 S^2 V_{SS}(S,t) + rSV_S(S,t) - rV(S,t) = 0,$$

where S is the price of the stock, σ is the volatility of the stock, r is the continuously compounded risk-free interest per annum, and time t is measured in years. This is known as the *Black–Scholes equation*.

Evolution of a quantum state. In quantum mechanics, the wave function $\psi(x,t)$ that characterizes the one-dimensional motion of a particle under the influence of a potential $V(x)$ satisfies the *Schrödinger equation*

$$i\hbar\psi_t(x,t) = -\frac{\hbar^2}{2m}\psi_{xx}(x,t) + \underline{V(x)}\underline{\psi(x,t)}.$$

This is a second-order parabolic PDE where $\hbar$ is the reduced Planck constant, m is the mass of the particle, and $i^2 = -1$.

Motion of a quantum scalar field. The wave function $\psi(x,y,z,t)$ of a free moving particle in relativistic quantum mechanics is the solution of the (second-order hyperbolic) *Klein–Gordon equation*

$$\psi_{tt}(x,y,z,t) = c^2\Delta\psi(x,y,z,t) - \frac{m^2c^4}{\hbar^2}\psi(x,y,z,t),$$

where c is the speed of light.

The one-dimensional version of this equation is of the form

$$u_{tt}(x,t) = c^2 u_{xx}(x,t) - au(x,t),$$

where $a = \text{const} > 0$.

Loss transmission line. The current and voltage in the propagation of signals on a telegraph line satisfy the second-order hyperbolic PDE

$$au_{tt}(x,t) + bu_t(x,t) + cu(x,t) = u_{xx}(x,t),$$

where the coefficients $a, b > 0$ and $c \geq 0$ are expressed in terms of the resistance, inductance, capacitance, and conductance characterizing the line. This is known as the *telegraph* (or *telegrapher's*) *equation*.

Dissipative waves. In the one-dimensional case, the propagation of such waves is governed by a second-order hyperbolic PDE of the form

$$u_{tt}(x,t) + au_t(x,t) + bu(x,t) = c^2 u_{xx}(x,t) - du_x(x,t),$$

where the coefficients $a, b, d \geq 0$ are not all zero and $c > 0$.

Transverse vibrations of a rod. The deflection $u(x,t)$ of a generic point in the rod satisfies the fourth-order PDE

$$u_{tt}(x,t) + c^2 u_{xxxx}(x,t) = 0,$$

where c is a physical constant related to the rigidity of the rod material.

Scattered waves. The *Helmholtz equation*

$$\Delta u(x,y,z) + k^2 u(x,y,z) = 0,$$

where $k = \text{const}$, plays an important role in the study of scattering of acoustic, electromagnetic, and elastic waves. It is a second-order elliptic PDE.

The equation

$$\Delta u(x,y,z) - k^2 u(x,y,z) = 0$$

is called the *modified Helmholtz equation.*

Steady-state convective heat. In the two-dimensional case, the temperature distribution $u(x,y)$ governing this process satisfies a second-order elliptic PDE of the form

$$\Delta u(x,y) - a u_x(x,y) - b u_y(x,y) + c u(x,y) = 0,$$

where the coefficients $a, b \geq 0$, not both zero, and c are related to the thermal properties of the medium, the heat flow rates in the x and y directions, and the strength of the heat source.

Plane problems in continuum mechanics. The Airy stress function in plane elasticity and the stream function in the slow flow of a viscous incompressible fluid are solutions of the elliptic fourth-order *biharmonic equation*

$$\Delta\Delta u(x,y) = u_{xxxx}(x,y) + 2u_{xxyy}(x,y) + u_{yyyy}(x,y) = 0.$$

Plane transonic flow. The transonic flow of a compressible gas is described by the *Euler–Tricomi equation*

$$u_{xx}(x,y) = x u_{yy}(x,y),$$

where $u(x,y)$ is a function of speed. This is a second-order PDE of mixed type: it is hyperbolic for $x > 0$, elliptic for $x < 0$, and parabolic for $x = 0$.

Exercises

(1) Show that Theorem 4.9 also holds for the IBVP in Example 4.4 with the BCs replaced by

(i) $u_x(0,t) - hu(0,t) = \alpha(t)$, $u(L,t) = \beta(t)$, $t > 0$;

(ii) $u_x(0,t) = \alpha(t)$, $u_x(L,t) + hu(L,t) = \beta(t)$, $t > 0$,

where α and β are given functions and $h = \text{const} > 0$.

(2) Verify the statements in Remarks 4.10(iv) and 4.25(iii).

(3) Show that any (classical) solution of the Poisson equation $\Delta u = q$ in D which is continuously differentiable, rather than merely continuous, up to the boundary ∂D satisfies

$$\int_D (u_x^2 + u_y^2)\,dA + \int_D uq\,dA = \int_{\partial D} uu_n\,ds.$$

Using this formula, show that

(i) each of the Dirichlet and Robin BVPs for the Poisson equation in D have at most one solution of this type;

(ii) any two such solutions of the Neumann BVP for the Poisson equation in D differ by a constant.

(4) Show that Theorem 4.24 also holds for the IBVP in Example 4.21 with $h = \text{const} > 0$ and the BCs replaced by

(i) $u_x(0,t) - hu(0,t) = 0$, $u(L,t) = 0$, $t > 0$;

(ii) $u_x(0,t) = 0$, $u_x(L,t) + hu(L,t) = 0$, $t > 0$.

In (5)–(8) find real numbers α and β such that the function substitution $u(x,t) = e^{\alpha x + \beta t}v(x,t)$ reduces the given diffusion–convection equation to the heat equation for v.

(5) $u_t(x,t) = 2u_{xx}(x,t) - 3u_x(x,t) - u(x,t)$.

(6) $u_t(x,t) = u_{xx}(x,t) - 6u_x(x,t) - 2u(x,t)$.

(7) $u_t(x,t) = \frac{1}{2}u_{xx}(x,t) - 4u_x(x,t) - 2u(x,t)$.

(8) $u_t(x,t) = 3u_{xx}(x,t) - 2u_x(x,t) - 3u(x,t)$.

In (9)–(12) determine if there are real numbers α and β such that the function substitution $u(x,t) = e^{\alpha x + \beta t} v(x,t)$ in the given dissipative wave equation eliminates (i) both first-order derivatives; (ii) the unknown function and its first-order t-derivative; (iii) the unknown function and its first-order x-derivative. (In all cases, the coefficients of the reduced equation must satisfy the restrictions mentioned in Section 4.4.) When such a substitution exists, write out the reduced equation.

(9) $u_{tt}(x,t) + u_t(x,t) + u(x,t) = u_{xx}(x,t) - 2u_x(x,t).$

(10) $u_{tt}(x,t) + 3u_t(x,t) + u(x,t) = 2u_{xx}(x,t) - u_x(x,t).$

(11) $u_{tt}(x,t) + 2u_t(x,t) + \frac{1}{2}u(x,t) = 2u_{xx}(x,t) - \sqrt{2}\,u_x(x,t).$

(12) $u_{tt}(x,t) + 2u_t(x,t) + 2u(x,t) = u_{xx}(x,t) - u_x(x,t).$

In (13)–(16) find real numbers α and β such that the function substitution $u(x,y) = e^{\alpha x + \beta y} v(x,y)$ reduces the given steady-state convective heat equation to the Helmholtz (modified Helmholtz) equation. In each case, write out the reduced equation.

(13) $u_{xx}(x,y) + u_{yy}(x,y) - 2u_x(x,y) - 4u_y(x,y) + 7u(x,y) = 0.$

(14) $u_{xx}(x,y) + u_{yy}(x,y) - u_x(x,y) - 2u_y(x,y) - u(x,y) = 0.$

(15) $u_{xx}(x,y) + u_{yy}(x,y) - 3u_x(x,y) - u_y(x,y) + 2u(x,y) = 0.$

(16) $u_{xx}(x,y) + u_{yy}(x,y) - 2u_x(x,y) - 3u_y(x,y) + 4u(x,y) = 0.$

Chapter 5
The Method of Separation of Variables

Separation of variables is one of the oldest and most efficient solution techniques for a certain class of PDE problems. Below we show it in application to initial boundary value problems for the heat and wave equations, and to boundary value problems for the Laplace equation.

5.1. The Heat Equation

Rod with zero temperature at the endpoints. Consider the IBVP

$$u_t(x,t) = ku_{xx}(x,t), \quad 0 < x < L, \ t > 0, \qquad \text{(PDE)}$$

$$u(0,t) = 0, \quad u(L,t) = 0, \quad t > 0, \qquad \text{(BCs)}$$

$$u(x,0) = f(x), \quad 0 < x < L, \qquad \text{(IC)}$$

where $f \neq 0$. We remark that the PDE and BCs are linear and homogeneous and seek a solution of the form

$$u(x,t) = X(x)T(t).$$

Substituting into the PDE, we obtain the equality

$$X(x)T'(t) = kX''(x)T(t).$$

It is clear that neither X nor T can be the zero function: if either of them were, then so would u, which is impossible, because the zero solution does not satisfy the nonhomogeneous IC. Hence, we may divide the above equality by $kX(x)T(t)$ to arrive at

$$\frac{1}{k}\frac{T'(t)}{T(t)} = \frac{X''(x)}{X(x)}.$$

Since the left-hand side above is a function of t and the right-hand side a function of x alone, this equality is possible if and only if both sides are equal to one and the same constant, say, $-\lambda$. Thus, we must have

$$\frac{1}{k}\frac{T'(t)}{T(t)} = \frac{X''(x)}{X(x)} = -\lambda,$$

where λ is called the *separation constant*. This leads to separate equations for the functions X and T:

$$X''(x) + \lambda X(x) = 0, \quad 0 < x < L, \tag{5.1}$$
$$T'(t) + \lambda k T(t) = 0, \quad t > 0. \tag{5.2}$$

From the first BC we see that

$$u(0,t) = X(0)T(t) = 0 \quad \text{for all } t > 0.$$

As $T \neq 0$, it follows that

$$X(0) = 0. \tag{5.3}$$

Similarly, the second BC yields

$$X(L) = 0. \tag{5.4}$$

We are now ready to find X and T. We want nonzero functions X that satisfy the regular Sturm–Liouville problem (5.1), (5.3), and (5.4), in other words, the eigenfunctions X_n corresponding to the eigenvalues λ_n, both computed in Example 3.16:

$$\lambda_n = \left(\frac{n\pi}{L}\right)^2, \quad X_n(x) = \sin\frac{n\pi x}{L}, \quad n = 1,2,\ldots. \tag{5.5}$$

For each λ_n, from (5.2) we find the corresponding time component

$$T_n(t) = e^{-k(n\pi/L)^2 t}. \tag{5.6}$$

We have taken the arbitrary constant of integration equal to 1 since these functions are used in the next stage with arbitrary numerical coefficients.

Combining (5.5) and (5.6), we conclude that all functions of the form

$$u_n(x,t) = X_n(x)T_n(t) = \sin\frac{n\pi x}{L} e^{-k(n\pi/L)^2 t}, \quad n = 1,2,\ldots,$$

satisfy both the PDE and the BCs. By the principle of superposition, so does any finite linear combination

$$\sum_{n=1}^{N} b_n u_n(x,t) = \sum_{n=1}^{N} b_n \sin \frac{n\pi x}{L} e^{-k(n\pi/L)^2 t}, \qquad (5.7)$$

where b_n are arbitrary numbers. The IC is satisfied by such an expression if

$$f(x) = \sum_{n=1}^{N} b_n \sin \frac{n\pi x}{L}.$$

But this is impossible unless f is a finite linear combination of the eigenfunctions, which, in general, is not the case. This implies that (5.7) is not a good representation for the solution $u(x,t)$ of the IBVP. However, we recall (see Section 2.2) that if f is piecewise smooth, then it can be written as an *infinite* linear combination of the eigenfunctions (its Fourier sine series):

$$f(x) = \sum_{n=1}^{\infty} b_n \sin \frac{n\pi x}{L}, \qquad 0 < x < L, \qquad (5.8)$$

where, by (2.10),

$$b_n = \frac{2}{L} \int_0^L f(x) \sin \frac{n\pi x}{L}\, dx, \qquad n = 1,2,\ldots. \qquad (5.9)$$

Therefore, the solution is given by the infinite series

$$u(x,t) = \sum_{n=1}^{\infty} b_n \sin \frac{n\pi x}{L} e^{-k(n\pi/L)^2 t} \qquad (5.10)$$

with the coefficients b_n computed from (5.9).

5.1. Example. Consider the IBVP

$$u_t(x,t) = u_{xx}(x,t), \qquad 0 < x < 1,\ t > 0,$$
$$u(0,t) = 0, \quad u(1,t) = 0, \qquad t > 0,$$
$$u(x,0) = \sin(3\pi x) - 2\sin(5\pi x), \qquad 0 < x < 1.$$

Here $k = 1$, $L = 1$, and the function on the right-hand side in the IC is a linear combination of the eigenfunctions. Using (5.8) and Theorem 3.20(ii), we find that

$$b_3 = 1, \quad b_5 = -2, \quad b_n = 0\ (n \neq 3,5);$$

so, by (5.10), the solution of the problem is

$$u(x,t) = \sin(3\pi x)e^{-9\pi^2 t} - 2\sin(5\pi x)e^{-25\pi^2 t}.$$

Equally, this result can be obtained from (5.9) and (5.10). ■

5.2. Example. To find the solution of the IBVP

$$u_t(x,t) = u_{xx}(x,t), \quad 0 < x < 1,\ t > 0,$$
$$u(0,t) = 0, \quad u(1,t) = 0, \quad t > 0,$$
$$u(x,0) = x, \quad 0 < x < 1,$$

we first use (5.9) with $f(x) = x$ and $L = 1$ and integration by parts to compute the coefficients b_n:

$$b_n = 2\int\limits_0^1 x\sin(n\pi x)\,dx = (-1)^{n+1}\frac{2}{n\pi}, \quad n = 1,2,\ldots;$$

then, by (5.10),

$$u(x,t) = \sum_{n=1}^\infty (-1)^{n+1}\frac{2}{n\pi}\sin(n\pi x)\,e^{-(n\pi)^2 t}$$

is the solution of the IBVP. ■

Rod with insulated endpoints. The heat conduction problem for a uniform rod with insulated endpoints is described by the IBVP

$$u_t(x,t) = ku_{xx}(x,t), \quad 0 < x < L,\ t > 0, \tag{PDE}$$
$$u_x(0,t) = 0, \quad u_x(L,t) = 0, \quad t > 0, \tag{BCs}$$
$$u(x,0) = f(x), \quad 0 < x < L. \tag{IC}$$

Since the PDE and BCs are linear and homogeneous, we again seek a solution of the form

$$u(x,t) = X(x)T(t).$$

Just as in the case of the rod with zero temperature at the endpoints, from the PDE and BCs we now find that X and T satisfy, respectively,

$$X''(x) + \lambda X(x) = 0, \quad 0 < x < L,$$
$$X'(0) = 0, \quad X'(L) = 0,$$

and

$$T'(t) + k\lambda T(t) = 0, \quad t > 0,$$

where λ is the separation constant.

The nonzero solutions X are the eigenfunctions of the above Sturm–Liouville problem, which were computed in Example 3.17:

$$\lambda_n = \left(\frac{n\pi}{L}\right)^2, \quad X_n(x) = \cos\frac{n\pi x}{L}, \quad n = 0,1,2,\ldots. \tag{5.11}$$

Integrating the equation satisfied by T with $\lambda = \lambda_n$, we obtain the associated time components

$$T_n(t) = e^{-k(n\pi/L)^2 t}. \tag{5.12}$$

In view of (5.11), (5.12), and the argument used in the preceding case, we now expect the solution of the IBVP to have the series representation

$$u(x,t) = \frac{1}{2}a_0 + \sum_{n=1}^{\infty} a_n \cos\frac{n\pi x}{L} e^{-k(n\pi/L)^2 t}, \tag{5.13}$$

where each term satisfies the PDE and the BCs. The IC is satisfied if

$$u(x,0) = f(x) = \frac{1}{2}a_0 + \sum_{n=1}^{\infty} a_n \cos\frac{n\pi x}{L}, \quad 0 < x < L.$$

This shows that $\frac{1}{2}a_0$ and a_n are the Fourier cosine series coefficients of f, given by (2.11):

$$a_n = \frac{2}{L}\int_0^L f(x)\cos\frac{n\pi x}{L}\,dx, \quad n = 0,1,2,\ldots. \tag{5.14}$$

Therefore, (5.13) with the a_n computed by means of (5.14) is the solution of the IBVP.

5.3. Example. The solution of the IBVP

$$u_t(x,t) = u_{xx}(x,t), \quad 0 < x < 1, \, t > 0,$$
$$u_x(0,t) = 0, \quad u_x(1,t) = 0, \quad t > 0,$$
$$u(x,0) = x, \quad 0 < x < 1,$$

is obtained from (5.13) and (5.14) with $L = 1$ and $f(x) = x$. Specifically,

$$a_0 = 2 \int_0^1 x \, dx = 1,$$

$$a_n = 2 \int_0^1 x \cos(n\pi x) \, dx = [(-1)^n - 1] \frac{2}{n^2 \pi^2}, \quad n = 1, 2, \ldots;$$

so, as in Example 5.2,

$$u(x,t) = \frac{1}{2} + \sum_{n=1}^{\infty} [(-1)^n - 1] \frac{2}{n^2 \pi^2} \cos(n\pi x) e^{-n^2 \pi^2 t}. \quad \blacksquare$$

Rod with mixed homogeneous boundary conditions. The method of separation of variables can also be used in the case where one endpoint is held at zero temperature while the other one is insulated.

5.4. Example. Consider the IBVP

$$u_t(x,t) = u_{xx}(x,t), \quad 0 < x < 1, \, t > 0,$$
$$u(0,t) = 0, \quad u_x(1,t) = 0, \quad t > 0,$$
$$u(x,0) = 1, \quad 0 < x < 1.$$

Since the PDE and BCs are linear and homogeneous, we seek a solution of the form

$$u(x,t) = X(x)T(t).$$

Proceeding as in the other cases, from the PDE and BCs we find that X and T satisfy, respectively,

$$X''(x) + \lambda X(x) = 0, \quad 0 < x < 1, \tag{5.15}$$
$$X(0) = 0, \quad X'(1) = 0,$$

and

$$T'(t) + \lambda T(t) = 0, \quad t > 0,$$

where λ is the separation constant.

To find X, we go back to the Rayleigh quotient (3.6) for the regular Sturm–Liouville problem (5.15). As in Example 3.16,

$$[p(x)f(x)f'(x)]_a^b = [X(x)X'(x)]_0^1 = 0,$$

and we arrive at inequality (3.7), from which, in view of the BCs in (5.15), we deduce that $\lambda > 0$. Hence, the general solution of the differential equation in (5.15) is

$$X(x) = C_1 \cos(\sqrt{\lambda}x) + C_2 \sin(\sqrt{\lambda}x), \quad C_1, C_2 = \text{const.}$$

Since $X(0) = 0$, it follows that $C_1 = 0$. Using the condition $X'(1) = 0$, we now find that $\cos\sqrt{\lambda} = 0$; therefore,

$$\sqrt{\lambda} = \frac{(2n-1)\pi}{2}, \quad n = 1, 2, \ldots.$$

This means that the eigenvalue-eigenfunction pairs of (5.15) are

$$\lambda_n = \frac{(2n-1)^2\pi^2}{4}, \quad X_n(x) = \sin\frac{(2n-1)\pi x}{2}, \quad n = 1, 2, \ldots,$$

and that the corresponding time components are

$$T_n(t) = e^{-(2n-1)^2\pi^2 t/4}.$$

Consequently, the functions

$$u_n(x,t) = X_n(x)T_n(t) = \sin\frac{(2n-1)\pi x}{2}e^{-(2n-1)^2\pi^2 t/4}, \quad n = 1, 2, \ldots,$$

satisfy the PDE and the BCs. Considering the usual arbitrary linear combination of all the (countably many) functions u_n, that is,

$$u(x,t) = \sum_{n=1}^{\infty} c_n u_n(x,t) = \sum_{n=1}^{\infty} c_n \sin\frac{(2n-1)\pi x}{2}e^{-(2n-1)^2\pi^2 t/4}, \quad (5.16)$$

we see that the IC is satisfied if

$$u(x,0) = 1 = \sum_{n=1}^{\infty} c_n \sin\frac{(2n-1)\pi x}{2}. \quad (5.17)$$

In other words, (5.16) is the solution of the given IBVP if c_n are the coefficients of the generalized Fourier series for the function $f(x) = 1$. These coefficients are computed by means of (3.10) with, according to (5.17),

$$u(x) = 1, \quad \sigma(x) = 1, \quad f_n(x) = \sin\frac{(2n-1)\pi x}{2}.$$

Direct calculation shows that

$$\int_0^1 \sin^2 \frac{(2n-1)\pi x}{2} \, dx = \frac{1}{2}, \quad n = 1,2,\ldots,$$

so

$$c_n = 2 \int_0^1 \sin \frac{(2n-1)\pi x}{2} \, dx = \frac{4}{(2n-1)\pi}, \quad n = 1,2,\ldots.$$

Thus, the solution of the IBVP has the series representation

$$u(x,t) = \sum_{n=1}^{\infty} \frac{4}{(2n-1)\pi} \sin \frac{(2n-1)\pi x}{2} e^{-(2n-1)^2\pi^2 t/4}. \quad \blacksquare$$

5.5. Example. The solution of the IBVP

$$u_t(x,t) = u_{xx}(x,t), \quad 0 < x < 1, \ t > 0,$$
$$u_x(0,t) = 0, \quad u(1,t) = 0, \quad t > 0,$$
$$u(x,0) = x, \quad 0 < x < 1,$$

is constructed just as in Example 5.4 except that here the BCs and the Rayleigh quotient yield the eigenvalue-eigenfunction pairs

$$\lambda_n = \frac{(2n-1)^2\pi^2}{4}, \quad X_n(x) = \cos \frac{(2n-1)\pi x}{2}, \quad n = 1,2,\ldots.$$

Since $\int_0^1 \cos^2\left((2n-1)\pi x/2\right) dx = 1/2$, $n = 1,2,\ldots$, the coefficients of the generalized Fourier series for the IC function, given by (3.10) with $u(x) = x$, $\sigma(x) = 1$, $k = 1$, and $L = 1$, are

$$c_n = 2 \int_0^1 x \cos \frac{(2n-1)\pi x}{2} \, dx$$

$$= (-1)^{n+1} \frac{4}{(2n-1)\pi} - \frac{8}{(2n-1)^2\pi^2}, \quad n = 1,2,\ldots.$$

Consequently, the solution of the IBVP is

$$u(x,t) = \sum_{n=1}^{\infty} \left[(-1)^{n+1} \frac{4}{(2n-1)\pi} - \frac{8}{(2n-1)^2\pi^2} \right]$$

$$\times \cos \frac{(2n-1)\pi x}{2} e^{-(2n-1)^2\pi^2 t/4}. \quad \blacksquare$$

Rod with an endpoint in a zero-temperature medium. Consider the problem of heat flow in a uniform rod without internal sources, when the near endpoint is kept at zero temperature and the far endpoint is kept in open air of zero temperature. The corresponding IBVP (see Section 4.1) is

$$u_t(x,t) = ku_{xx}(x,t), \quad 0 < x < L, \ t > 0, \tag{PDE}$$

$$u(0,t) = 0, \quad u_x(L,t) + hu(L,t) = 0, \quad t > 0, \tag{BCs}$$

$$u(x,0) = f(x), \quad 0 < x < L, \tag{IC}$$

where $h = \text{const} > 0$.

Since the PDE and BCs are linear and homogeneous, we use separation of variables and seek a solution of the form

$$u(x,t) = X(x)T(t).$$

In the usual way, from the PDE and BCs we find that X satisfies the regular S–L problem

$$X''(x) + \lambda X(x) = 0, \quad 0 < x < L,$$
$$X(0) = 0, \quad X'(L) + hX(L) = 0, \tag{5.18}$$

and that T is the solution of the equation

$$T'(t) + k\lambda T(t) = 0, \quad t > 0.$$

The eigenvalue-eigenfunction pairs of (5.18) were computed in Example 3.18:

$$\lambda_n = \left(\frac{\zeta_n}{L}\right)^2, \quad X_n(x) = \sin\frac{\zeta_n x}{L}, \quad n = 1, 2, \ldots,$$

where ζ_n are the positive roots of the equation

$$\tan\zeta = -\zeta/(hL).$$

Hence,

$$T_n(t) = e^{-k(\zeta_n/L)^2 t},$$

so we expect the solution of the IBVP to have a series representation of the form

$$u(x,t) = \sum_{n=1}^{\infty} c_n X_n(x) T_n(t) = \sum_{n=1}^{\infty} c_n \sin\frac{\zeta_n x}{L} e^{-k(\zeta_n/L)^2 t}. \tag{5.19}$$

The coefficients c_n are found from the IC, according to which

$$u(x,0) = f(x) = \sum_{n=1}^{\infty} c_n \sin \frac{\zeta_n x}{L}.$$

Since the eigenfunctions of (5.18) are orthogonal on $[0, L]$, the c_n are computed by means of (3.10):

$$c_n = \frac{\int\limits_0^L f(x) \sin(\zeta_n x/L) \, dx}{\int\limits_0^L \sin^2(\zeta_n x/L) \, dx}, \qquad n = 1, 2, \dots. \tag{5.20}$$

Therefore, the solution of the IBVP is (5.19) with the c_n given by (5.20).

5.6. Example. Consider the IBVP

$$u_t(x,t) = u_{xx}(x,t), \quad 0 < x < 1, \ t > 0,$$
$$u(0,t) = 0, \quad u_x(1,t) + u(1,t) = 0, \quad t > 0,$$
$$u(x,0) = \begin{cases} 0, & 0 < x < 1/2, \\ 1, & 1/2 < x < 1. \end{cases}$$

From Examples 3.18 and 3.22 with $L = h = 1$ we know that the eigenvalue-eigenfunction pairs associated with this problem are

$$\lambda_n = \zeta_n^2, \quad X_n(x) = \sin(\zeta_n x), \quad n = 1, 2, \dots,$$

where, to four decimal places,

$$\zeta_1 = 2.0288, \quad \zeta_2 = 4.9132, \quad \zeta_3 = 7.9787, \quad \zeta_4 = 11.0855, \quad \zeta_5 = 14.2074.$$

Using (5.20) and $f(x) = \begin{cases} 0, & 0 < x < 1/2, \\ 1, & 1/2 < x < 1, \end{cases}$ we now compute the coefficients c_n to the same degree of accuracy:

$$c_1 = 0.8001, \quad c_2 = -0.3813, \quad c_3 = -0.1326, \quad c_4 = 0.1160, \quad c_5 = 0.1053.$$

Consequently, by (5.19), we obtain the approximate solution

$$u(x,t) = 0.8001 \sin(2.0288x) e^{-4.1160t} - 0.3813 \sin(4.9132x) e^{-24.1395t}$$
$$- 0.1326 \sin(7.9787x) e^{-63.6597t} + 0.1160 \sin(11.0855x) e^{-122.8883t}$$
$$+ 0.1053 \sin(14.2074x) e^{-201.8502t} + \cdots. \quad \blacksquare$$

Heat conduction in a thin uniform circular ring. Physically, the ring of circumference $2L$ shown in Fig. 5.1 can be regarded as a rod of length $2L$ that, for continuity reasons, has the same temperature and heat flux at the two endpoints $x = -L$ and $x = L$.

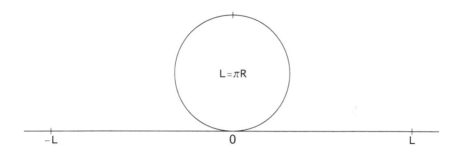

Fig. 5.1. Representation of a circular ring as a rod.

Thus, the corresponding IBVP is

$$u_t(x,t) = ku_{xx}(x,t), \quad -L < x < L,\ t > 0, \tag{PDE}$$

$$u(-L,t) = u(L,t), \quad u_x(-L,t) = u_x(L,t), \quad t > 0, \tag{BCs}$$

$$u(x,0) = f(x), \quad -L < x < L. \tag{IC}$$

The PDE and BCs are homogeneous, so, as before, we seek a solution of the form $u(x,t) = X(x)T(t)$. Then from the PDE and BCs we find by the standard argument that

$$X''(x) + \lambda X(x) = 0, \quad -L < x < L,$$
$$X(-L) = X(L), \quad X'(-L) = X'(L), \tag{5.21}$$

and

$$T'(t) + \lambda kT(t) = 0, \quad t > 0. \tag{5.22}$$

For a nonzero solution u of the IBVP, X must be a nonzero solution of the periodic Sturm–Liouville problem (5.21), which was discussed in Example 3.26. The eigenvalues of this S–L problem are

$$\lambda_0 = 0, \quad \lambda_n = \left(\frac{n\pi}{L}\right)^2, \quad n = 1,2,\ldots,$$

with corresponding eigenfunctions

$$X_0(x) = \frac{1}{2}, \quad X_{1n}(x) = \cos\frac{n\pi x}{L}, \quad X_{2n}(x) = \sin\frac{n\pi x}{L}, \quad n = 1,2,\ldots.$$

Therefore, the time components given by (5.22) with $\lambda = \lambda_n$ are

$$T_n(t) = Ce^{-k(n\pi/L)^2 t}, \quad C = \text{const},$$

so we consider a series solution of the form

$$u(x,t) = \frac{1}{2}a_0 + \sum_{n=1}^{\infty}\left[a_n X_{1n}(x) + b_n X_{2n}(x)\right]T_n(t)$$

$$= \frac{1}{2}a_0 + \sum_{n=1}^{\infty}\left(a_n\cos\frac{n\pi x}{L} + b_n\sin\frac{n\pi x}{L}\right)e^{-k(n\pi/L)^2 t}. \quad (5.23)$$

The IC requires that

$$u(x,0) = f(x)$$

$$= \frac{1}{2}a_0 + \sum_{n=1}^{\infty}\left(a_n\cos\frac{n\pi x}{L} + b_n\sin\frac{n\pi x}{L}\right), \quad -L < x < L.$$

As mentioned in Example 3.26, the coefficients of this full Fourier series of f are computed by means of formulas (2.7)–(2.9):

$$a_n = \frac{1}{L}\int_{-L}^{L} f(x)\cos\frac{n\pi x}{L}\,dx, \quad n = 0,1,2,\ldots,$$

$$(5.24)$$

$$b_n = \frac{1}{L}\int_{-L}^{L} f(x)\sin\frac{n\pi x}{L}\,dx, \quad n = 1,2,\ldots.$$

Consequently, the solution of the IBVP is given by (5.23) and (5.24).

5.7. Example. Consider the IBVP

$$u_t(x,t) = u_{xx}(x,t), \quad -1 < x < 1,\ t > 0,$$
$$u(-1,t) = u(1,t), \quad u_x(-1,t) = u_x(1,t), \quad t > 0,$$
$$u(x,0) = x + 1, \quad -1 < x < 1.$$

Here $L = 1$ and $f(x) = x + 1$, so, by (5.24),

$$a_0 = \int_{-1}^{1} (x + 1)\, dx = 2,$$

$$a_n = \int_{-1}^{1} (x + 1) \cos(n\pi x)\, dx = 0,$$

$$b_n = \int_{-1}^{1} (x + 1) \sin(n\pi x)\, dx = (-1)^{n+1} \frac{2}{n\pi}.$$

Hence, by (5.23), the solution of the IBVP is

$$u(x,t) = 1 + \sum_{n=1}^{\infty} (-1)^{n+1} \frac{2}{n\pi} \sin(n\pi x)\, e^{-n^2 \pi^2 t}. \quad \blacksquare$$

5.2. The Wave Equation

Generally speaking, the method of separation of variables is applied to IBVPs for the wave equation in much the same way as for the heat equation. Here, however, the time component satisfies a second-order ODE, which introduces a second IC.

String with fixed endpoints. We know from Section 4.3 that the corresponding IBVP is

$$u_{tt}(x,t) = c^2 u_{xx}(x,t), \quad 0 < x < L,\ t > 0, \qquad \text{(PDE)}$$
$$u(0,t) = 0, \quad u(L,t) = 0, \quad t > 0, \qquad \text{(BCs)}$$
$$u(x,0) = f(x), \quad u_t(x,0) = g(x), \quad 0 < x < L. \qquad \text{(ICs)}$$

Since, as in the earlier problems, the PDE and BCs are linear and homogeneous, we seek a solution of the form

$$u(x,t) = X(x)T(t),$$

where, as in the case of the heat equation, neither X nor T can be the zero function. Replacing this in the PDE, we obtain

$$X(x)T''(t) = c^2 X''(x)T(t),$$

which, divided by $c^2 X(x)T(t)$, yields

$$\frac{1}{c^2}\frac{T''(t)}{T(t)} = \frac{X''(x)}{X(x)} = -\lambda, \tag{5.25}$$

where $\lambda = $ const is the separation constant. At the same time, it can be seen that the BCs lead to

$$X(0) = 0, \quad X(L) = 0;$$

hence, X satisfies

$$X''(x) + \lambda X(x) = 0, \quad 0 < x < L,$$
$$X(0) = 0, \quad X(L) = 0.$$

This regular S–L problem was solved in Example 3.16, where it was shown that its eigenvalue-eigenfunction pairs are

$$\lambda_n = \left(\frac{n\pi}{L}\right)^2, \quad X_n(x) = \sin\frac{n\pi x}{L}, \quad n = 1,2,\dots. \tag{5.26}$$

From (5.25) we deduce that for each eigenvalue λ_n we obtain a function T_n that satisfies the equation

$$T_n''(t) + c^2\lambda_n T_n(t) = 0, \quad t > 0,$$

whose general solution (since $\lambda_n > 0$) is

$$T_n(t) = b_{1n}\cos\frac{n\pi ct}{L} + b_{2n}\sin\frac{n\pi ct}{L}, \quad b_{1n}, b_{2n} = \text{const}. \tag{5.27}$$

In view of (5.26) and (5.27), we assume that the solution of the IBVP has the series representation

$$u(x,t) = \sum_{n=1}^{\infty} X_n(x)T_n(t)$$

$$= \sum_{n=1}^{\infty}\sin\frac{n\pi x}{L}\left(b_{1n}\cos\frac{n\pi ct}{L} + b_{2n}\sin\frac{n\pi ct}{L}\right). \tag{5.28}$$

Each term of this series satisfies the PDE and the BCs. To satisfy the ICs, we set $t = 0$ in (5.28) and in the expression of $u_t(x,t)$ obtained by differentiating (5.28) term by term:

$$u(x,0) = f(x) = \sum_{n=1}^{\infty} b_{1n} \sin \frac{n\pi x}{L},$$

$$u_t(x,0) = g(x) = \sum_{n=1}^{\infty} b_{2n} \frac{n\pi c}{L} \sin \frac{n\pi x}{L}.$$

We see that the numbers b_{1n} and $b_{2n}(n\pi c)/L$ are the Fourier sine series coefficients of f and g, respectively; so, by (2.10),

$$b_{1n} = \frac{2}{L} \int_0^L f(x) \sin \frac{n\pi x}{L}\, dx, \quad n = 1,2,\ldots,$$

$$(5.29)$$

$$b_{2n} = \frac{2}{n\pi c} \int_0^L g(x) \sin \frac{n\pi x}{L}\, dx, \quad n = 1,2,\ldots.$$

Therefore, the solution of the IBVP is given by (5.28) with the b_{1n} and b_{2n} computed from (5.29).

5.8. Remark. The terms in the expansion of u are called *normal modes of vibration*. Solutions of this type are called *standing waves*. ■

5.9. Example. We notice that in the IBVP

$$u_{tt}(x,t) = u_{xx}(x,t), \quad 0 < x < 1, \ t > 0,$$
$$u(0,t) = 0, \quad u(1,t) = 0, \quad t > 0,$$
$$u(x,0) = 3\sin(\pi x) - 2\sin(3\pi x),$$
$$u_t(x,0) = 4\pi \sin(2\pi x), \quad 0 < x < 1,$$

the functions f and g are linear combinations of the eigenfunctions. Setting $c = 1$ and $L = 1$ in their eigenfunction expansions above, from Theorem 3.20(ii) we deduce that

$$b_{11} = 3, \quad b_{13} = -2, \quad b_{1n} = 0 \ (n \neq 1,3), \quad b_{22} = 2, \quad b_{2n} = 0 \ (n \neq 2).$$

Hence, by (5.28), the solution of the IBVP is

$$u(x,t) = 3\sin(\pi x)\cos(\pi t) - 2\sin(3\pi x)\cos(3\pi t) + 2\sin(2\pi x)\sin(2\pi t). \ ■$$

5.10. Example. If the initial conditions in the IBVP in Example 5.9 are replaced by

$$u(x,0) = \begin{cases} x, & 0 < x \le 1/2, \\ 1 - x, & 1/2 < x < 1, \end{cases} \quad u_t(x,0) = 0, \quad 0 < x < 1,$$

then, using (5.29) (with $c = 1$ and $L = 1$) and integration by parts, we find that

$$b_{1n} = 2\left\{ \int_0^{1/2} f(x)\sin(n\pi x)\,dx + \int_{1/2}^1 f(x)\sin(n\pi x)\,dx \right\}$$

$$= 2\left\{ \int_0^{1/2} x\sin(n\pi x)\,dx + \int_{1/2}^1 (1 - x)\sin(n\pi x)\,dx \right\} = \frac{4}{n^2\pi^2}\sin\frac{n\pi}{2},$$

$$b_{2n} = \frac{2}{n\pi} \int_0^1 g(x)\sin(n\pi x)\,dx = 0, \quad n = 1, 2, \ldots;$$

so, by (5.28), the solution of the IBVP is

$$u(x,t) = \sum_{n=1}^\infty \frac{4}{n^2\pi^2}\sin\frac{n\pi}{2}\sin(n\pi x)\cos(n\pi t). \quad \blacksquare$$

Vibrating string with free endpoints. The IBVP corresponding to this case is (see Section 4.3)

$$u_{tt}(x,t) = c^2 u_{xx}(x,t), \quad 0 < x < L, \ t > 0, \tag{PDE}$$

$$u_x(0,t) = 0, \quad u_x(L,t) = 0, \quad t > 0, \tag{BCs}$$

$$u(x,0) = f(x), \quad u_t(x,0) = g(x), \quad 0 < x < L. \tag{ICs}$$

Separating the variables, we are led to the eigenvalue-eigenfunction pairs

$$\lambda_n = \left(\frac{n\pi}{L}\right)^2, \quad X_n(x) = \cos\frac{n\pi x}{L}, \quad n = 0, 1, 2, \ldots.$$

For $\lambda_0 = 0$, the general solution of the equation satisfied by T is

$$T_0(t) = \frac{1}{2}a_{10} + \frac{1}{2}a_{20}t, \quad a_{10}, a_{20} = \text{const},$$

while for λ_n, $n \geq 1$, the functions T_n are given by (5.27). Hence, we expect the solution of the IBVP to be of the form

$$u(x,t) = \frac{1}{2}a_{10} + \frac{1}{2}a_{20}t$$

$$+ \sum_{n=1}^{\infty} \cos\frac{n\pi x}{L}\left(a_{1n}\cos\frac{n\pi ct}{L} + a_{2n}\sin\frac{n\pi ct}{L}\right), \qquad (5.30)$$

where the a_{1n} and a_{2n}, $n = 0,1,2,\ldots$, are constants. Every term in (5.30) satisfies the PDE and the BCs.

To satisfy the ICs, we require that

$$u(x,0) = f(x) = \frac{1}{2}a_{10} + \sum_{n=1}^{\infty}a_{1n}\cos\frac{n\pi x}{L},$$

$$u_t(x,0) = g(x) = \frac{1}{2}a_{20} + \sum_{n=1}^{\infty}a_{2n}\frac{n\pi c}{L}\cos\frac{n\pi x}{L};$$

consequently, by (2.11),

$$a_{10} = \frac{2}{L}\int_0^L f(x)\,dx,$$

$$a_{1n} = \frac{2}{L}\int_0^L f(x)\cos\frac{n\pi x}{L}\,dx, \quad n = 1,2,\ldots,$$

$$\tag{5.31}$$

$$a_{20} = \frac{2}{L}\int_0^L g(x)\,dx,$$

$$a_{2n} = \frac{2}{n\pi c}\int_0^L g(x)\cos\frac{n\pi x}{L}\,dx, \quad n = 1,2,\ldots,$$

and we conclude that the solution of the IBVP is given by (5.30) with the a_{1n} and a_{2n} computed by means of (5.31).

5.11. Example. To find the solution of the IBVP

$$u_{tt}(x,t) = u_{xx}(x,t), \quad 0 < x < 1,\ t > 0,$$

$$u_x(0,t) = 0, \quad u_x(1,t) = 0, \quad t > 0,$$

$$u(x,0) = \cos(2\pi x), \quad u_t(x,0) = -2\pi\cos(\pi x), \quad 0 < x < 1,$$

we remark that $c = 1$ and $L = 1$, and that the initial data functions f and g are linear combinations of the eigenfunctions; thus,

$$a_{12} = 1, \quad a_{1n} = 0 \ (n \neq 2), \quad a_{21} = -2, \quad a_{2n} = 0 \ (n \neq 1),$$

so, by (5.30), the solution of the IBVP is

$$u(x,t) = -2\cos(\pi x)\sin(\pi t) + \cos(2\pi x)\cos(2\pi t). \quad \blacksquare$$

5.12. Example. Consider the IBVP

$$u_{tt}(x,t) = u_{xx}(x,t), \quad 0 < x < 1, \ t > 0,$$

$$u_x(0,t) = 0, \quad u_x(1,t) = 0, \quad t > 0,$$

$$u(x,0) = 0, \quad u_t(x,0) = \begin{cases} -1, & 1/4 \le x \le 3/4, \\ 0 & \text{otherwise.} \end{cases}$$

Again, here we have $c = 1$ and $L = 1$; so, by (5.31),

$$a_{1n} = 0, \quad n = 0,1,2,\ldots,$$

$$a_{20} = 2\left\{ \int_0^{1/4} g(x)\,dx + \int_{1/4}^{3/4} g(x)\,dx + \int_{3/4}^{1} g(x)\,dx \right\} = 2\int_{1/4}^{3/4}(-1)\,dx = -1,$$

$$a_{2n} = \frac{2}{n\pi}\left\{ \int_0^{1/4} g(x)\cos(n\pi x)\,dx + \int_{1/4}^{3/4} g(x)\cos(n\pi x)\,dx \right.$$

$$\left. + \int_{3/4}^{1} g(x)\cos(n\pi x)\,dx \right\}$$

$$= \frac{2}{n\pi}\int_{1/4}^{3/4}(-1)\cos(n\pi x)\,dx = -\frac{4}{n^2\pi^2}\sin\frac{n\pi}{4}\cos\frac{n\pi}{2}, \quad n = 1,2,\ldots.$$

Hence, by (5.30), the solution of the IBVP is

$$u(x,t) = -\frac{1}{2}t - \sum_{n=1}^{\infty} \frac{4}{n^2\pi^2}\sin\frac{n\pi}{4}\cos\frac{n\pi}{2}\cos(n\pi x)\sin(n\pi t). \quad \blacksquare$$

5.13. Remark. IBVPs with mixed BCs are handled similarly. $\quad \blacksquare$

5.3. The Laplace Equation

In this section we discuss the solution of two types of problems: one formulated in terms of Cartesian coordinates, and one where the use of polar coordinates is more appropriate.

The Laplace equation in a rectangle. Consider the equilibrium temperature in a uniform rectangle

$$D = \{(x,y) : 0 \le x \le L, \ 0 \le y \le K\}, \ L, K = \text{const}\}$$

with no sources and with temperature prescribed on the boundary.

The corresponding problem is (see Section 4.2)

$$u_{xx}(x,y) + u_{yy}(x,y) = 0, \quad 0 < x < L, \ 0 < y < K,$$
$$u(0,y) = f_1(y), \quad u(L,y) = f_2(y), \quad 0 < y < K,$$
$$u(x,0) = g_1(x), \quad u(x,K) = g_2(x), \quad 0 < x < L.$$
functions

Although this is a boundary value problem, it turns out that the method of separation of variables can still be applied. However, since here the BCs are nonhomogeneous, we need to do some additional preliminary work.

By the principle of superposition (see Theorem 1.19), we can seek the solution of the above BVP as

$$u(x,y) = \overset{X}{u_1(x,y)} + \overset{Y}{u_2(x,y)},$$

where both u_1 and u_2 satisfy the PDE, u_1 satisfies the BCs with $f_1(y) = 0$, $f_2(y) = 0$, and u_2 satisfies the BCs with $g_1(x) = 0$, $g_2(x) = 0$; thus, the BVP for u_1 is

$$(u_1)_{xx}(x,y) + (u_1)_{yy}(x,y) = 0, \quad 0 < x < L, \ 0 < y < K, \quad \text{(PDE)}$$
$$u_1(0,y) = 0, \quad u_1(L,y) = 0, \quad 0 < y < K,$$
$$u_1(x,0) = g_1(x), \quad u_1(x,K) = g_2(x), \quad 0 < x < L. \quad \text{(BCs)}$$

We seek a solution of this BVP of the form

$$u_1(x,y) = X(x)Y(y).$$

Clearly, for the same reason as in the case of the heat and wave equations, neither X nor Y can be the zero function. Replacing in the PDE, we obtain

$$X''(x)Y(y) + X(x)Y''(y) = 0,$$

from which, on division by $X(x)Y(y)$, we find that

$$\frac{X''(x)}{X(x)} = -\frac{Y''(y)}{Y(y)} = -\lambda,$$

where $\lambda = \text{const}$ is the separation constant.

The two homogeneous BCs yield

$$X(0)Y(y) = 0, \quad X(L)Y(y) = 0, \quad 0 < y < K.$$

Since, as already mentioned, $Y(y)$ cannot be identically zero, we conclude that

$$X(0) = 0, \quad X(L) = 0.$$

Hence, X is the solution of the regular Sturm–Liouville problem

$$X''(x) + \lambda X(x) = 0, \quad 0 < x < L,$$
$$X(0) = 0, \quad X(L) = 0.$$

From Example 3.16 we know that the eigenvalue-eigenfunction pairs of this problem are

$$\lambda_n = \left(\frac{n\pi}{L}\right)^2, \quad X_n(x) = \sin\frac{n\pi x}{L}, \quad n = 1, 2, \ldots. \tag{5.32}$$

Then for each $n = 1, 2, \ldots$ we seek functions Y_n that satisfy

$$Y_n''(y) - \left(\frac{n\pi}{L}\right)^2 Y_n(y) = 0, \quad 0 < y < K.$$

By Remark 1.4, the general solution of the equation for Y_n can be written as

$$Y_n(y) = C_{1n} \sinh\frac{n\pi(y - K)}{L} + C_{2n} \sinh\frac{n\pi y}{L}, \quad C_{1n}, C_{2n} = \text{const}. \tag{5.33}$$

In view of (5.32) and (5.33), it seems reasonable to expect u_1 to be of the form

$$u_1(x,y) = \sum_{n=1}^{\infty} X_n(x) Y_n(y)$$

$$= \sum_{n=1}^{\infty} \sin \frac{n\pi x}{L} \left(C_{1n} \sinh \frac{n\pi(y-K)}{L} + C_{2n} \sinh \frac{n\pi y}{L} \right), \quad (5.34)$$

where each term satisfies the PDE and the two homogeneous BCs. To satisfy the two remaining, nonhomogeneous BCs, we must have

$$u_1(x,0) = g_1(x)$$

$$= \sum_{n=1}^{\infty} \left(- C_{1n} \sinh \frac{n\pi K}{L} \right) \sin \frac{n\pi x}{L} = \sum_{n=1}^{\infty} b_{1n} \sin \frac{n\pi x}{L},$$

$$u_1(x,K) = g_2(x)$$

$$= \sum_{n=1}^{\infty} \left(C_{2n} \sinh \frac{n\pi K}{L} \right) \sin \frac{n\pi x}{L} = \sum_{n=1}^{\infty} b_{2n} \sin \frac{n\pi x}{L};$$

in other words, we need the Fourier sine series of g_1 and g_2. By (2.10), for $n = 1,2,\ldots,$

$$b_{1n} = -C_{1n} \sinh \frac{n\pi K}{L} = \frac{2}{L} \int_0^L g_1(x) \sin \frac{n\pi x}{L} \, dx,$$

$$b_{2n} = C_{2n} \sinh \frac{n\pi K}{L} = \frac{2}{L} \int_0^L g_2(x) \sin \frac{n\pi x}{L} \, dx;$$

therefore,

$$C_{1n} = -\frac{b_{1n}}{\sinh(n\pi K/L)} = -\frac{2}{L} \operatorname{csch}(n\pi K/L) \int_0^L g_1(x) \sin \frac{n\pi x}{L} \, dx, \quad (5.35)$$

$$C_{2n} = \frac{b_{2n}}{\sinh(n\pi K/L)} = \frac{2}{L} \operatorname{csch}(n\pi K/L) \int_0^L g_2(x) \sin \frac{n\pi x}{L} \, dx. \quad (5.36)$$

This means that the function $u_1(x,y)$ is given by (5.34) with the coefficients C_{1n} and C_{2n} computed from (5.35) and (5.36).

The procedure for finding u_2 is similar, with the obvious modifications.

5.14. Example. For the BVP

$$u_{xx}(x,y) + u_{yy}(x,y) = 0, \quad 0 < x < 1, \ 0 < y < 2,$$
$$u(0,y) = 0, \quad u(1,y) = 0, \quad 0 < y < 2,$$
$$u(x,0) = x, \quad u(x,2) = 3\sin(\pi x), \quad 0 < x < 1,$$

we have

$$L = 1, \ K = 2, \ f_1(y) = 0, \ f_2(y) = 0, \ g_1(x) = x, \ g_2(x) = 3\sin(\pi x).$$

First, integrating by parts, we see that

$$\int_0^1 x\sin(n\pi x)\,dx = (-1)^{n+1}\frac{1}{n\pi};$$

so, by (5.35),

$$C_{1n} = (-1)^n \frac{2}{n\pi}\operatorname{csch}(2n\pi), \quad n = 1,2,\ldots.$$

Second, using (5.36), we easily convince ourselves that

$$C_{21} = 3\operatorname{csch}(2\pi), \quad C_{2n} = 0, \quad n = 2,3,\ldots.$$

Consequently, by (5.34), the solution of the given BVP is

$$u(x,y) = \sum_{n=1}^{\infty} (-1)^n \frac{2}{n\pi}\operatorname{csch}(2n\pi)\sin(n\pi x)\sinh\big(n\pi(y-2)\big)$$
$$+ 3\operatorname{csch}(2\pi)\sin(\pi x)\sinh(\pi y). \quad \blacksquare$$

5.15. Remark. The general solution of Y_n should be expressed in the most convenient form for the given nonhomogeneous BCs. Thus,

(i) if $(u_1)_y(x,0) = g_1(x)$, $(u_1)_y(x,K) = g_2(x)$, then

$$Y_n(y) = C_{1n}\cosh\frac{n\pi(y-K)}{L} + C_{2n}\cosh\frac{n\pi y}{L};$$

(ii) if $u_1(x,0) = g_1(x)$, $(u_1)_y(x,K) = g_2(x)$, then

$$Y_n(y) = C_{1n}\cosh\frac{n\pi(y-K)}{L} + C_{2n}\sinh\frac{n\pi y}{L};$$

(iii) if $(u_1)_y(x,0) = g_1(x), \; u_1(x,K) = g_2(x)$, then

$$Y_n(y) = C_{1n}\cosh\frac{n\pi y}{L} + C_{2n}\sinh\frac{n\pi(y-K)}{L}. \quad \blacksquare$$

The Laplace equation in a circular disk. Consider the equilibrium temperature in a uniform circular disk

$$D = \{(r,\theta) : 0 \leq r < \alpha, \; -\pi < \theta \leq \pi\}$$

with no sources, under the condition of prescribed temperature on the boundary $r = \alpha = $ const. Because of the geometry of the body, it is advisable that we write the modeling BVP in terms of polar coordinates r and θ with the pole at the center of the disk. Thus, recalling the form of the Laplacian in polar coordinates given in Remark 4.11(ii), we have

$$u_{rr}(r,\theta) + r^{-1}u_r(r,\theta) + r^{-2}u_{\theta\theta}(r,\theta) = 0,$$

$$0 < r < \alpha, \; -\pi < \theta < \pi, \qquad \text{(PDE)}$$

$$u(\alpha,\theta) = f(\theta), \quad -\pi < \theta \leq \pi. \qquad \text{(BC)}$$

Unlike the BVP discussed earlier in terms of Cartesian coordinates, where the boundary values of the solution proved sufficient for us to solve the problem, the form of the PDE in polar coordinates requires us to consider additional "boundary" conditions. These conditions are imposed at the remaining endpoints of the intervals on which the variables r and θ are defined, namely, at $r = 0$, $\theta = -\pi$, and $\theta = \pi$, and their form is suggested by analytic requirements based on physical considerations: u (the temperature) and u_r, u_θ (the heat flux) in a problem in a bounded region with "reasonably behaved" boundary data are expected to be continuous (therefore bounded) everywhere in that region. Thus, our additional conditions can be formulated as follows:

$$u(r,\theta), \; u_r(r,\theta) \text{ bounded as } r \to 0+, \quad -\pi < \theta < \pi,$$

$$u(r,-\pi) = u(r,\pi), \quad u_\theta(r,-\pi) = u_\theta(r,\pi), \quad 0 < r < \alpha.$$

The PDE and all the BCs except $u(\alpha,\theta) = f(\theta)$ are homogeneous; consequently, it seems advisable to seek a solution of the form

$$u(r,\theta) = R(r)\Theta(\theta). \qquad (5.37)$$

Then the last two (periodic) BCs become

$$\Theta(-\pi)R(r) = \Theta(\pi)R(r), \quad \Theta'(-\pi)R(r) = \Theta'(\pi)R(r), \quad 0 < r < \alpha,$$

from which, to avoid the identically zero—hence, unacceptable—solution, we deduce that

$$\Theta(-\pi) = \Theta(\pi), \quad \Theta'(-\pi) = \Theta'(\pi).$$

In view of (5.37), the PDE can be written as

$$\left[R''(r) + r^{-1}R'(r)\right]\Theta(\theta) + r^{-2}R(r)\Theta''(\theta) = 0,$$

which, on division by $r^{-2}R(r)\Theta(\theta)$ and in accordance with the usual argument, yields

$$\frac{\Theta''(\theta)}{\Theta(\theta)} = -\frac{r^2 R''(r) + rR'(r)}{R(r)} = -\lambda, \tag{5.38}$$

where $\lambda = $ const is the separation constant. Then Θ is a solution of the periodic Sturm–Liouville problem

$$\Theta''(\theta) + \lambda\Theta(\theta) = 0, \quad -\pi < \theta < \pi,$$
$$\Theta(-\pi) = \Theta(\pi), \quad \Theta'(-\pi) = \Theta'(\pi).$$

This problem was discussed in Example 3.26. Hence, for $L = \pi$, the eigenvalues and their corresponding eigenfunctions are, respectively,

$$\lambda_0 = 0, \quad \lambda_n = n^2, \quad n = 1, 2, \ldots,$$
$$\Theta_0(\theta) = 1, \quad \Theta_{1n}(\theta) = \cos(n\theta), \quad \Theta_{2n}(\theta) = \sin(n\theta), \quad n = 1, 2, \ldots.$$

From the second equality in (5.38) we see that for each $n = 0, 1, 2, \ldots$ the correspondent radial component R_n satisfies

$$r^2 R_n''(r) + rR_n'(r) - n^2 R_n = 0, \quad 0 < r < \alpha. \tag{5.39}$$

This is a Cauchy–Euler equation, for which, if $n \neq 0$, we seek solutions of the form (see Section 1.4)

$$R_n(r) = r^p, \quad p = \text{const}.$$

Replacing in (5.39), we find that

$$[p(p-1)+p-n^2]r^p = 0,$$

or $p^2 - n^2 = 0$, with roots $p_1 = n$ and $p_2 = -n$. Therefore, the general solution of (5.39) in this case is

$$R_n(r) = C_1 r^n + C_2 r^{-n}, \quad C_1, C_2 = \text{const.} \tag{5.40}$$

If $n = 0$, then (5.39) reduces to $(rR_0'(r))' = 0$, so

$$rR_0'(r) = \bar{C}_2 = \text{const};$$

in turn, this leads to $R_0'(r) = r^{-1}\bar{C}_2$ and then to the general solution

$$R_0(r) = \bar{C}_1 + \bar{C}_2 \ln r, \quad \bar{C}_1, \bar{C}_2 = \text{const.} \tag{5.41}$$

From (5.37) and the condition of boundedness at $r = 0$ we deduce that all the numbers $R_n(0)$ must be finite. As (5.40) and (5.41) indicate, this happens only if $C_2 = 0$ and $\bar{C}_2 = 0$; in other words,

$$R_0(r) = \bar{C}_1, \quad R_n(r) = C_1 r^n, \quad n = 1, 2, \ldots.$$

Taking the arbitrary constant equal to 1, we express this by the single formula

$$R_n(r) = r^n, \quad n = 0, 1, 2, \ldots.$$

In accordance with the separation of variables procedure, we expect the solution of the IBVP to have a series representation of the form

$$u(r, \theta) = \frac{1}{2} a_0 R_0(r)\Theta_0(\theta) + \sum_{n=1}^{\infty} R_n(r)\left[a_n \Theta_{1n}(\theta) + b_n \Theta_{2n}(\theta)\right]$$

$$= \frac{1}{2} a_0 + \sum_{n=1}^{\infty} r^n \left[a_n \cos(n\theta) + b_n \sin(n\theta)\right], \tag{5.42}$$

where each term satisfies the PDE and the three homogeneous BCs. The unknown numerical coefficients $\frac{1}{2} a_0$, a_n, and b_n are determined from the remaining (nonhomogeneous) condition:

$$u(\alpha,\theta) = f(\theta) = \tfrac{1}{2}a_0 + \sum_{n=1}^{\infty} \alpha^n \big[a_n \cos(n\theta) + b_n \sin(n\theta)\big],$$
$$-\pi < \theta \le \pi. \qquad (5.43)$$

This is the full Fourier series for f with $L = \pi$ and coefficients $\tfrac{1}{2}a_0$, $\alpha^n a_n$, and $\alpha^n b_n$; so, by (2.7)–(2.9),

$$a_0 = \frac{1}{\pi} \int_{-\pi}^{\pi} f(\theta)\,d\theta,$$

$$a_n = \frac{1}{\pi\alpha^n} \int_{-\pi}^{\pi} f(\theta)\cos(n\theta)\,d\theta, \quad n = 1,2,\ldots, \qquad (5.44)$$

$$b_n = \frac{1}{\pi\alpha^n} \int_{-\pi}^{\pi} f(\theta)\sin(n\theta)\,d\theta, \quad n = 1,2,\ldots.$$

Consequently, the solution of the BVP is given by (5.42) with the a_0, a_n, and b_n computed by means of (5.44).

5.16. Remark. From (5.42) and (5.44) it follows that

$$u(0,\theta) = \frac{1}{2}a_0 = \frac{1}{2\pi} \int_{-\pi}^{\pi} f(\theta)\,d\theta;$$

in other words, the temperature at the center of the disk is equal to the average of the temperature on the boundary circle. This result, already used in the proof of Theorem 4.14, is called the *mean value property* for the Laplace equation. ■

5.17. Example. In the BVP

$$(\Delta u)(r,\theta) = 0, \quad 0 < r < 2, \ -\pi < \theta < \pi,$$
$$u(2,\theta) = 1 + 8\sin\theta - 32\cos(4\theta), \quad -\pi < \theta < \pi,$$

we have

$$\alpha = 2, \quad f(\theta) = 1 + 8\sin\theta - 32\cos(4\theta).$$

Since f is a linear combination of the eigenfunctions associated with this problem, we use (5.43) and Theorem 3.20(ii) to deduce that $\frac{1}{2}a_0 = 1$, $2^4 a_4 = -32$, and $2b_1 = 8$, while the remaining coefficients are zero; therefore,

$$a_0 = 2, \quad a_4 = -2, \quad a_n = 0 \ (n \neq 0,4), \quad b_1 = 4, \quad b_n = 0 \ (n \neq 1),$$

which means that, by (5.42), the solution of the given BVP is

$$u(r,\theta) = 1 + 4r\sin\theta - 2r^4\cos(4\theta). \quad \blacksquare$$

5.4. Other Equations

The method of separation of variables can also be applied to problems involving some of the equations mentioned in Section 4.4.

5.18. Example. Consider the diffusion–convection IBVP

$$u_t(x,t) = 2u_{xx}(x,t) - u_x(x,t), \quad 0 < x < 1, \ t > 0, \qquad \text{(PDE)}$$
$$u(0,t) = 0, \quad u(1,t) = 0, \quad t > 0, \qquad \text{(BCs)}$$
$$u(x,0) = -2e^{x/4}\sin(3\pi x), \quad 0 < x < 1. \qquad \text{(IC)}$$

Applying the arguments developed earlier in this chapter, we seek the solution in the form $u(x,t) = X(x)T(t)$ and, from the PDE and BCs, deduce that the components X and T satisfy

$$2X''(x) - X'(x) + \lambda X(x) = 0, \quad 0 < x < 1,$$
$$X(0) = 0, \quad X(1) = 0, \quad t > 0, \tag{5.45}$$

and

$$T'(t) + \lambda T(t) = 0, \quad t > 0. \tag{5.46}$$

The regular Sturm–Liouville problem (5.45) is solved by the method used in Example 3.19, and its eigenvalues and eigenfunctions are found to be

$$\lambda_n = \tfrac{1}{8}(16n^2\pi^2 + 1), \quad X_n(x) = e^{x/4}\sin(n\pi x), \quad n = 1,2,\ldots.$$

Then (5.46) becomes

$$T_n(t) + \tfrac{1}{8}(16n^2\pi^2 + 1)T_n(t) = 0,$$

with general solution

$$T_n(t) = c_n e^{-(16n^2\pi^2+1)t/8}, \quad c_n = \text{const.}$$

Combining the X_n and T_n in the usual way, we find that the function

$$u(x,t) = \sum_{n=1}^{\infty} c_n e^{x/4-(16n^2\pi^2+1)t/8} \sin(n\pi x) \qquad (5.47)$$

satisfies both the PDE and the BCs.

The coefficients c_n are found from the IC; thus, for $t = 0$ we must have

$$u(x,0) = \sum_{n=1}^{\infty} c_n e^{x/4} \sin(n\pi x) = -2e^{x/4} \sin(3\pi x),$$

from which

$$c_3 = -2, \quad c_n = 0 \ (n \neq 3),$$

so the solution, given by (5.47), of the IBVP is

$$u(x,t) = -2e^{x/4-(144\pi^2+1)t/8} \sin(3\pi x). \ \blacksquare$$

5.19. Example. Consider the dissipative wave propagation problem

$$u_{tt}(x,t) + 2u_t(x,t) = u_{xx}(x,t) - 3u_x(x,t),$$
$$0 < x < 1, \ t > 0,$$
$$u(0,t) = 0, \quad u(1,t) = 0, \quad t > 0,$$
$$u(x,0) = -e^{3x/2} \sin(2\pi x), \quad u_t(x,0) = 2e^{3x/2} \sin(\pi x),$$
$$0 < x < 1.$$

Writing $u(x,t) = X(x)T(t)$ and proceeding as in Example 5.18, we find that

$$X''(x) - 3X'(x) + \lambda X(x) = 0, \quad 0 < x < 1,$$
$$X(0) = 0, \quad X(1) = 0, \qquad (5.48)$$

and

$$T''(t) + 2T' + \lambda T(t) = 0. \qquad (5.49)$$

The regular S–L problem (5.48), solved as in Example 3.19, yields the eigenvalues and eigenfunctions

$$\lambda_n = \tfrac{1}{4}(n^2\pi^2 + 9), \quad X_n(x) = e^{3x/2}\sin(n\pi x), \quad n = 1,2,\ldots.$$

Then, from (5.49) with λ replaced by $\lambda_n = \tfrac{1}{4}(n^2\pi^2 + 9)$, we find that

$$T_n(t) = e^{-t/2}\big[c_{1n}\cos(\alpha_n t) + c_{2n}\sin(\alpha_n t)\big], \quad c_{1n}, c_{2n} = \text{const},$$

where

$$\alpha_n = \tfrac{1}{2}\sqrt{n^2\pi^2 + 5}, \quad n = 1,2,\ldots.$$

Putting together the X_n and T_n, we now see that the function

$$u(x,t) = \sum_{n=1}^{\infty} e^{3x/2 - t/2}\sin(n\pi x)\big[c_{1n}\cos(\alpha_n t) + c_{2n}\sin(\alpha_n t)\big] \tag{5.50}$$

satisfies the PDE and the BCs. We now set $t = 0$ in (5.50) and in its t-derivative and compute the coefficients c_{1n} and c_{2n} from the ICs:

$$u(x,0) = \sum_{n=1}^{\infty} c_{1n}e^{3x/2}\sin(n\pi x) = -e^{3x/2}\sin(2\pi x),$$

$$u_t(x,0) = \sum_{n=1}^{\infty}\big(-\tfrac{1}{2}c_{1n} + \alpha_n c_{2n}\big)e^{3x/2}\sin(n\pi x) = 2e^{3x/2}\sin(\pi x);$$

hence,

$$c_{12} = -1, \quad c_{1n} = 0 \ (n \neq 2),$$

$$-\tfrac{1}{2}c_{11} + \alpha_1 c_{21} = 2, \quad -\tfrac{1}{2}c_{1n} + \alpha_n c_{2n} = 0 \ (n \neq 1).$$

From these equalities we find that

$$c_{21} = \frac{2}{\alpha_1}, \quad c_{22} = -\frac{1}{2\alpha_2}, \quad c_{2n} = 0 \ (n \neq 1,2),$$

so the solution of the IBVP, obtained from (5.50), is

$$u(x,t) = e^{(3x-t)/2}\bigg[\frac{2}{\alpha_1}\sin(\alpha_1 t)\sin(\pi x) - \cos(\alpha_1 t)\sin(2\pi x)$$

$$-\frac{1}{2\alpha_2}\sin(\alpha_2 t)\sin(2\pi x)\bigg],$$

where $\alpha_1 = \tfrac{1}{2}\sqrt{\pi^2 + 5}$ and $\alpha_2 = \tfrac{1}{2}\sqrt{4\pi^2 + 5}$. ∎

5.20. Example. Consider the two-dimensional steady state diffusion–convection problem

$$u_{xx}(x,y) + u_{yy}(x,y) - 2u_x(x,y) = 0, \quad 0 < x < 1, \ 0 < y < 2, \quad \text{(PDE)}$$

$$u(0,y) = 0, \quad u(1,y) = -3\sin(\pi y), \quad 0 < y < 2,$$
$$u(x,0) = 0, \quad u(x,2) = 0, \quad 0 < x < 1. \qquad \text{(BCs)}$$

Seeking the solution in the form $u(x,y) = X(x)Y(y)$, from the PDE and the last two (homogeneous) BCs it follows that

$$X''(x) - 2X'(x) - \lambda X(x) = 0, \quad 0 < x < 1, \qquad (5.51)$$

and
$$Y''(y) + \lambda Y(y) = 0, \quad 0 < y < 2,$$
$$Y(0) = 0, \quad Y(1) = 0.$$

The regular S–L problem for Y has been solved in Example 3.16 and has the eigenvalues and eigenfunctions

$$\lambda_n = \frac{n^2\pi^2}{4}, \quad Y_n(y) = \sin\frac{n\pi y}{2}, \quad n = 1,2,\ldots,$$

so (5.51) becomes

$$X_n''(x) - 2X_n'(x) - \frac{n^2\pi^2}{4}X_n(x) = 0,$$

with general solution

$$X_n(x) = c_{1n}e^{s_{1n}x} + c_{2n}e^{s_{2n}x},$$

where

$$s_{1n} = 1 + \tfrac{1}{2}\sqrt{n^2\pi^2 + 4}, \quad s_{2n} = 1 - \tfrac{1}{2}\sqrt{n^2\pi^2 + 4}, \quad c_{1n}, c_{2n} = \text{const.}$$

Consequently, the function

$$u(x,y) = \sum_{n=1}^{\infty} \sin\frac{n\pi y}{2}\left(c_{1n}e^{s_{1n}x} + c_{2n}e^{s_{2n}x}\right) \qquad (5.52)$$

satisfies the PDE and the last two BCs. The coefficients c_{1n} and c_{2n} are determined from the first two BCs; thus, setting $x = 0$ and then $x = 1$ in (5.52), we find that

$$\sum_{n=1}^{\infty} (c_{1n} + c_{2n}) \sin \frac{n\pi y}{2} = 0,$$

$$\sum_{n=1}^{\infty} \left(c_{1n} e^{s_{1n}} + c_{2n} e^{s_{2n}} \right) \sin \frac{n\pi y}{2} = -3\sin(\pi y),$$

so

$$c_{1n} + c_{2n} = 0, \quad n = 1, 2, \ldots,$$

$$c_{12} e^{s_{12}} + c_{22} e^{s_{22}} = -3, \quad c_{1n} e^{s_{1n}} + c_{2n} e^{s_{2n}} = 0 \ \ (n \neq 2).$$

Then

$$c_{12} = -c_{22} = \frac{3}{e^{s_{22}} - e^{s_{12}}}, \quad c_{1n} = c_{2n} = 0 \ \ (n \neq 2),$$

which, substituted in (5.52), yield the solution of the given BVP:

$$u(x,y) = \frac{3}{e^{s_{22}} - e^{s_{12}}} \left(e^{s_{12}x} - e^{s_{22}x} \right) \sin(\pi y),$$

where

$$s_{12} = 1 + \sqrt{\pi^2 + 1}, \quad s_{22} = 1 - \sqrt{\pi^2 + 1}. \ \blacksquare$$

5.5. Equations with More than Two Variables

In Section 4.2 we derived a model for heat conduction in a uniform plate, where the unknown temperature function depends on the time variable and two space variables. Here we consider the two-dimensional analog of the wave equation and solve it by the method of separation of variables in terms of both Cartesian and polar coordinates.

Vibrating rectangular membrane. A vibrating rectangular membrane with negligible body force and clamped edges is modeled by the IBVP

$$u_{tt}(x,t) = c^2 (\Delta u)(x,y) = c^2 \left[u_{xx}(x,y) + u_{yy}(x,y) \right],$$

$$0 < x < L, \ 0 < y < K, \ t > 0, \quad \text{(PDE)}$$

$$u(0,y,t) = 0, \quad u(L,y,t) = 0, \quad 0 < y < K, \ t > 0,$$

$$u(x,0,t) = 0, \quad u(x,K,t) = 0, \quad 0 < x < L, \ t > 0, \quad \text{(BCs)}$$

$$u(x,y,0) = f(x,y), \quad u_t(x,y,0) = g(x,y),$$

$$0 < x < L, \ 0 < y < K. \quad \text{(ICs)}$$

Since the PDE and BCs are linear and homogeneous, we seek the solution as a product of a function of the space variables and a function of the time variable:

$$u(x,y,t) = S(x,y)T(t). \tag{5.53}$$

Reasoning as in previous similar situations, from the PDE and BCs we find that T satisfies the ODE

$$T''(t) + \lambda c^2 T(t) = 0, \quad t > 0,$$

and that S is a solution of the BVP

$$S_{xx}(x,y) + S_{yy}(x,y) + \lambda S(x,y) = 0,$$
$$0 < x < L, \ 0 < y < K,$$
$$S(0,y) = 0, \quad S(L,y) = 0, \quad 0 < y < K,$$
$$S(x,0) = 0, \quad S(x,K) = 0, \quad 0 < x < L, \tag{5.54}$$

where $\lambda = \mathrm{const}$ is the separation constant.

We notice that the PDE and BCs for S are themselves linear and homogeneous, so we use separation of variables once more and seek a solution for (5.54) of the form

$$S(x,y) = X(x)Y(y). \tag{5.55}$$

The PDE in (5.54) then becomes

$$X''(x)Y(y) + X(x)Y''(y) = -\lambda X(x)Y(y),$$

which, on division by $X(x)Y(y)$, yields

$$\frac{X''(x)}{X(x)} = -\lambda - \frac{Y''(y)}{Y(y)} = -\mu,$$

where $\mu = \mathrm{const}$ is a second separation constant.

From the above chain of equalities and the BCs in (5.54) we find in the usual way that X and Y are, respectively, solutions of the regular Sturm–Liouville problems

$$X''(x) + \mu X(x) = 0, \quad 0 < x < L,$$
$$X(0) = 0, \quad X(L) = 0, \tag{5.56}$$

and

$$Y''(y) + (\lambda - \mu)Y(y) = 0, \quad 0 < y < K,$$
$$Y(0) = 0, \quad Y(K) = 0. \tag{5.57}$$

The nonzero solutions that we require for (5.56) are the eigenfunctions of this S–L problem, which were computed in Example 3.16 together with the corresponding eigenvalues:

$$\mu_n = \left(\frac{n\pi}{L}\right)^2, \quad X_n(x) = \sin\frac{n\pi x}{L}, \quad n = 1, 2, \dots.$$

For each fixed value of n, the function Y satisfies the S–L problem (5.57) with $\mu = \mu_n$. Since this problem is similar to (5.56), its eigenvalues and eigenfunctions (nonzero solutions) are

$$\lambda_{nm} - \mu_n = \left(\frac{m\pi}{K}\right)^2, \quad Y_{nm}(y) = \sin\frac{m\pi y}{K}, \quad m = 1, 2, \dots.$$

We remark that (5.54) may be regarded as a two-dimensional eigenvalue problem with eigenvalues

$$\lambda_{nm} = \mu_n + \left(\frac{m\pi}{K}\right)^2 = \left(\frac{n\pi}{L}\right)^2 + \left(\frac{m\pi}{K}\right)^2, \quad n, m = 1, 2, \dots,$$

and corresponding eigenfunctions

$$S_{nm}(x, y) = X_n(x)Y_{nm}(y) = \sin\frac{n\pi x}{L} \sin\frac{m\pi y}{K}, \quad n, m = 1, 2, \dots.$$

Going back to the equation satisfied by T, for $\lambda = \lambda_{nm}$ we obtain the functions

$$T_{nm}(t) = a_{nm} \cos\left(\sqrt{\lambda_{nm}}\, ct\right) + b_{nm} \sin\left(\sqrt{\lambda_{nm}}\, ct\right), \quad a_{nm}, b_{nm} = \text{const.}$$

From this, (5.53), and (5.55) we now conclude that the solution of the original IBVP should have a representation of the form

$$u(x, y, t) = \sum_{n=1}^{\infty} \sum_{m=1}^{\infty} S_{nm}(x, y) T_{nm}(t)$$
$$= \sum_{n=1}^{\infty} \sum_{m=1}^{\infty} \sin\frac{n\pi x}{L} \sin\frac{m\pi y}{K}$$
$$\times \left[a_{nm} \cos\left(\sqrt{\lambda_{nm}}\, ct\right) + b_{nm} \sin\left(\sqrt{\lambda_{nm}}\, ct\right)\right]. \tag{5.58}$$

Each term in (5.58) satisfies the PDE and the BCs. To satisfy the first IC, we must have

$$u(x,y,0) = f(x,y) = \sum_{m=1}^{\infty} \left(\sum_{n=1}^{\infty} a_{nm} \sin \frac{n\pi x}{L} \right) \sin \frac{m\pi y}{K}.$$

Multiplying the second equality by $\sin(p\pi y/K)$, $p = 1,2,\ldots$, integrating the result with respect to y over $[0,K]$, and taking into account formula (2.5) with L replaced by K, we find that

$$\int_0^K f(x,y) \sin \frac{p\pi y}{K}\, dy = \frac{K}{2} \sum_{n=1}^{\infty} a_{np} \sin \frac{n\pi x}{L}.$$

Multiplying this new equality by $\sin(q\pi x/L)$, $q = 1,2,\ldots$, and integrating with respect to x over $[0,L]$, we arrive at

$$\int_0^L \int_0^K f(x,y) \sin \frac{p\pi y}{K} \sin \frac{q\pi x}{L}\, dy\, dx = \frac{LK}{4} a_{qp},$$

from which, replacing q by n and p by m, we obtain

$$a_{nm} = \frac{4}{LK} \int_0^L \int_0^K f(x,y) \sin \frac{n\pi x}{L} \sin \frac{m\pi y}{K}\, dy\, dx, \quad n,m = 1,2,\ldots. \tag{5.59}$$

The second IC is satisfied if

$$u_t(x,y,0) = g(x,y) = \sum_{m=1}^{\infty} \left(\sum_{n=1}^{\infty} \sqrt{\lambda_{nm}}\, cb_{nm} \sin \frac{n\pi x}{L} \right) \sin \frac{m\pi y}{K};$$

as above, this yields

$$b_{nm} = \frac{4}{LK\sqrt{\lambda_{nm}}\, c} \int_0^L \int_0^K g(x,y) \sin \frac{n\pi x}{L} \sin \frac{m\pi y}{K}\, dy\, dx,$$

$$n,m = 1,2,\ldots. \tag{5.60}$$

In conclusion, the solution of the IBVP is given by series (5.58) with the coefficients a_{nm} and b_{nm} computed by means of (5.59) and (5.60).

Other homogeneous BCs are treated similarly.

5.21. Example. In the IBVP

$$u_{tt}(x,y,t) = u_{xx}(x,y,t) + u_{yy}(x,y,t),$$
$$0 < x < 1, \ 0 < y < 2, \ t > 0,$$
$$u(0,y,t) = 0, \quad u(1,y,t) = 0, \quad 0 < y < 2, \ t > 0,$$
$$u(x,0,t) = 0, \quad u(x,2,t) = 0, \quad 0 < x < 1, \ t > 0,$$
$$u(x,y,0) = \sin(2\pi x)\left[\sin(\tfrac{1}{2}\pi y) - 2\sin(\pi y)\right],$$
$$0 < x < 1, \ 0 < y < 2,$$
$$u_t(x,y,0) = 2\sin(2\pi x)\sin(\tfrac{1}{2}\pi y), \quad 0 < x < 1, \ 0 < y < 2,$$

we have $c = 1$, $L = 1$, and $K = 2$; thus, comparing the initial data functions with the double Fourier series expansions of f and g, we see that

$$a_{21} = 1, \quad a_{22} = -2, \quad a_{2m} = 0 \ (m \neq 1,2),$$
$$\sqrt{\lambda_{21}}\, b_{21} = 2, \quad \sqrt{\lambda_{2m}}\, b_{2m} = 0 \ (m \neq 1).$$

Since $\lambda_{21} = (2\pi)^2 + (\pi/2)^2 = 17\pi^2/4$, formula (5.58) yields the solution

$$u(x,y,t) = \sin(2\pi x)\sin\left(\tfrac{1}{2}\pi y\right)\cos\left(\tfrac{1}{2}\sqrt{17}\,\pi t\right)$$
$$- 2\sin(2\pi x)\sin(\pi y)\cos\left(\sqrt{5}\,\pi t\right)$$
$$+ \left(4\pi/\sqrt{17}\right)\sin(2\pi x)\sin\left(\tfrac{1}{2}\pi y\right)\sin\left(\tfrac{1}{2}\sqrt{17}\,\pi t\right). \ \blacksquare$$

Vibrating circular membrane. A two-dimensional body of this type is defined in polar coordinates by $0 \leq r \leq \alpha$, $-\pi < \theta \leq \pi$. If the body force is negligible and the boundary $r = \alpha$ is clamped (fixed), then the vertical vibrations of the membrane are described by the IBVP

$$u_{tt}(r,\theta,t) = c^2(\Delta u)(r,\theta,t), \quad 0 < r < \alpha, \ -\pi < \theta < \pi, \ t > 0, \quad \text{(PDE)}$$
$$u(\alpha,\theta,t) = 0, \quad u(r,\theta,t), \ u_r(r,\theta,t) \text{ bounded as } r \to 0+,$$
$$-\pi < \theta < \pi, \ t > 0,$$
$$u(r,-\pi,t) = u(r,\pi,t), \quad u_\theta(r,-\pi,t) = u_\theta(r,\pi,t),$$
$$0 < r < \alpha, \ t > 0, \quad \text{(BCs)}$$
$$u(r,\theta,0) = f(r,\theta), \quad u_t(r,\theta,0) = g(r,\theta),$$
$$0 < r < \alpha, \ -\pi < \theta < \pi. \quad \text{(ICs)}$$

As in Section 5.3, the BCs at $r = 0$ and $\theta = -\pi, \pi$ express the continuity (therefore boundedness) of the displacement (u) and tension (u_r, u_θ) in the membrane for "reasonably behaved" functions f and g.

Since the PDE and BCs are linear and homogeneous, we try the method of separation of variables and seek a solution of the form

$$u(r,\theta,t) = S(r,\theta)T(t). \tag{5.61}$$

Then, following the standard procedure, from the PDE and BCs we find that S is a solution of the problem

$$(\Delta S)(r,\theta) + \lambda S(r,\theta) = 0, \quad 0 < r < \alpha, \quad -\pi < \theta < \pi,$$
$$S(\alpha,\theta) = 0, \quad S(r,\theta), \ S_r(r,\theta) \text{ bounded as } r \to 0+, \quad -\pi < \theta < \pi, \tag{5.62}$$
$$S(r,-\pi) = S(r,\pi), \quad S_\theta(r,-\pi) = S_\theta(r,\pi), \quad 0 < r < \alpha,$$

where $\lambda = \text{const}$ is the separation constant, while T satisfies

$$T''(t) + \lambda c^2 T(t) = 0, \quad t > 0.$$

Since the PDE and BCs for S are linear and homogeneous, we seek S of the form

$$S(r,\theta) = R(r)\Theta(\theta). \tag{5.63}$$

Next, recalling the expression of the Laplacian in polar coordinates (see Remark 4.11(ii)), we write the PDE in (5.62) as

$$\left[R''(r) + r^{-1}R'(r)\right]\Theta(\theta) + r^{-2}R(r)\Theta''(\theta) + \lambda R(r)\Theta(\theta) = 0,$$

which, on division by $r^{-2}R(r)\Theta(\theta)$, yields

$$\frac{\Theta''(\theta)}{\Theta(\theta)} = -\frac{r^2 R''(r) + rR'(r)}{R(r)} - \lambda r^2 = -\mu,$$

where $\mu = \text{const}$ is a second separation constant.

Applying the usual argument to the BCs, we now conclude that R and Θ are solutions of Sturm–Liouville problems. First, Θ satisfies the periodic S–L problem

$$\Theta''(\theta) + \mu\Theta(\theta) = 0, \quad -\pi < \theta < \pi,$$
$$\Theta(-\pi) = \Theta(\pi), \quad \Theta'(-\pi) = \Theta'(\pi),$$

whose eigenvalues and eigenfunctions, computed in Example 3.26, are

$$\mu_0 = 0, \quad \Theta_0(\theta) = \tfrac{1}{2},$$
$$\mu_n = n^2, \quad \Theta_{1n}(\theta) = \cos(n\theta), \quad \Theta_{2n}(\theta) = \sin(n\theta), \quad n = 1,2,\ldots.$$

For each fixed value of n, we compute a solution R_n of the singular S–L problem

$$r^2 R_n''(r) + r R_n'(r) + (\lambda r^2 - n^2) R_n(r) = 0, \quad 0 < r < \alpha,$$
$$R_n(\alpha) = 0, \quad R_n(r), \; R_n'(r) \text{ bounded as } r \to 0+.$$

This is the singular Sturm–Liouville problem (3.12), (3.13) (with m replaced by n) discussed in Section 3.3, which has countably many eigenvalues and a complete set of corresponding eigenfunctions

$$\lambda_{nm} = \left(\frac{\xi_{nm}}{\alpha}\right)^2, \quad R_{nm}(r) = J_n\left(\frac{\xi_{nm} r}{\alpha}\right), \quad n = 0,1,2,\ldots, \; m = 1,2,\ldots,$$

where J_n is the Bessel function of the first kind and order n and ξ_{nm} are the zeros of J_n.

Returning to the ODE satisfied by T, for $\lambda = \lambda_{nm} = (\xi_{nm}/\alpha)^2$ we now obtain the solutions

$$T_{nm}(t) = A_{nm} \cos\frac{\xi_{nm} ct}{\alpha} + B_{nm} \sin\frac{\xi_{nm} ct}{\alpha}, \quad A_{nm}, B_{nm} = \text{const.}$$

Suppose, for simplicity, that $g = 0$; then it is clear that the sine term in T_{nm} must vanish. Hence, if we piece together the various components of u according to (5.61) and (5.63), we conclude that the solution of the IBVP should be of the form

$$u(r,\theta,t) = \sum_{n=0}^{\infty}\sum_{m=1}^{\infty} \left[a_{nm}\Theta_{1n}(\theta) + b_{nm}\Theta_{2n}(\theta)\right] R_{nm}(r) T_{nm}(t)$$
$$= \sum_{n=0}^{\infty}\sum_{m=1}^{\infty} \left[a_{nm}\cos(n\theta) + b_{nm}\sin(n\theta)\right] J_n\left(\frac{\xi_{nm} r}{\alpha}\right) \cos\frac{\xi_{nm} ct}{\alpha}.$$

Each term in this sum satisfies the PDE, the BCs, and the second IC. To satisfy the first IC, we need to have

$$u(r,\theta,0) = f(r,\theta) = \sum_{m=1}^{\infty}\left\{\sum_{n=0}^{\infty}\left[a_{nm}\cos(n\theta) + b_{nm}\sin(n\theta)\right]\right\} J_n\left(\frac{\xi_{nm} r}{\alpha}\right).$$

The a_{nm} and b_{nm} are now computed as in other problems of this type, by means of the orthogonality properties of $\cos(n\theta)$, $\sin(n\theta)$, and $J_n(\xi_{nm}r/\alpha)$ expressed by (2.4)–(2.6) with $L = \pi$ and by (3.16).

5.22. Remark. In the circularly symmetric case (that is, when the data functions are independent of θ) we use the solution split $u(r,t) = R(r)T(t)$. Here the problem reduces to solving Bessel's equation of order zero. In the end, we obtain

$$u(r,t) = \sum_{m=1}^{\infty} \left[a_m \cos\left(\sqrt{\lambda_m}\, ct\right) + b_m \sin\left(\sqrt{\lambda_m}\, ct\right) \right] J_0\left(\sqrt{\lambda_m}\, r\right),$$

where λ_m is the same as $\lambda_{0m} = (\xi_{0m}/\alpha)^2$ in the general problem and the a_m and b_m are determined from the property of orthogonality with weight r on $[0,\alpha]$ of the functions $J_0\left(\sqrt{\lambda_m}\, r\right)$. ■

5.23. Example. With $\alpha = 1$, $c = 1$, $f(r,\theta) = 1$, and $g(r,\theta) = 0$, the formula in Remark 5.22 yields

$$u(r,0) = \sum_{m=1}^{\infty} a_m J_0\left(\sqrt{\lambda_m}\, r\right) = 1,$$

$$u_t(r,0) = \sum_{m=1}^{\infty} b_m \sqrt{\lambda_m}\, J_0\left(\sqrt{\lambda_m}\, r\right) = 0,$$

so the a_m are the coefficients of the expansion of the constant function 1 in the functions $J_0\left(\sqrt{\lambda_m}\, r\right)$ on $(0,1)$ and $b_m = 0$. Since, with a computational approximation to four decimal places,

$$1 = 1.6020 J_0(2.4048r) - 1.0463 J_0(5.5201r) + 0.8514 J_0(8.6537r) + \cdots,$$

it follows that

$$a_1 = 1.6020, \quad a_2 = -1.0463, \quad a_3 = 0.8514, \quad \ldots,$$

so the solution of the problem is

$$u(r,t) = 1.6020 \cos(2.4048t) J_0(2.4048r)$$
$$- 1.0463 \cos(5.5201t) J_0(5.5201r)$$
$$+ 0.8514 \cos(8.6537t) J_0(8.6537r) + \cdots. \quad ■$$

Equilibrium temperature in a solid sphere. The steady state distribution of heat inside a homogeneous sphere of radius α when the (time-independent) temperature is prescribed on the surface and no sources are present is the solution of the BVP

$$\Delta u(r,\theta,\varphi) = 0, \quad 0 < r < \alpha, \ 0 < \theta < 2\pi, \ 0 < \varphi < \pi, \qquad \text{(PDE)}$$

$$u(\alpha,\theta,\varphi) = f(\theta,\varphi), \quad u(r,\theta,\varphi), \ u_r(r,\theta,\varphi) \text{ bounded as } r \to 0+,$$

$$0 < \theta < 2\pi, \ 0 < \varphi < \pi,$$

$$u(r,0,\varphi) = u(r,2\pi,\varphi), \quad u_\theta(r,0,\varphi) = u_\theta(r,2\pi,\varphi),$$

$$0 < r < \alpha, \ 0 < \varphi < \pi, \qquad \text{(BCs)}$$

$$u(r,\theta,\varphi), \ u_\varphi(r,\theta,\varphi) \text{ bounded as } \varphi \to 0+ \text{ and as } \varphi \to \pi-,$$

$$0 < r < \alpha, \ 0 < \theta < 2\pi,$$

where f is a known function. The first boundary condition represents the given surface temperature; the remaining ones have been adjoined on the basis of arguments similar to those used in the case of a uniform circular disk.

In terms of the spherical coordinates r, θ, φ, the above PDE is written as (see Remark 4.11(v))

$$\frac{1}{r^2}(r^2 u_r)_r + \frac{1}{r^2 \sin^2\varphi} u_{\theta\theta} + \frac{1}{r^2 \sin\varphi}((\sin\varphi)u_\varphi)_\varphi = 0.$$

Since the PDE and all but one of the BCs are homogeneous, we try a solution of the form

$$u(r,\theta,\varphi) = R(r)\Theta(\theta)\Phi(\varphi).$$

As before, the nonhomogeneous BC implies that none of R, Θ, Φ can be the zero function. Replacing in the PDE and multiplying every term by $(r^2 \sin^2\varphi)/(R\Theta\Phi)$, we find that

$$\frac{\Theta''}{\Theta} = -(\sin^2\varphi)\frac{(r^2 R')'}{R} - (\sin\varphi)\frac{((\sin\varphi)\Phi')'}{\Phi},$$

where the left-hand side is a function of θ and the right-hand side is a function of r and φ. Consequently, both sides must be equal to one and the same constant, which, for convenience, we denote by $-\mu$. This leads to the pair of equations

$$\Theta''(\theta) + \mu\Theta(\theta) = 0, \quad 0 < \theta < 2\pi, \tag{5.64}$$

$$\frac{(r^2 R')'}{R} + \frac{1}{\sin\varphi}\frac{((\sin\varphi)\Phi')'}{\Phi} - \frac{\mu}{\sin^2\varphi} = 0,$$
$$0 < r < \alpha, \ 0 < \varphi < \pi. \tag{5.65}$$

(In (5.65) we have also divided by $\sin^2\varphi$.)

To avoid the identically zero solution, from the two periodicity conditions in the BVP we deduce in the usual way that

$$\Theta(0) = \Theta(2\pi), \quad \Theta_\theta(0) = \Theta_\theta(2\pi). \tag{5.66}$$

The regular S–L problem (5.64), (5.66) for Θ is the same as in the case of the circular membrane; it has the eigenvalues $\mu = m^2$, $m = 0, 1, 2, \ldots$, and corresponding eigenfunctions

$$\Theta_m(\theta) = C_{1m}\cos(m\theta) + C_{2m}\sin(m\theta),$$

which, using Euler's formula, we can rewrite as

$$\Theta_m(\theta) = C'_{1m}e^{im\theta} + C'_{12}e^{-im\theta}. \tag{5.67}$$

We now operate a second separation of variables, this time in (5.65):

$$\frac{1}{\sin\varphi}\frac{((\sin\varphi)\Phi')'}{\Phi} - \frac{m^2}{\sin^2\varphi} = \frac{(r^2 R')'}{R} = -\lambda = \text{const};$$

this yields the equations

$$\frac{1}{\sin\varphi}((\sin\varphi)\Phi'(\varphi))' + \left(\lambda - \frac{m^2}{\sin^2\varphi}\right)\Phi(\varphi) = 0, \quad 0 < \varphi < \pi, \tag{5.68}$$

$$(r^2 R')' - \lambda R = 0, \quad 0 < r < \alpha. \tag{5.69}$$

In (5.68) we substitute

$$\xi = \cos\varphi, \quad d\xi = -\sin\varphi\, d\varphi, \quad \frac{d}{d\xi} = -\frac{1}{\sin\varphi}\frac{d}{d\varphi},$$
$$\sin^2\varphi = 1 - \cos^2\varphi = 1 - \xi^2, \quad \Phi(\varphi) = \Psi(\xi)$$

and bring the equation to the form

$$((1-\xi^2)\Psi'(\xi))' + \left(\lambda - \frac{m^2}{1-\xi^2}\right)\Psi(\xi) = 0, \quad -1 < \xi < 1. \tag{5.70}$$

The boundary conditions for Ψ are obtained from those at $\varphi = 0$ and $\varphi = \pi$ in the original problem by recalling that Φ cannot be the zero function:

$$\Psi(\xi), \ \Psi'(\xi) \text{ bounded as } \xi \to -1+ \text{ and as } \xi \to 1-. \tag{5.71}$$

We remark that (5.70), (5.71) is the singular S–L problem for the associated Legendre equation mentioned in Section 3.5, with eigenvalues and eigenfunctions

$$\lambda_n = n(n+1), \quad \Psi(\xi) = P_n^m(\xi), \quad n = m, m+1, \ldots,$$

where P_n^m are the associated Legendre functions; consequently,

$$\Phi_n(\varphi) = P_n^m(\cos\varphi). \tag{5.72}$$

Turning to the Cauchy–Euler equation (5.69) with $\lambda = n(n+1)$, we find that its general solution (see Section 1.4) is

$$R_n(r) = C_1 r^n + C_2 r^{-(n+1)}, \quad C_1, C_2 = \text{const.}$$

When the condition that $R(r)$ and $R'(r)$ be bounded as $r \to 0+$, which follows from the second BC in the given BVP, is applied, we find that $C_2 = 0$, so

$$R_n(r) = r^n. \tag{5.73}$$

Here, as usual, we have taken $C_1 = 1$.

Finally, we combine (5.67), (5.72), (5.73), (3.31), and (3.32) and write the solution of the original problem in the form

$$\begin{aligned}
u(r,\theta,\varphi) &= \sum_{\substack{m=0,1,2,\ldots \\ n=m,m+1,\ldots}} R_n(r)\Theta_m(\theta)\Phi_{nm}(\varphi) \\
&= \sum_{\substack{m=0,1,2,\ldots \\ n=m,m+1,\ldots}} r^n \left(C'_{1m} e^{im\theta} + C'_{2m} e^{-im\theta} \right) P_n^m(\cos\varphi) \\
&= \sum_{n=0}^{\infty} \sum_{m=-n}^{m=n} C_{nm} r^n e^{im\theta} P_n^m(\cos\varphi) \\
&= \sum_{n=0}^{\infty} \sum_{m=-n}^{m=n} c_{n,m} r^n Y_{n,m}(\theta,\varphi),
\end{aligned} \tag{5.74}$$

where $Y_{n,m}$ are the orthonormalized spherical harmonics defined by (3.32).

According to the nonhomogeneous BC in the given BVP, we must have

$$u(\alpha,\theta,\varphi) = \sum_{n=1}^{\infty} \sum_{m=-n}^{m=n} c_{n,m} \alpha^n Y_{n,m}(\theta,\varphi) = f(\theta,\varphi).$$

The coefficients $c_{n,m}$ are now determined in the standard way, by means of the orthonormality relations (3.33):

$$c_{n,m} = \frac{1}{\alpha^n} \int_0^{2\pi} \int_0^\pi f(\theta,\varphi) \bar{Y}_{n,m}(\theta,\varphi) \sin\varphi \, d\varphi \, d\theta. \qquad (5.75)$$

5.24. Example. If in the preceding problem the upper and lower hemispheres of the sphere with the center at the origin and radius $\alpha = 1$ are kept at constant temperatures of 2 and -1, respectively, then

$$f(\theta,\varphi) = \begin{cases} 2, & 0 < \theta < 2\pi, \ 0 < \varphi < \pi/2, \\ -1, & 0 < \theta < 2\pi, \ \pi/2 < \varphi < \pi, \end{cases}$$

so, using (5.75), we find that the first two nonzero coefficients $c_{n,m}$ are $c_{0,0} = \sqrt{\pi}$ and $c_{1,0} = 3\sqrt{3\pi}/2$. By (5.74) and the expressions of the spherical harmonics given in Example 3.36, the solution of the problem is

$$u(r,\theta,\varphi) = \sqrt{\pi}\, Y_{0,0}(\theta,\varphi) + \tfrac{3}{2}\sqrt{3\pi}\, r Y_{1,0}(\theta,\varphi) + \cdots = \tfrac{1}{2} + \tfrac{9}{4} r\cos\varphi + \cdots. \quad \blacksquare$$

Exercises

In (1)–(24) use separation of variables to solve the PDE

$$u_t(x,t) = u_{xx}(x,t), \quad 0 < x < 1, \ t > 0,$$

with the BCs and IC as indicated.

(1) $u(0,t) = 0, \ u(1,t) = 0, \ u(x,0) = \sin(2\pi x) - 3\sin(6\pi x).$
(2) $u(0,t) = 0, \ u(1,t) = 0, \ u(x,0) = 3\sin(\pi x) - \sin(4\pi x).$
(3) $u(0,t) = 0, \ u(1,t) = 0, \ u(x,0) = -2.$
(4) $u(0,t) = 0, \ u(1,t) = 0, \ u(x,0) = 1 - 2x.$
(5) $u(0,t) = 0, \ u(1,t) = 0, \ u(x,0) = 2x + 1.$
(6) $u(0,t) = 0, \ u(1,t) = 0, \ u(x,0) = \begin{cases} 0, & 0 < x \le 1/2, \\ 2, & 1/2 < x < 1. \end{cases}$
(7) $u(0,t) = 0, \ u(1,t) = 0, \ u(x,0) = \begin{cases} x, & 0 < x \le 1/2, \\ 0, & 1/2 < x < 1. \end{cases}$

(8) $u(0,t) = 0$, $u(1,t) = 0$, $u(x,0) = \begin{cases} 2, & 0 < x \le 1/2, \\ x - 1, & 1/2 < x < 1. \end{cases}$

(9) $u_x(0,t) = 0$, $u_x(1,t) = 0$, $u(x,0) = 3 - 2\cos(4\pi x)$.

(10) $u_x(0,t) = 0$, $u_x(1,t) = 0$, $u(x,0) = \cos(2\pi x) - 3\cos(3\pi x)$.

(11) $u_x(0,t) = 0$, $u_x(1,t) = 0$, $u(x,0) = 2 - 3x$.

(12) $u_x(0,t) = 0$, $u_x(1,t) = 0$, $u(x,0) = 3 - 2x$.

(13) $u_x(0,t) = 0$, $u_x(1,t) = 0$, $u(x,0) = \begin{cases} -2, & 0 < x \le 1/2, \\ 0, & 1/2 < x < 1. \end{cases}$

(14) $u_x(0,t) = 0$, $u_x(1,t) = 0$, $u(x,0) = \begin{cases} 3, & 0 < x \le 1/2, \\ -1, & 1/2 < x < 1. \end{cases}$

(15) $u_x(0,t) = 0$, $u_x(1,t) = 0$, $u(x,0) = \begin{cases} 0, & 0 < x \le 1/2, \\ 2x, & 1/2 < x < 1. \end{cases}$

(16) $u_x(0,t) = 0$, $u_x(1,t) = 0$, $u(x,0) = \begin{cases} 1, & 0 < x \le 1/2, \\ x + 1, & 1/2 < x < 1. \end{cases}$

(17) $u(0,t) = 0$, $u_x(1,t) = 0$, $u(x,0) = 3\sin(\pi x/2) - \sin(5\pi x/2)$.

(18) $u(0,t) = 0$, $u_x(1,t) = 0$, $u(x,0) = -3$.

(19) $u(0,t) = 0$, $u_x(1,t) = 0$, $u(x,0) = 2 + x$.

(20) $u(0,t) = 0$, $u_x(1,t) = 0$, $u(x,0) = \begin{cases} 0, & 0 < x \le 1/2, \\ -1, & 1/2 < x < 1. \end{cases}$

(21) $u_x(0,t) = 0$, $u(1,t) = 0$, $u(x,0) = 2\cos(5\pi x/2)$.

(22) $u_x(0,t) = 0$, $u(1,t) = 0$, $u(x,0) = 4$.

(23) $u_x(0,t) = 0$, $u(1,t) = 0$, $u(x,0) = 2x - 3$.

(24) $u_x(0,t) = 0$, $u(1,t) = 0$, $u(x,0) = \begin{cases} 2 - x, & 0 < x \le 1/2, \\ 1, & 1/2 < x < 1. \end{cases}$

In (25)–(28) use separation of variables to solve the IBVP

$$u_t(x,t) = u_{xx}(x,t), \quad -1 < x < 1, \ t > 0,$$
$$u(-1,t) = u(1,t), \quad u_x(-1,t) = u_x(1,t), \quad t > 0,$$
$$u(x,0) = f(x), \quad 0 < x < 1,$$

with the function f as indicated.

(25) $f(x) = 2\sin(2\pi x) - \cos(5\pi x)$.

(26) $f(x) = 3x - 2$.

(27) $f(x) = \begin{cases} 3, & -1 < x \le 0, \\ 0, & 0 \le x < 1. \end{cases}$

(28) $f(x) = \begin{cases} 2, & -1 < x \le 0, \\ x - 1, & 0 < x < 1. \end{cases}$

In (29)–(46) use separation of variables to solve the PDE

$$u_{tt}(x,t) = u_{xx}(x,t), \quad 0 < x < 1, \ t > 0,$$

with BCs and ICs as indicated.

(29) $u(0,t) = 0, \quad u(1,t) = 0,$
$\quad u(x,0) = -3\sin(2\pi x) + 4\sin(7\pi x), \quad u_t(x,0) = \sin(3\pi x).$

(30) $u(0,t) = 0, \quad u(1,t) = 0, \quad u(x,0) = -1, \quad u_t(x,0) = 3\sin(\pi x).$

(31) $u(0,t) = 0, \quad u(1,t) = 0, \quad u(x,0) = 2\sin(3\pi x), \quad u_t(x,0) = 2.$

(32) $u(0,t) = 0, \quad u(1,t) = 0, \quad u(x,0) = 1, \quad u_t(x,0) = x.$

(33) $u(0,t) = 0, \quad u(1,t) = 0,$
$\quad u(x,0) = \begin{cases} 1, & 0 < x \le 1/2, \\ 2, & 1/2 < x < 1, \end{cases} \quad u_t(x,0) = 3\sin(2\pi x).$

(34) $u(0,t) = 0, \quad u(1,t) = 0,$
$\quad u(x,0) = 1, \quad u_t(x,0) = \begin{cases} 2, & 0 < x \le 1/2, \\ -1, & 1/2 < x < 1. \end{cases}$

(35) $u(0,t) = 0, \quad u(1,t) = 0,$
$\quad u(x,0) = \begin{cases} x, & 0 < x \le 1/2, \\ 1, & 1/2 < x < 1, \end{cases} \quad u_t(x,0) = -1.$

(36) $u(0,t) = 0, \quad u(1,t) = 0,$
$\quad u(x,0) = x + 1, \quad u_t(x,0) = \begin{cases} -1, & 0 < x \le 1/2, \\ 2x, & 1/2 < x < 1. \end{cases}$

(37) $u_x(0,t) = 0, \quad u_x(1,t) = 0,$
$\quad u(x,0) = 2 - 3\cos(4\pi x), \quad u_t(x,0) = 2\cos(3\pi x).$

(38) $u_x(0,t) = 0, \quad u_x(1,t) = 0, \quad u(x,0) = x - 1, \quad u_t(x,0) = 2 - \cos(\pi x).$

(39) $u_x(0,t) = 0, \quad u_x(1,t) = 0, \quad u(x,0) = -3\cos(2\pi x), \quad u_t(x,0) = 2x - 1.$

(40) $u_x(0,t) = 0, \quad u_x(1,t) = 0, \quad u(x,0) = x, \quad u_t(x,0) = 2x - 1.$

(41) $u_x(0,t) = 0, \quad u_x(1,t) = 0,$
$\quad u(x,0) = \begin{cases} 2, & 0 < x \le 1/2, \\ 3, & 1/2 < x < 1, \end{cases} \quad u_t(x,0) = -\cos(3\pi x).$

(42) $u_x(0,t) = 0, \quad u_x(1,t) = 0,$
$\quad u(x,0) = x, \quad u_t(x,0) = \begin{cases} 2, & 0 < x \le 1/2, \\ -1, & 1/2 < x < 1. \end{cases}$

(43) $u_x(0,t) = 0, \quad u_x(1,t) = 0,$
$\quad u(x,0) = \begin{cases} x + 1, & 0 < x \le 1/2, \\ 2, & 1/2 < x < 1, \end{cases} \quad u_t(x,0) = 1 - x.$

(44) $u_x(0,t) = 0$, $u_x(1,t) = 0$,

$$u(x,0) = 2x, \quad u_t(x,0) = \begin{cases} 2x + 1, & 0 < x \le 1/2, \\ x + 2, & 1/2 < x < 1. \end{cases}$$

(45) $u(0,t) = 0$, $u_x(1,t) = 0$,

$$u(x,0) = \begin{cases} 0, & 0 < x \le 1/2, \\ 2, & 1/2 < x < 1, \end{cases} \quad u_t(x,0) = 3\sin(5\pi x/2).$$

(46) $u_x(0,t) = 0$, $u(1,t) = 0$,

$$u(x,0) = x - 1, \quad u_t(x,0) = \begin{cases} 1, & 0 < x \le 1/2, \\ 2x - 1, & 1/2 < x < 1. \end{cases}$$

In (47)–(52) use separation of variables to solve the PDE

$$u_{xx}(x,y) + u_{yy}(x,y) = 0, \quad 0 < x < 1, \ 0 < y < 2,$$

with the BCs as indicated.

(47) $u(0,y) = 0$, $u(1,y) = 0$,
  $u(x,0) = 3\sin(2\pi x)$, $u(x,2) = \sin(3\pi x)$.

(48) $u_x(0,y) = 2\sin(\pi y)$, $u_x(1,y) = -\sin(2\pi y)$,
  $u(x,0) = 0$, $u(x,2) = 0$.

(49) $u_x(0,y) = y - 2$, $u(1,y) = 3\cos(2\pi y)$,
  $u_y(x,0) = 0$, $u_y(x,2) = 0$.

(50) $u_x(0,y) = 0$, $u_x(1,y) = 0$,
  $u(x,0) = 2 - \cos(\pi x)$, $u_y(x,2) = x$.

(51) $u_x(0,y) = 0$, $u(1,y) = 0$,

$$u(x,0) = \begin{cases} 1, & 0 < x \le 1/2, \\ 2, & 1/2 < x < 1, \end{cases} \quad u_y(x,2) = 3\cos(\pi x/2).$$

(52) $u_x(0,y) = -\sin(3\pi y/4)$, $u_x(1,y) = \begin{cases} 0, & 0 < y \le 1, \\ 1, & 1 < y < 2, \end{cases}$
  $u(x,0) = 0$, $u_y(x,2) = 0$.

In (53)–(56) use separation of variables to solve the BVP

$$u_{rr}(r,\theta) + r^{-1}u_r(r,\theta) + r^{-2}u_{\theta\theta}(r,\theta) = 0, \quad 0 < r < \alpha, \ -\pi < \theta < \pi,$$

$$u(r,\theta), \ u_r(r,\theta) \text{ bounded as } r \to 0+, \quad u(\alpha,\theta) = f(\theta), \quad -\pi < \theta < \pi,$$

$$u(r,-\pi) = u(r,\pi), \quad u_\theta(r,-\pi) = u_\theta(r,\pi), \quad 0 < r < \alpha,$$

with the constant α and the function f as indicated.

(53) $\alpha = 1/2$, $f(\theta) = 3 - 4\cos(3\theta)$.

(54) $\alpha = 3$, $f(\theta) = 2\cos(4\theta) + \sin(2\theta)$.

(55) $\alpha = 2$, $f(\theta) = \begin{cases} -1, & -\pi < \theta \le 0, \\ 3, & 0 < \theta < \pi. \end{cases}$

(56) $\alpha = 1$, $f(\theta) = \begin{cases} 1, & -\pi < \theta \le 0, \\ 0, & 0 < \theta < \pi. \end{cases}$

In (57)–(60) use separation of variables to solve the IBVP

$$u_t(x,t) = u_{xx}(x,t) - u_x(x,t), \quad 0 < x < 1, \ t > 0,$$
$$u(0,t) = 0, \quad u(1,t) = 0, \quad t > 0,$$
$$u(x,0) = f(x), \quad 0 < x < 1,$$

with the function f as indicated.

(57) $f(x) = 3e^{x/2}\sin(2\pi x)$.

(58) $f(x) = 1$.

(59) $f(x) = \begin{cases} 0, & 0 < x \le 1/2, \\ 1, & 1/2 < x < 1. \end{cases}$

(60) $f(x) = \begin{cases} 1, & 0 < x \le 1/2, \\ -2, & 1/2 < x < 1. \end{cases}$

In (61)–(64) use separation of variables to solve the IBVP

$$u_{tt}(x,t) + u_t(x,t) + u(x,t) = 4u_{xx}(x,t) - 2u_x(x,t),$$
$$0 < x < 1, \ t > 0,$$
$$u(0,t) = 0, \quad u(1,t) = 0, \quad t > 0,$$
$$u(x,0) = f(x), \quad u_t(x,0) = g(x), \quad 0 < x < 1,$$

with the functions f and g as indicated.

(61) $f(x) = -2e^{x/4}\sin(3\pi x)$, $g(x) = e^{x/4}\sin(2\pi x)$.

(62) $f(x) = -3$, $g(x) = 1$.

(63) $f(x) = \begin{cases} -1, & 0 < x \le 1/2, \\ 0, & 1/2 < x < 1, \end{cases}$ $g(x) = 4e^{x/4}\sin(\pi x)$.

(64) $f(x) = -3e^{x/4}\sin(4\pi x)$, $g(x) = \begin{cases} 3, & 0 < x \le 1/2, \\ -1, & 1/2 < x < 1. \end{cases}$

In (65)–(68) use separation of variables to solve the PDE

$$u_{xx}(x,y) + u_{yy}(x,y) - 2u_y(x,y) = 0, \quad 0 < x < 2, \ 0 < y < 1,$$

with the BCs as indicated.

(65) $u(0,y) = 0, \quad u(2,y) = 0,$
   $u(x,0) = -\sin(\pi x/2), \quad u(x,1) = 2\sin(\pi x).$

(66) $u(0,y) = 0, \quad u(2,y) = 0,$
   $u(x,0) = 1, \quad u(x,1) = -2\sin(3\pi x/2).$

(67) $u(0,y) = 3e^y \sin(2\pi y), \quad u(2,y) = -2e^y \sin(\pi y),$
   $u(x,0) = 0, \quad u(x,1) = 0.$

(68) $u(0,y) = 4e^y \sin(3\pi y), \quad u(2,y) = \begin{cases} 1, & 0 < y \le 1/2, \\ 0, & 1/2 < y < 1, \end{cases}$
   $u(x,0) = 0, \quad u(x,1) = 0.$

In (69)–(74) use separation of variables to solve the PDE

$$u_{tt}(x,y,t) = u_{xx}(x,y,t) + u_{yy}(x,y,t),$$
$$0 < x < 1, \ 0 < y < 1, \ t > 0,$$

with the BCs and ICs as indicated.

(69) $u(0,y,t) = 0, \quad u(1,y,t) = 0, \quad u(x,0,t) = 0, \quad u(x,1,t) = 0,$
   $u(x,y,0) = \sin(\pi x)\sin(2\pi y), \quad u_t(x,y,0) = -2\sin(2\pi x)\sin(\pi y).$

(70) $u(0,y,t) = 0, \quad u(1,y,t) = 0, \quad u(x,0,t) = 0, \quad u(x,1,t) = 0,$
   $u(x,y,0) = 3\sin(2\pi x)\sin(\pi y),$
   $u_t(x,y,0) = \sin(3\pi x)\big[\sin(2\pi y) - 2\sin(3\pi y)\big].$

(71) $u(0,y,t) = 0, \quad u(1,y,t) = 0, \quad u_y(x,0,t) = 0, \quad u_y(x,1,t) = 0,$
   $u(x,y,0) = -\sin(\pi x)\cos(2\pi y), \quad u_t(x,y,0) = 3\sin(2\pi x).$

(72) $u_x(0,y,t) = 0, \quad u_x(1,y,t) = 0, \quad u(x,0,t) = 0, \quad u(x,1,t) = 0,$
   $u(x,y,0) = \cos(2\pi x), \quad u_t(x,y,0) = -2\cos(\pi x)\sin(3\pi y).$

(73) $u(0,y,t) = 0, \quad u(1,y,t) = 0, \quad u(x,0,t) = 0, \quad u(x,1,t) = 0,$
   $u(x,y,0) = 1, \quad u_t(x,y,0) = xy.$

(74) $u(0,y,t) = 0, \quad u(1,y,t) = 0, \quad u_y(x,0,t) = 0, \quad u_y(x,1,t) = 0,$
   $u(x,y,0) = (x-1)y, \quad u_t(x,y,0) = 2xy.$

In (75)–(78), use separation of variables to compute the first five terms of the series solution of the IBVP

$$u_{tt}(r,\theta,t) = u_{rr}(r,\theta,t) + r^{-1}u_r(r,\theta,t) + r^{-2}u_{\theta\theta}(r,\theta,t),$$
$$0 < r < 1, \ -\pi < \theta < \pi, \ t > 0,$$
$$u(1,\theta,t) = 0, \quad u(r,\theta,t), \ u_r(r,\theta,t) \text{ bounded as } r \to 0+,$$
$$-\pi < \theta < \pi, \ t > 0,$$
$$u(r,-\pi,t) = u(r,\pi,t), \quad u_\theta(r,-\pi,t) = u_\theta(r,\pi,t),$$
$$0 < r < 1, \ t > 0,$$
$$u(r,\theta,0) = f(r,\theta), \quad u_t(r,\theta,0) = g(r,\theta),$$
$$0 < r < 1, \ -\pi < \theta < \pi,$$

with the functions f and g as indicated.

(75) $f(r,\theta) = 2\sin(2\theta), \ g(r,\theta) = 0.$
(76) $f(r,\theta) = 0, \ g(r,\theta) = -\cos\theta.$
(77) $f(r,\theta) = r\sin\theta, \ g(r,\theta) = 0.$
(78) $f(r,\theta) = 0, \ g(r,\theta) = (r-1)\cos(2\theta).$

In (79) and (80) use separation of variables to compute, up to r^2-terms, the series solution of the BVP

$$\frac{1}{r^2}(r^2 u_r)_r + \frac{1}{r^2 \sin^2\varphi} u_{\theta\theta} + \frac{1}{r^2 \sin\varphi}\left((\sin\varphi)u_\varphi\right)_\varphi = 0,$$
$$0 < r < 1, \ 0 < \theta < 2\pi, \ 0 < \varphi < \pi,$$
$$u(1,\theta,\varphi) = f(\theta,\varphi), \quad u(r,\theta,\varphi), \ u_r(r,\theta,\varphi) \text{ bounded as } r \to 0+,$$
$$0 < \theta < 2\pi, \ 0 < \varphi < \pi,$$
$$u(r,0,\varphi) = u(r,2\pi,\varphi), \quad u_\theta(r,0,\varphi) = u_\theta(r,2\pi,\varphi),$$
$$0 < r < 1, \ 0 < \varphi < \pi,$$
$$u(r,\theta,\varphi), \ u_\varphi(r,\theta,\varphi) \text{ bounded as } \varphi \to 0+ \text{ and as } \varphi \to \pi-,$$
$$0 < r < 1, \ 0 < \theta < 2\pi,$$

with the function f as indicated.

(79) $f(\theta,\varphi) = 2\cos^2\theta\sin^2\varphi - \sin\theta\sin\varphi\cos\varphi - \cos^2\varphi.$

(80) $f(\theta,\varphi) = \begin{cases} 1, & 0 < \varphi \leq \pi/2, \\ \sin\theta\sin\varphi\cos\varphi, & \pi/2 < \varphi < \pi. \end{cases}$

Chapter 6
Linear Nonhomogeneous Problems

The method of separation of variables can be applied only if the PDE and BCs are homogeneous. However, a mathematical model may include nonhomogeneous terms (that is, terms not containing the unknown function or its derivatives) in the equation, and/or may have nonhomogeneous (nonzero) data prescribed at its boundary points. In this chapter we show how, in certain cases, such problems can be reduced to their corresponding homogeneous versions.

6.1. Equilibrium Solutions

We have already discussed equilibrium (or steady-state, or time-independent) solutions for the higher-dimensional heat equation (see Section 4.2). Below we consider equilibrium temperature distributions for a uniform rod.

Temperature prescribed at the endpoints. The general IBVP for a rod with internal sources in this case is of the form

$$u_t(x,t) = ku_{xx}(x,t) + q(x,t), \quad 0 < x < L, \ t > 0,$$
$$u(0,t) = \alpha(t), \quad u(L,t) = \beta(t), \quad t > 0,$$
$$u(x,0) = f(x), \quad 0 < x < L,$$

where α, β, and f are given functions. If α and β are constant and $q = q(x)$, then we may have an equilibrium solution $u_\infty = u_\infty(x)$, which satisfies the BVP

$$u_\infty''(x) + q(x) = 0, \quad 0 < x < L,$$
$$u_\infty(0) = \alpha, \quad u_\infty(L) = \beta,$$

with q incorporating the factor $1/k$. As can be seen, here the IC plays no role in the computation of u_∞. For a time-dependent problem, it seems physically reasonable to expect that if an equilibrium solution exists, then

$$\lim_{t \to \infty} u(x,t) = u_\infty(x).$$

This justifies the use of the subscript ∞ in the symbol of the equilibrium solution.

6.1. Example. The equilibrium temperature for the IBVP

$$u_t(x,t) = 4u_{xx}(x,t) + 2x - 1, \quad 0 < x < 2, \ t > 0,$$

$$u(0,t) = 1, \quad u(2,t) = -2, \quad t > 0,$$

$$u(x,0) = \sin\frac{\pi x}{2}, \quad 0 < x < 2,$$

is obtained by solving the BVP

$$4u_\infty''(x) + 2x - 1 = 0, \quad 0 < x < 2,$$

$$u_\infty(0) = 1, \quad u_\infty(2) = -2.$$

Integrating the ODE twice, we find that

$$u_\infty(x) = -\tfrac{1}{12}x^3 + \tfrac{1}{8}x^2 + C_1 x + C_2, \quad C_1, C_2 = \text{const.}$$

The constants C_1 and C_2 are now easily found from the BCs. The equilibrium solution is

$$u_\infty(x) = -\tfrac{1}{12}x^3 + \tfrac{1}{8}x^2 - \tfrac{17}{12}x + 1. \quad \blacksquare$$

Flux prescribed at the endpoints. The general IBVP is

$$u_t(x,t) = ku_{xx}(x,t) + q(x,t), \quad 0 < x < L, \ t > 0,$$

$$u_x(0,t) = \alpha(t), \quad u_x(L,t) = \beta(t), \quad t > 0,$$

$$u(x,0) = f(x), \quad 0 < x < L.$$

As above, an equilibrium solution may exist if α and β are constant and q is independent of t. Such a solution satisfies the BVP

$$u_\infty''(x) + q(x) = 0, \quad 0 < x < L,$$

$$u_\infty'(0) = \alpha, \quad u_\infty'(L) = \beta,$$

with q incorporating the factor $1/k$, and, if it exists, may once again be regarded as the limit, as $t \to \infty$, of the solution $u(x,t)$ of the IBVP. It turns out that in this case the IC cannot be set aside. In fact, this condition now plays an essential role in the computation of the equilibrium temperature. Its input comes through the law of conservation of heat energy in the full

$$-2 = \frac{1}{8}(2)^2 - \frac{1}{12}(2)^3 + C_1(2) + 1 \qquad \frac{-17}{12} = C_1$$

$$-3 = \frac{1}{2} - \frac{2}{3} + 2C_1 \qquad \longrightarrow \quad \frac{-18}{6} = -\frac{1}{6} + 2C_1$$

time-dependent problem, which is equivalent to the PDE itself. Thus, under the assumption made above on q, α, and β, integrating the PDE term by term from 0 to L and using the BCs, we obtain

$$\frac{d}{dt}\int_0^L u(x,t)\,dx = k\int_0^L u_{xx}(x,t)\,dx + \int_0^L q(x)\,dx$$

$$= k\big[u_x(L,t) - u_x(0,t)\big] + \int_0^L q(x)\,dx$$

$$= k\big[\beta(t) - \alpha(t)\big] + \int_0^L q(x)\,dx.$$

Given that the lateral (cylindrical) surface of the rod is insulated, an equilibrium temperature cannot physically exist unless the total contribution of the sources in the rod and of the heat flux through its endpoints is zero; that is,

$$k\big[\beta(t) - \alpha(t)\big] + \int_0^L q(x,t)\,dx = 0.$$

If this happens, then, as the above equality shows, $\int_0^L u(x,t)\,dx$ is constant for all $t > 0$, and so,

$$\int_0^L u(x,0)\,dx = \lim_{t\to\infty}\int_0^L u(x,t)\,dx;$$

in other words,

$$\int_0^L u_\infty(x)\,dx = \int_0^L f(x)\,dx. \tag{6.1}$$

6.2. Example. The equilibrium temperature for the IBVP

$$u_t(x,t) = 4u_{xx}(x,t) + \gamma x + 24, \quad 0 < x < 2, \ t > 0,$$
$$u_x(0,t) = 2, \quad u_x(2,t) = 14, \quad t > 0,$$
$$u(x,0) = \pi\sin\frac{\pi x}{2} + 1, \quad 0 < x < 2,$$

where $\gamma = \text{const}$, satisfies the BVP

$$4u_\infty''(x) + \gamma x + 24 = 0, \quad 0 < x < 2,$$
$$u_\infty'(0) = 2, \quad u_\infty'(2) = 14.$$

Handwritten annotations: $U''(x) = \dfrac{-\lambda x + 24}{4} = \dfrac{xx}{4} + 6$; $U'(x) = \dfrac{\lambda x^2}{8} + 6x + C_1$

The ODE yields

$$u'_\infty(x) = -\frac{1}{8}\gamma x^2 - 6x + C_1 = 2$$

and the BCs imply that $C_1 = 2$ and $\gamma = -48$. This is the only value of γ for which an equilibrium temperature exists. Integrating again, we get

$$u_\infty(x) = 2x^3 - 3x^2 + 2x + C_2.$$

Since

$$\int_0^L f(x)\,dx = \int_0^2 \left(\pi \sin \frac{\pi x}{2} + 1\right) dx = 6,$$

$$\int_0^L u(x)\,dx = \int_0^2 \left(2x^3 - 3x^2 + 2x + C_2\right) dx = 2C_2 + 4,$$

it follows that, by (6.1), $C_2 = 1$; hence,

$$u_\infty(x) = 2x^3 - 3x^2 + 2x + 1.$$

It is easily verified that, as expected, the total contribution of the source term $\gamma x + 24 = -48x + 24$ over the length of the rod and of the heat flux at the endpoints is zero. ■

Mixed boundary conditions. The same method can be applied to compute the equilibrium solution in the case of mixed BCs.

6.3. Example. To find the equilibrium temperature for the IBVP

$$u_t(x,t) = 4u_{xx}(x,t) - 16, \quad 0 < x < 2,\ t > 0,$$
$$u(0,t) = -5, \quad u_x(2,t) = 4, \quad t > 0,$$
$$u(x,0) = \sin \frac{\pi x}{2}, \quad 0 < x < 2,$$

we need to solve the BVP

$$u''_\infty(x) - 4 = 0, \quad 0 < x < L,$$
$$u_\infty(0) = -5, \quad u'_\infty(2) = 4.$$

Direct integration shows that $u_\infty(x) = 2x^2 - 4x - 5$. ■

6.4. Example. The equilibrium temperature for the IBVP

$$u_t(x,t) = 4u_{xx}(x,t) + 32, \quad 0 < x < 2, \ t > 0,$$

$$u_x(0,t) = 2\left[u(0,t) - \tfrac{13}{2}\right], \quad u(2,t) = 1, \quad t > 0,$$

$$u(x,0) = \sin\frac{\pi x}{2}, \quad 0 < x < 2,$$

satisfies

$$u_\infty''(x) + 32 = 0, \quad 0 < x < 1,$$

$$u_\infty'(0) = 2\left[u_\infty(0) - \tfrac{13}{2}\right], \quad u_\infty(2) = 1,$$

so $u_\infty(x) = -4x^2 + 7x + 3.$ ∎

Equilibrium temperatures can sometimes also be computed explicitly for heat IBVPs in higher dimensions.

Circular annulus with prescribed temperature on the boundary. In view of the geometry of the domain, in this case it is advisable to use the Laplacian in polar coordinates (see Remark 4.11(ii)).

6.5. Example. The IBVP

$$u_t(r,t) = k(\Delta u)(r,t) = kr^{-1}(ru_r(r,t))_r, \quad 1 < r < 2, \ t > 0,$$

$$u(1,t) = 3, \quad u(2,t) = -1, \quad t > 0,$$

$$u(r,0) = f(r), \quad 1 < r < 2,$$

models heat conduction in a uniform circular annulus with no sources when the inner and outer boundary circles $r = 1$ and $r = 2$ are held at two different constant temperatures and the IC function depends only on r. (Since the body and the IC have circular symmetry, we may assume that the solution u is independent of the polar angle θ.) Then the equilibrium temperature $u_\infty(r)$ satisfies the BVP

$$(ru_\infty'(r))' = 0, \quad 1 < r < 2,$$

$$u_\infty(1) = 3, \quad u_\infty(2) = -1.$$

Integrating the ODE once, we find that

$$ru_\infty'(r) = C_1 = \text{const},$$

while a second integration yields the general solution

$$u_\infty(r) = C_1 \ln r + C_2, \quad C_2 = \text{const.}$$

Using the BC at $r = 1$, we see that $C_2 = 3$; then the condition at $r = 2$ leads to $C_1 = 4/(\ln 2)$, so

$$u_\infty(r) = \frac{4}{\ln 2} \ln r + 3. \quad \blacksquare$$

6.2. Nonhomogeneous Problems

Time-independent boundary conditions. Consider the IBVP

$$
\begin{aligned}
u_t(x,t) &= k u_{xx}(x,t), \quad 0 < x < L,\ t > 0, \\
u(0,t) &= \alpha, \quad u(L,t) = \beta, \quad t > 0, \\
u(x,0) &= f(x), \quad 0 < x < L,
\end{aligned}
\tag{6.2}
$$

where $k, \alpha, \beta = $ const and f is a given function. To be able to use the method of separation of variables, we need to make the BCs homogeneous, while keeping the PDE also homogeneous.

Proceeding as in Section 6.1, it is easily seen that the equilibrium solution for this problem is

$$u_\infty(x) = \alpha + \frac{\beta - \alpha}{L} x, \quad 0 < x < L. \tag{6.3}$$

Then the function $v = u - u_\infty$ is a solution of the IBVP

$$
\begin{aligned}
v_t(x,t) &= k v_{xx}(x,t), \quad 0 < x < L,\ t > 0, \\
v(0,t) &= 0, \quad v(L,t) = 0, \quad t > 0, \\
v(x,0) &= f(x) - u_\infty(x), \quad 0 < x < L.
\end{aligned}
$$

This new IBVP was solved in Section 5.1 and, by (5.10), its solution is

$$v(x,t) = \sum_{n=1}^{\infty} b_n \sin \frac{n\pi x}{L} e^{-k(n^2 \pi^2/L)t},$$

where, by (5.9) with f replaced by $f - u_\infty$,

$$b_n = \frac{2}{L} \int_0^L [f(x) - u_\infty(x)] \sin \frac{n\pi x}{L} \, dx, \quad n = 1, 2, \ldots. \tag{6.4}$$

Consequently, the solution of (6.2) is

$$u(x,t) = u_\infty(x) + v(x,t) = u_\infty(x) + \sum_{n=1}^{\infty} b_n \sin \frac{n\pi x}{L} e^{-k(n\pi/L)^2 t},$$

with u_∞ and the b_n given by (6.3) and (6.4), respectively.

Time-independent sources and boundary conditions. The above method also works for an IBVP of the form

$$u_t(x,t) = ku_{xx}(x,t) + q(x), \quad 0 < x < L, \ t > 0,$$
$$u(0,t) = \alpha, \quad u(L,t) = \beta, \quad t > 0,$$
$$u(x,0) = f(x), \quad 0 < x < L,$$

since, as seen in Section 6.1, the computation of the equilibrium temperature distribution takes the source term into account.

Other types of BCs are treated similarly if a steady-state solution exists.

6.6. Example. The equilibrium solution u_∞ of the IBVP

$$u_t(x,t) = u_{xx}(x,t) + \pi^2 \sin(\pi x), \quad 0 < x < 1, \ t > 0,$$
$$u(0,t) = 1, \quad u(1,t) = -3, \quad t > 0,$$
$$u(x,0) = x^2, \quad 0 < x < 1,$$

satisfies

$$u_\infty''(x) + \pi^2 \sin(\pi x) = 0, \quad 0 < x < 1,$$
$$u_\infty(0) = 1, \quad u_\infty(1) = -3.$$

The solution of this BVP is $u_\infty(x) = \sin(\pi x) - 4x + 1$. Then the substitution $v(x) = u(x) - \sin(\pi x) + 4x - 1$ reduces the given IBVP to

$$v_t(x,t) = v_{xx}(x,t), \quad 0 < x < 1, \ t > 0,$$
$$v(0,t) = 0, \quad v(1,t) = 0, \quad t > 0,$$
$$v(x,0) = x^2 + 4x - 1 - \sin(\pi x), \quad 0 < x < 1,$$

which can be solved by the method of separation of variables. ∎

6.7. Example. If the IBVP

$$u_t(x,t) = u_{xx}(x,t) + x - \gamma, \quad 0 < x < 1, \ t > 0,$$
$$u_x(0,t) = 0, \quad u_x(1,t) = 0, \quad t > 0,$$
$$u(x,0) = -\tfrac{1}{2} + x + \tfrac{1}{4}x^2 - \tfrac{1}{6}x^3, \quad 0 < x < 1,$$

where $\gamma = \text{const}$, has a steady-state solution u_∞, then this solution must satisfy

$$u_\infty''(x) + x - \gamma = 0, \quad 0 < x < 1,$$
$$u_\infty'(0) = 0, \quad u_\infty'(1) = 0.$$

[handwritten marginalia:]
$U''(x) = \lambda - x$
$U'(x) = \lambda x - \frac{1}{2}x^2 + C$

Integrating the equation once, we find that

$$u_\infty'(x) = -\tfrac{1}{2}x^2 + \gamma x + C_1, \quad C_1 = \text{const},$$

[handwritten marginalia:] $\lambda = \frac{1}{2}$ $U(x) = \frac{1}{4}x^2 - \frac{1}{6}x^3 + C_x$

which satisfies both boundary conditions only if $\gamma = 1/2$ and $C_1 = 0$. Another integration yields

$$u_\infty(x) = -\tfrac{1}{6}x^3 + \tfrac{1}{4}x^2 + C_2, \quad C_2 = \text{const}.$$

Following the procedure in Example 6.2, we deduce that $C_2 = 0$. Then the substitution $v(x) = u(x) + \tfrac{1}{6}x^3 - \tfrac{1}{4}x^2$ reduces the original problem to the IBVP

$$v_t(x,t) = v_{xx}(x,t), \quad 0 < x < 1, \ t > 0,$$
$$v_x(0,t) = 0, \quad v_x(1,t) = 0, \quad t > 0,$$
$$v(x,0) = x - \tfrac{1}{2}, \quad 0 < x < 1,$$

to which we can apply the method of separation of variables. ∎

The general case. Consider the IBVP

$$u_t(x,t) = ku_{xx}(x,t) + q(x,t), \quad 0 < x < L, \ t > 0,$$
$$u(0,t) = \alpha(t), \quad u(L,t) = \beta(t), \quad t > 0,$$
$$u(x,0) = f(x), \quad 0 < x < L,$$

where α, β, f, and q are given functions. Also, let

$$p(x,t) = C_1(t) + C_2(t)x$$

be a linear polynomial in x satisfying the BCs; that is, $p(0,t) = \alpha(t)$ and $p(L,t) = \beta(t)$. Then $C_1 = \alpha$ and $C_2 = (\beta - \alpha)/L$, and the substitution $u = v + p$ reduces the given problem to the new IBVP

$$v_t(x,t) = kv_{xx}(x,t) + \left[q(x,t) - p_t(x,t)\right], \quad 0 < x < L, \ t > 0,$$
$$v(0,t) = 0, \quad v(L,t) = 0, \quad t > 0,$$
$$v(x,0) = f(x) - p(x,0), \quad 0 < x < L,$$

which has homogeneous BCs but a nonhomogeneous PDE. Such problems, and similar ones with other types of BCs, are solved by the method of eigenfunction expansion, discussed in the next chapter.

6.8. Example. The IBVP

$$u_t(x,t) = u_{xx}(x,t) + xt, \quad 0 < x < 1, \ t > 0,$$
$$u(0,t) = 1 - t, \quad u(1,t) = t^2, \quad t > 0,$$
$$u(x,0) = x, \quad 0 < x < 1,$$

is of the above form, with $L = 1$, $\alpha(t) = 1 - t$, $\beta(t) = t^2$, and $q(x,t) = xt$; therefore,

$$p(x,t) = 1 - t + (t^2 + t - 1)x.$$

The substitution

$$u(x,t) = v(x,t) + 1 - t + (t^2 + t - 1)x$$

now reduces the given IBVP to the problem

$$v_t(x,t) = v_{xx}(x,t) + 1 - xt - x, \quad 0 < x < 1, \ t > 0,$$
$$v(0,t) = 0, \quad v(1,t) = 0, \quad t > 0,$$
$$v(x,0) = 2x - 1, \quad 0 < x < 1,$$

with homogeneous BCs. ∎

6.9. Example. The IBVP

$$u_t(x,t) = u_{xx}(x,t) + xt, \quad 0 < x < 1, \ t > 0,$$
$$u_x(0,t) = t, \quad u_x(1,t) = t^2, \quad t > 0,$$
$$u(x,0) = x + 1, \quad 0 < x < 1,$$

needs slightly different handling. Here we take p_x instead of p to be a linear polynomial in x satisfying the BCs; that is,

$$p_x(x,t) = C_1(t) + C_2(t)x, \quad p_x(0,t) = t, \quad p_x(1,t) = t^2.$$

Then $p_x(x,t) = t + (t^2 - t)x$, from which, by integration, we obtain

$$p(x,t) = xt + \tfrac{1}{2}x^2(t^2 - t).$$

(Since we need only one such function p, we have taken the constant of integration to be zero.) The substitution

$$u(x,t) = v(x,t) + xt + \tfrac{1}{2}x^2(t^2 - t)$$

now reduces the original problem to the IBVP

$$v_t(x,t) = v_{xx}(x,t) + t^2 - t + xt - \tfrac{1}{2}x^2(2t - 1) - x,$$
$$0 < x < 1, \ t > 0,$$
$$v_x(0,t) = 0, \quad v_x(1,t) = 0, \quad t > 0,$$
$$v(x,0) = x + 1, \quad 0 < x < 1. \ \blacksquare$$

6.10. Remark. This technique can also be applied to reduce other types of IBVPs to simpler versions where the PDE is nonhomogeneous but the BCs are homogeneous. In the case of BVPs, we can similarly make two of the BCs homogeneous by means of a suitably chosen linear polynomial. ∎

Exercises

In (1)–(6) find the equilibrium solution for the IBVP consisting of the PDE

$$u_t(x,t) = u_{xx}(x,t) + q(x), \quad 0 < x < 1, \ t > 0,$$

and the function q, BCs, and IC as indicated. (In (5) and (6) also find the value of γ for which such a solution exists.)

(1) $q(x) = 6 - 12x, \quad u(0,t) = -1, \quad u(1,t) = 2, \quad u(x,0) = f(x)$.

(2) $q(x) = 4 - 6x, \quad u(0,t) = -3, \quad u_x(1,t) = -1, \quad u(x,0) = f(x)$.

(3) $q(x) = -4 - 12x^2, \quad u_x(0,t) = 3, \quad u_x(1,t) = 3[u(1,t) - \tfrac{1}{3}],$
   $u(x,0) = f(x)$.

(4) $q(x) = 24x - 12x^2, \quad u_x(0,t) = -3[u(0,t) - \tfrac{1}{3}], \quad u(1,t) = -2,$
   $u(x,0) = f(x)$.

(5) $q(x) = \gamma x - 4, \quad u_x(0,t) = -3, \quad u_x(1,t) = 4, \quad u(x,0) = -3x - \tfrac{1}{12}$.

(6) $q(x) = \gamma + 6x - 12x^2, \quad u_x(0,t) = -2, \quad u_x(1,t) = 3,$
   $u(x,0) = \tfrac{157}{120}\pi\sin(\pi x)$.

In (7)–(12) find the equilibrium solution of the IBVP consisting of the PDE and IC

$$u_t(x,t) = u_{xx}(x,t) - 3u_x(x,t) + 2u(x,t) + q(x), \quad 0 < x < 1, \ t > 0,$$
$$u(x,0) = f(x), \quad 0 < x < 1,$$

and the function q and BCs as indicated.

(7) $q(x) = -2x^2 + 2x + 6, \quad u(0,t) = -1, \quad u(1,t) = 2.$

(8) $q(x) = -2x + 7, \quad u(0,t) = 0, \quad u_x(1,t) = 1 + 2e.$

(9) $q(x) = 2x - 3, \quad u_x(0,t) = -4, \quad u_x(1,t) = -1 + e - 4e^2.$

(10) $q(x) = 8 - 4x, \quad u_x(0,t) = 8, \quad u_x(1,t) = 6e^2 + 2.$

(11) $q(x) = -6x^2 + 28x - 17, \quad u(0,t) = -2, \quad u_x(1,t) = 2[u(1,t) - \frac{9}{2}].$

(12) $q(x) = 2x - 9, \quad u_x(0,t) = -2[u(0,t) - \frac{11}{2}], \quad u_x(1,t) = 2e - 1.$

In (13)–(18) find a substitution that reduces the IBVP consisting of the PDE and IC

$$u_t(x,t) = u_{xx}(x,t) + x + t - 2, \quad 0 < x < 1, \ t > 0,$$
$$u(x,0) = 2x, \quad 0 < x < 1,$$

and the BCs as indicated, to an equivalent IBVP with homogeneous BCs. In each case compute the nonhomogeneous term $q(x,t)$ in the transformed PDE and the transformed IC function $f(x)$.

(13) $u(0,t) = t + 1, \quad u(1,t) = 2t - t^2.$

(14) $u_x(0,t) = 2 - t^2, \quad u_x(1,t) = t - 3.$

(15) $u_x(0,t) = 3t, \quad u_x(1,t) = t^2 + t.$

(16) $u(0,t) = t + 2, \quad u_x(1,t) = 1 - 2t.$

(17) $u_x(0,t) = -2[u(0,t) - t], \quad u_x(1,t) = 2 + t.$

(18) $u(0,t) = t^2, \quad u_x(1,t) = 2[u(1,t) - 3].$

In (19)–(22) find a substitution that reduces the IBVP consisting of the PDE and ICs

$$u_{tt}(x,t) = u_{xx}(x,t) - xt + 2x, \quad 0 < x < 1, \ t > 0,$$
$$u(x,0) = x + 1, \quad u_t(x,0) = x, \quad 0 < x < 1,$$

and the BCs as indicated, to an equivalent IBVP with homogeneous BCs. In each case compute the nonhomogeneous term $q(x,t)$ in the transformed PDE and the transformed IC functions $f(x)$ and $g(x)$.

(19) $u(0,t) = 2t + 3, \quad u(1,t) = t^2 - 4.$

(20) $u(0,t) = t - t^2, \quad u(1,t) = 2t.$

(21) $u_x(0,t) = t - 1, \quad u_x(1,t) = t^2 + 2.$

(22) $u_x(0,t) = t^2 + 2t, \quad u_x(1,t) = 3 - 2t.$

In (23)–(26) find a substitution that reduces the BVP consisting of the PDE

$$u_{xx}(x,y) + u_{yy}(x,y) + 2x - y = 0, \quad 0 < x, y < 1,$$

the two boundary conditions

$$u(x,0) = 3x, \quad u(x,1) = 1 - 2x, \quad 0 < x < 1,$$

and the other two BCs as indicated, to an equivalent BVP with homogeneous BCs at $x = 0$ and $x = 1$. In each case compute the nonhomogeneous term $q(x,y)$ in the transformed PDE and the transformed BC functions $f(x)$ and $g(x)$ at $y = 0$ and $y = 1$.

(23) $u(0,y) = 2y^2 - 1, \quad u(1,y) = y - y^2.$
(24) $u(0,y) = y^2 + 2y, \quad u(1,y) = 3y^2.$
(25) $u_x(0,y) = 2 - y, \quad u_x(1,y) = 1 + 3y.$
(26) $u_x(0,y) = y^2 + 1, \quad u_x(1,y) = 3y^2 - 2.$

In (27)–(30) find a substitution that reduces the IBVP consisting of the PDE and IC

$$u_t(x,t) = u_{xx}(x,t) - 2u_x(x,t) + xt - x + 1, \quad 0 < x < 1, \ t > 0,$$
$$u(x,0) = x + 2, \quad 0 < x < 1,$$

and the BCs as indicated, to an equivalent IBVP with homogeneous BCs. In each case compute the nonhomogeneous term $q(x,t)$ in the transformed PDE and the transformed IC function $f(x)$.

(27) $u(0,t) = 2t + 3, \quad u(1,t) = 4 - 3t.$
(28) $u(0,t) = t + 4, \quad u(1,t) = 2t - 1.$
(29) $u_x(0,t) = 1 - 2t, \quad u_x(1,t) = 3 + t.$
(30) $u_x(0,t) = t^2, \quad u_x(1,t) = 3t + 2.$

Chapter 7
The Method of
Eigenfunction Expansion

Separation of variables cannot be performed if the PDE and/or BCs are not homogeneous. As we saw in Chapter 6, there are particular situations where we can reduce the problem to an equivalent one with homogeneous PDE and BCs, but this is not always possible. In the general case, the best we can do is make the boundary conditions homogeneous. The eigenfunction expansion technique is designed for IBVPs with a nonhomogeneous equation and homogeneous BCs.

7.1. The Heat Equation

Consider the IBVP

$$u_t(x,t) = ku_{xx}(x,t) + q(x,t), \quad 0 < x < L,\ t > 0, \qquad \text{(PDE)}$$

$$u(0,t) = 0, \quad u(L,t) = 0, \quad t > 0, \qquad \text{(BCs)}$$

$$u(x,0) = f(x), \quad 0 < x < L. \qquad \text{(IC)}$$

The eigenvalues and eigenfunctions of the corresponding homogeneous problem $(q = 0)$ are, respectively (see Section 5.1),

$$\lambda_n = \left(\frac{n\pi}{L}\right)^2, \quad X_n(x) = \sin\frac{n\pi x}{L}, \quad n = 1, 2, \ldots.$$

Since $\{X_n\}_{n=1}^\infty$ is a complete set, we may consider for the solution an expansion of the form

$$u(x,t) = \sum_{n=1}^\infty c_n(t) X_n(x). \qquad (7.1)$$

Differentiating series (7.1) term by term and recalling that $X_n'' + \lambda_n X_n = 0$, from the PDE we obtain

$$\sum_{n=1}^\infty c_n'(t) X_n(x) = k \sum_{n=1}^\infty c_n(t) X_n''(x) + q(x,t)$$

$$= -k \sum_{n=1}^\infty c_n(t) \lambda_n X_n(x) + q(x,t),$$

or

$$\sum_{n=1}^{\infty} \left[c_n'(t) + k\lambda_n c_n(t)\right] X_n(x) = q(x,t).$$

Multiplying this equality by $X_m(x)$, integrating from 0 to L, and taking the orthogonality of the X_n on $[0,L]$ (see Theorem 3.8(iii)) into account, we find that

$$\left[c_m'(t) + k\lambda_m c_m(t)\right] \int_0^L X_m^2(x)\,dx = \int_0^L q(x,t) X_m(x)\,dx,$$

which (with m replaced by n) yields the equations

$$c_n'(t) + k\lambda_n c_n(t) = \frac{\int_0^L q(x,t) X_n(x)\,dx}{\int_0^L X_n^2(x)\,dx}, \qquad t > 0, \quad n = 1,2,\ldots. \qquad (7.2)$$

The BCs are automatically satisfied since each of the X_n in (7.1) satisfies them.

From (7.1) and the IC we see that

$$u(x,0) = f(x) = \sum_{n=1}^{\infty} c_n(0) X_n(x).$$

Proceeding as above, we arrive at the initial conditions

$$c_n(0) = \frac{\int_0^L f(x) X_n(x)\,dx}{\int_0^L X_n^2(x)\,dx}, \qquad n = 1,2,\ldots \qquad (7.3)$$

Clearly, this is the same as finding formal expansions

$$q(x,t) = \sum_{n=1}^{\infty} q_n(t) X_n(x), \qquad f(x) = \sum_{n=1}^{\infty} f_n X_n(x),$$

where the $q_n(t)$ and f_n are given by the right-hand sides in (7.2) and (7.3), respectively. Then the solution of the IBVP is of the form (7.1) with the coefficients $c_n(t)$ computed from (7.2) and (7.3), which, for our problem, reduce to

X_n

$$c_n'(t) + k\left(\frac{n\pi}{L}\right)^2 c_n(t) = q_n(t) = \frac{2}{L}\int_0^L q(x,t)\sin\frac{n\pi x}{L}\,dx,$$

$$c_n(0) = f_n = \frac{2}{L}\int_0^L f(x)\sin\frac{n\pi x}{L}\,dx.$$

IBVPs with other types of BCs are treated similarly.

7.1. Example. The eigenvalues and eigenfunctions for the IBVP

$q(x,t)$

$$u_t(x,t) = u_{xx}(x,t) + \pi^2 e^{-24\pi^2 t}\sin(5\pi x), \quad 0 < x < 1,\ t > 0,$$

$$u(0,t) = 0, \quad u(1,t) = 0, \quad t > 0,$$

$f(x) =$ $u(x,0) = 3\sin(4\pi x), \quad 0 < x < 1,$

are (here $k = 1$ and $L = 1$) $\lambda_n = n^2\pi^2$ and $X_n(x) = \sin(n\pi x)$, $n = 1,2,\ldots$.
Since $q(x,t) = \pi^2 e^{-24\pi^2 t}\sin(5\pi x)$ and $f(x) = 3\sin(4\pi x)$ coincide with their
eigenfunction expansions, the above formulas yield the IVPs

$$c_n'(t) + n^2\pi^2 c_n(t) = \begin{cases} \pi^2 e^{-24\pi^2 t}, & n = 5, \\ 0, & n \neq 5, \end{cases} \quad t > 0,$$

$$c_n(0) = \begin{cases} 3, & n = 4, \\ 0, & n \neq 4. \end{cases}$$

Hence, for $n = 4$ we have

$$c_4'(t) + 16\pi^2 c_4(t) = 0, \quad t > 0,$$

$$c_4(0) = 3,$$

with solution $c_4(t) = 3e^{-16\pi^2 t}$ (obtained, for example, by means of the
integrating factor $e^{25\pi^2 t}$); for $n = 5$,

$$c_5'(t) + 25\pi^2 c_5(t) = \pi^2 e^{-24\pi^2 t}, \quad t > 0,$$

$$c_5(0) = 0,$$

with solution $c_5(t) = e^{-25\pi^2 t}(e^{\pi^2 t} - 1)$; and for $n \neq 4,5$,

$$c_n'(t) + n^2\pi^2 c_n(t) = 0, \quad t > 0,$$

$$c_n(0) = 0,$$

with solution $c_n(t) = 0$. Consequently, by (7.1), the solution of the IBVP is

$$u(x,t) = c_4(t)X_4(x) + c_5(t)X_5(x)$$

$$= 3e^{-16\pi^2 t}\sin(4\pi x) + e^{-25\pi^2 t}(e^{\pi^2 t} - 1)\sin(5\pi x). \quad \blacksquare$$

7.2. Example. The IBVP

$$u_t(x,t) = u_{xx}(x,t) + xe^{-t}, \quad 0 < x < 1, \, t > 0,$$
$$u(0,t) = 0, \quad u(1,t) = 0, \quad t > 0,$$
$$u(x,0) = x - 1, \quad 0 < x < 1,$$

has the same eigenvalues and eigenfunctions as that in Example 7.1. Here $q(x,t) = xe^{-t}$ and $f(x) = x - 1$. Hence, by (7.2), (7.3), and (2.5), and using integration by parts, we find that

$$c'_n(t) + n^2\pi^2 c_n(t) = 2e^{-t} \int_0^1 x\sin(n\pi x)\,dx$$

$$= (-1)^{n+1}\frac{2}{n\pi}e^{-t}, \quad t > 0, \tag{7.4}$$

and

$$c_n(0) = 2\int_0^1 (x-1)\sin(n\pi x)\,dx = -\frac{2}{n\pi}. \tag{7.5}$$

An integrating factor for (7.4) is $e^{n^2\pi^2 t}$, so the solution of (7.4), (7.5) is

$$c_n(t) = (-1)^{n+1}\frac{2}{n\pi}\left\{\frac{1}{n^2\pi^2 - 1}e^{-t} + \left[(-1)^n - \frac{1}{n^2\pi^2 - 1}\right]e^{-n^2\pi^2 t}\right\},$$

which means that, by (7.1), the solution of the given IBVP is

$$u(x,t) = \sum_{n=1}^{\infty}(-1)^{n+1}\frac{2}{n\pi}\left\{\frac{1}{n^2\pi^2 - 1}e^{-t}\right.$$

$$\left. + \left[(-1)^n - \frac{1}{n^2\pi^2 - 1}\right]e^{-n^2\pi^2 t}\right\}\sin(n\pi x). \quad \blacksquare$$

7.3. Example. The eigenvalues and eigenfunctions associated with the IBVP

$$u_t(x,t) = u_{xx}(x,t) + 2t + \cos(2\pi x), \quad 0 < x < 1, \, t > 0,$$
$$u_x(0,t) = 0, \quad u_x(1,t) = 0, \quad t > 0,$$
$$u(x,0) = \frac{1}{2\pi^2}\cos(2\pi x), \quad 0 < x < 1,$$

are (see Section 5.1) $\lambda_n = n^2\pi^2$ and $X_n(x) = \cos(n\pi x)$, $n = 0, 1, 2, \ldots$ (for convenience, here we have taken $\lambda_0 = 1$ instead of $1/2$). Reasoning just as in the case of the sine eigenfunctions, we deduce that (7.1)–(7.3) still hold except that now $n = 0,1,2,\dots$ and the X_n are cosines. Since $q(x,t) = 2t + \cos(2\pi x)$ and $f(x) = (2\pi^2)^{-1}\cos(2\pi x)$ are linear combinations of the eigenfunctions, (7.2) (with $k = 1$) and (7.3) lead to the IVPs

$$c_n'(t) + n^2\pi^2 c_n(t) = \begin{cases} 2t, & n = 0, \\ 1, & n = 2, \\ 0, & n \neq 0,2, \end{cases} \quad t > 0,$$

$$c_n(0) = \begin{cases} \dfrac{1}{2\pi^2}, & n = 2, \\ 0, & n \neq 2. \end{cases}$$

Thus, for $n = 0$ we have

$$c_0'(t) = 2t, \quad t > 0,$$
$$c_0(0) = 0,$$

with solution $c_0(t) = t^2$; for $n = 2$,

$$c_2'(t) + 4\pi^2 c_2(t) = 1, \quad t > 0,$$
$$c_2(0) = \frac{1}{2\pi^2},$$

with solution $c_2(t) = (4\pi^2)^{-1}(1 + e^{-4\pi^2 t})$; and for $n \neq 0,2$,

$$c_n'(t) + n^2\pi^2 c_n(t) = 0, \quad t > 0,$$
$$c_n(0) = 0,$$

with solution $c_n(t) = 0$. Hence, the solution of the given IBVP is

$$u(x,t) = c_0(t)X_0(x) + c_2(t)X_2(x) = t^2 + \frac{1}{4\pi^2}(1 + e^{-4\pi^2 t})\cos(2\pi x). \quad \blacksquare$$

7.4. Example. The eigenvalues and eigenfunctions for the IBVP

$$u_t(x,t) = u_{xx}(x,t) - \sin\left(\tfrac{3}{2}\pi x\right) + t\sin\left(\tfrac{5}{2}\pi x\right), \quad 0 < x < 1, \, t > 0,$$
$$u(0,t) = 0, \quad u_x(1,t) = 0, \quad t > 0,$$
$$u(x,0) = \sin\left(\tfrac{1}{2}\pi x\right) + 2\sin\left(\tfrac{3}{2}\pi x\right), \quad 0 < x < 1,$$

are (see Section 5.1)

$$\lambda_n = \frac{(2n-1)^2\pi^2}{4}, \quad X_n(x) = \sin\frac{(2n-1)\pi x}{2}, \quad n = 1,2,\dots.$$

By direct calculation, we can show that the general scheme also works in this case, with $k = 1$, $L = 1$, and the above λ_n and X_n. We could therefore find the equations and ICs for the c_n from (7.2) and (7.3). However, as before, we notice that the functions $q(x,t) = -\sin(\frac{3}{2}\pi x) + t\sin(\frac{5}{2}\pi x)$ and $f(x) = \sin(\frac{1}{2}\pi x) + 2\sin(\frac{3}{2}\pi x)$ are already linear combinations of the X_n, so we deduce that

$$c_n'(t) + \tfrac{1}{4}(2n-1)^2\pi^2 c_n(t) = \begin{cases} -1, & n = 2, \\ t, & n = 3, \\ 0, & n \neq 2,3, \end{cases} \quad t > 0,$$

$$c_n(0) = \begin{cases} 1, & n = 1, \\ 2, & n = 2, \\ 0, & n \neq 1,2. \end{cases}$$

Thus, for $n = 1$ we have

$$c_1'(t) + \tfrac{1}{4}\pi^2 c_1(t) = 0, \quad t > 0,$$
$$c_1(0) = 2,$$

with solution $c_1(t) = e^{-\pi^2 t/4}$; for $n = 2$,

$$c_2'(t) + \tfrac{9}{4}\pi^2 c_2(t) = -1, \quad t > 0,$$
$$c_2(0) = 2,$$

with solution $c_2(t) = (2 + \frac{4}{9}\pi^{-2})e^{-9\pi^2 t/4} - \frac{4}{9}\pi^{-2}$; for $n = 3$,

$$c_3'(t) + \tfrac{25}{4}\pi^2 c_3(t) = t, \quad t > 0,$$
$$c_3(0) = 0,$$

with solution $c_3(t) = \frac{16}{625}\pi^{-4}e^{-25\pi^2 t/4} + \frac{4}{25}\pi^{-2}t - \frac{16}{625}\pi^{-4}$; finally, for $n \neq 1,2,3$,

$$c_n(t) + \tfrac{1}{4}(2n-1)^2\pi^2 c_n(t) = 0, \quad t > 0,$$
$$c_n(0) = 0,$$

with solution $c_n(t) = 0$. Hence, by (7.1), the solution of the given IBVP is

$$u(x,t) = e^{-\pi^2 t/4}\sin\left(\tfrac{1}{2}\pi x\right) + \left(\frac{18\pi^2 + 4}{9\pi^2}e^{-9\pi^2 t/4} - \frac{4}{9\pi^2}\right)\sin\left(\tfrac{3}{2}\pi x\right)$$

$$+ \left(\frac{4}{25\pi^2}t - \frac{16}{625\pi^4} + \frac{16}{625\pi^4}e^{-25\pi^2 t/4}\right)\sin\left(\tfrac{5}{2}\pi x\right). \quad \blacksquare$$

7.2. The Wave Equation

Consider the IBVP

$$u_{tt}(x,t) = c^2 u_{xx}(x,t) + q(x,t), \quad 0 < x < L, \ t > 0, \qquad \text{(PDE)}$$

$$u(0,t) = 0, \quad u(L,t) = 0, \quad t > 0, \qquad \text{(BCs)}$$

$$u(x,0) = f(x), \quad u_t(x,0) = g(x), \quad 0 < x < L. \qquad \text{(ICs)}$$

The eigenvalues and eigenfunctions associated with this problem are (see Section 5.2)

$$\lambda_n = \left(\frac{n\pi}{L}\right)^2, \quad X_n(x) = \sin\frac{n\pi x}{L}, \quad n = 1,2,\dots.$$

Starting with an expansion of the form (7.1) and following the same general procedure as for the heat equation, this time also including the second IC, we find that (7.2) is replaced by

$$c_n''(t) + c^2 \lambda_n c_n(t) = \frac{\int\limits_0^L q(x,t) X_n(x)\, dx}{\int\limits_0^L X_n^2(x)\, dx}, \quad t > 0, \quad n = 1,2,\dots, \qquad (7.6)$$

that (7.3) remains unchanged, and that

$$c_n'(0) = \frac{\int\limits_0^L g(x) X_n(x)\, dx}{\int\limits_0^L X_n^2(x)\, dx}, \quad n = 1,2,\dots. \qquad (7.7)$$

The solution of the IBVP is then given by (7.1) with the c_n computed by means of (7.6), (7.3), and (7.7), which in our case take the form

$$c_n''(t) + c^2 \left(\frac{n\pi}{L}\right)^2 c_n(t) = \frac{2}{L} \int\limits_0^L q(x,t) \sin\frac{n\pi x}{L}\, dx,$$

$$c_n(0) = \frac{2}{L} \int\limits_0^L f(x) \sin\frac{n\pi x}{L}\, dx, \quad c_n'(0) = \frac{2}{L} \int\limits_0^L g(x) \sin\frac{n\pi x}{L}\, dx.$$

Other BCs can be treated in similar fashion.

7.5. Example. For the IBVP

$$u_{tt}(x,t) = u_{xx}(x,t) + \pi^2 \sin(\pi x), \quad 0 < x < 1, \ t > 0,$$

$$u(0,t) = 0, \quad u(1,t) = 0, \quad t > 0,$$

$$u(x,0) = \pi, \quad u_t(x,0) = 2\pi \sin(2\pi x), \quad 0 < x < 1,$$

the above formulas with $c^2 = 1$, $L = 1$, $q(x,t) = \pi^2 \sin(\pi x)$, $f(x) = \pi$, and $g(x) = 2\pi \sin(2\pi x)$ yield

$$c_n''(t) + n^2 \pi^2 c_n(t) = \begin{cases} \pi^2, & n = 1, \\ 0, & n \neq 1, \end{cases} \quad t > 0,$$

$$c_n(0) = 2 \int_0^1 \pi \sin(n\pi x) \, dx = \left[1 - (-1)^n\right] \frac{2}{n}, \quad n = 1, 2, \ldots,$$

$$c_n'(0) = \begin{cases} 2\pi, & n = 2, \\ 0, & n \neq 2. \end{cases}$$

Hence, for $n = 1$,

$$c_1''(t) + \pi^2 c_1(t) = \pi^2, \quad t > 0,$$

$$c_1(0) = 4, \quad c_1'(0) = 0,$$

with solution $c_1(t) = 3\cos(\pi t) + 1$; for $n = 2$,

$$c_2''(t) + 4\pi^2 c_2(t) = 0, \quad t > 0,$$

$$c_2(0) = 0, \quad c_2'(0) = 2\pi,$$

with solution $c_2(t) = \sin(2\pi t)$; and for $n \neq 1, 2$,

$$c_n''(t) + n^2 \pi^2 c_n(t) = 0, \quad t > 0,$$

$$c_n(0) = \left[1 - (-1)^n\right] \frac{2}{n}, \quad c_n'(0) = 0,$$

with solution $c_n(t) = \left[1 - (-1)^n\right](2/n)\cos(n\pi t)$. Consequently, by (7.1),

$$u(x,t) = \left[3\cos(\pi t) + 1\right]\sin(\pi x) + \sin(2\pi t)\sin(2\pi x)$$

$$+ \sum_{n=3}^{\infty} \left[1 - (-1)^n\right] \frac{2}{n} \cos(n\pi t)\sin(n\pi x). \quad \blacksquare$$

7.6. Example. The IBVP

$$u_{tt}(x,t) = u_{xx}(x,t) + 1 + t\cos(\pi x), \quad 0 < x < 1, \ t > 0,$$

$$u_x(0,t) = 0, \quad u_x(1,t) = 0, \quad t > 0,$$

$$u(x,0) = 2, \quad u_t(x,0) = -2\cos(2\pi x), \quad 0 < x < 1,$$

gives rise to the eigenvalues and eigenfunctions (see Section 5.3)

$$\lambda_n = n^2\pi^2, \quad X_n(x) = \cos(n\pi x), \quad n = 0,1,2,\dots.$$

The above general procedure accommodates this case as well, with the obvious modifications. Taking $c^2 = 1$ and $L = 1$ and seeing that $q(x,t) = 1 + t\cos(\pi x)$, $f(x) = 2$, and $g(x) = -2\cos(2\pi x)$ coincide with their eigenfunction expansions, we find that (7.6), (7.3), and (7.7) lead to the IVPs

$$c_n''(t) + n^2\pi^2 c_n(t) = \begin{cases} 1, & n = 0, \\ t, & n = 1, \\ 0, & n \neq 0,1, \end{cases} \quad t > 0,$$

$$c_n(0) = \begin{cases} 2, & n = 0, \\ 0, & n \neq 0, \end{cases} \quad c_n'(0) = \begin{cases} -2, & n = 2, \\ 0, & n \neq 2. \end{cases}$$

So for $n = 0$ we have

$$c_0''(t) = 1, \quad t > 0,$$
$$c_0(0) = 2, \quad c_0'(0) = 0,$$

with solution $c_0(t) = \frac{1}{2}t^2 + 2$; for $n = 1$,

$$c_1''(t) + \pi^2 c_1(t) = t, \quad t > 0,$$
$$c_1(0) = 0, \quad c_1'(0) = 0,$$

with solution $c_1(t) = -\pi^{-3}\sin(\pi t) + \pi^{-2}t$; for $n = 2$,

$$c_2''(t) + 4\pi^2 c_2(t) = 0, \quad t > 0,$$
$$c_2(0) = 0, \quad c_2'(0) = -2,$$

with solution $c_2(t) = -\pi^{-1}\sin(2\pi t)$; and for $n \neq 0,1,2$,

$$c_n''(t) + n^2\pi^2 c_n(t) = 0, \quad t > 0,$$
$$c_n(0) = 0, \quad c_n'(0) = 0,$$

with solution $c_n(t) = 0$. Consequently, by (7.1), the solution of the IBVP is

$$u(x,t) = \frac{1}{2}t^2 + 2 + \left[\frac{1}{\pi^2}t - \frac{1}{\pi^3}\sin(\pi t)\right]\cos(\pi x)$$

$$- \frac{1}{\pi}\sin(2\pi t)\cos(2\pi x). \quad \blacksquare$$

7.3. The Laplace Equation

Consider the BVP

$$u_{xx}(x,y) + u_{yy}(x,y) = q(x,y), \quad 0 < x < L, \, 0 < y < K,$$
$$u(0,y) = 0, \quad u(L,y) = 0, \quad 0 < y < K,$$
$$u(x,0) = f_1(x), \quad u(x,K) = f_2(x), \quad 0 < x < L,$$

where, for convenience, the source term has been shifted to the right-hand side of the PDE. The eigenvalues and eigenfunctions associated with this problem for the nonhomogeneous Laplace equation (Poisson equation) were found in Section 5.3; they are, respectively,

$$\lambda_n = \left(\frac{n\pi}{L}\right)^2, \quad X_n(x) = \sin\frac{n\pi x}{L}, \quad n = 1,2,\ldots.$$

Seeking a solution of the form

$$u(x,y) = \sum_{n=1}^{\infty} c_n(y) \sin\frac{n\pi x}{L},$$

we deduce as in Sections 7.1 and 7.2 that the coefficients c_n, $n = 1,2,\ldots$, satisfy the ODE boundary value problems

$$c_n''(y) - n^2\pi^2 c_n(y) = \frac{2}{L}\int_0^L q(x,y)\sin\frac{n\pi x}{L}\,dx, \quad 0 < y < K,$$

$$c_n(0) = \frac{2}{L}\int_0^L f_1(x)\sin\frac{n\pi x}{L}\,dx, \tag{7.8}$$

$$c_n(K) = \frac{2}{L}\int_0^L f_2(x)\sin\frac{n\pi x}{L}\,dx.$$

7.7. Example. In the BVP

$$u_{xx}(x,y) + u_{yy}(x,y) = \pi^2\sin(\pi x), \quad 0 < x < 1, \, 0 < y < 2,$$
$$u(0,y) = 0, \quad u(1,y) = 0, \quad 0 < y < 2,$$
$$u(x,0) = 2\sin(3\pi x), \quad u(x,2) = -\sin(\pi x), \quad 0 < x < 1,$$

we have $L = 1$, $K = 2$, $q(x,y) = \pi^2 \sin(\pi x)$, $f_1(x) = 2\sin(3\pi x)$, and $f_2(x) = -\sin(\pi x)$. Since q, f_1, and f_2 are linear combinations of the eigenfunctions, we can bypass (7.8) and see directly that c_n satisfies

$$c_n''(y) - n^2\pi^2 c_n(y) = \begin{cases} \pi^2, & n = 1, \\ 0, & n \neq 1, \end{cases} \quad 0 < y < 2,$$

$$c_n(0) = \begin{cases} 2, & n = 3, \\ 0, & n \neq 3, \end{cases}$$

$$c_n(2) = \begin{cases} -1, & n = 1, \\ 0, & n \neq 1. \end{cases}$$

Then for $n = 1$,

$$c_1''(y) - \pi^2 c_1(y) = \pi^2, \quad 0 < y < 2,$$
$$c_1(0) = 0, \quad c_1(2) = -1,$$

with solution $c_1(y) = -\operatorname{csch}(2\pi)\sinh\left(\pi(y-2)\right) - 1$; for $n = 3$,

$$c_3''(y) - 9\pi^2 c_3(y) = 0, \quad 0 < y < 2,$$
$$c_3(0) = 2, \quad c_3(2) = 0,$$

with solution $c_3(y) = -2\operatorname{csch}(6\pi)\sinh\left(3\pi(y-2)\right)$; and for $n \neq 1, 3$,

$$c_n''(y) - n^2\pi^2 c_n(y) = 0, \quad 0 < y < 2,$$
$$c_n(0) = 0, \quad c_n(2) = 0,$$

with solution $c_n(y) = 0$. Hence, the solution of the BVP is

$$u(x,y) = -\left[\operatorname{csch}(2\pi)\sinh\left(\pi(y-2)\right) + 1\right]\sin(\pi x)$$
$$- 2\operatorname{csch}(6\pi)\sinh\left(3\pi(y-2)\right)\sin(3\pi x). \quad \blacksquare$$

The same method can be applied to the nonhomogeneous Laplace equation in polar coordinates.

7.8. Example. In Section 5.3 we established that the eigenvalues and eigenfunctions associated with a BVP such as

$$u_{rr}(r,\theta) + r^{-1}u_r(r,\theta) + r^{-2}u_{\theta\theta}(r,t) = 4,$$
$$0 < r < 1, \ -\pi < \theta < \pi,$$
$$u(1,\theta) = 2\cos\theta - \sin(2\theta), \quad -\pi < \theta < \pi,$$

are

$$\lambda_0 = 0, \quad \Theta_0(\theta) = 1,$$

$$\lambda_n = n^2, \quad \Theta_{1n}(\theta) = \cos(n\theta), \quad \Theta_{2n}(\theta) = \sin(n\theta), \quad n = 1, 2, \ldots.$$

Seeking a solution of the form

$$u(r, \theta) = c_0(r) + \sum_{n=1}^{\infty} \left[c_{1n}(r) \Theta_{1n}(\theta) + c_{2n}(r) \Theta_{2n}(\theta) \right]$$

and reasoning just as in Example 7.7, we deduce that the coefficients c_0, c_{1n}, and c_{2n}, $n = 1, 2, \ldots$, satisfy

$$c_0''(r) + r^{-1} c_0'(r) = 4, \quad 0 < r < 1,$$
$$c_0(1) = 0,$$

$$c_{1n}''(r) + r^{-1} c_{1n}'(r) - n^2 r^{-2} c_{1n}(r) = 0, \quad 0 < r < 1,$$
$$c_{1n}(1) = \begin{cases} 2, & n = 1, \\ 0, & n \neq 1, \end{cases}$$

$$c_{2n}''(r) + r^{-1} c_{2n}'(r) - n^2 r^{-2} c_{2n}(r) = 0, \quad 0 < r < 1,$$
$$c_{2n}(1) = \begin{cases} -1, & n = 2, \\ 0, & n \neq 2. \end{cases}$$

Since, according to the explanation given in Section 5.3, $u(r, \theta)$, $u_r(r, \theta)$, and $u_\theta(r, \theta)$ must be continuous (hence, bounded) for $0 \le r \le 1$, $-\pi < \theta \le \pi$, we look for solutions $c_0(r)$, $c_{1n}(r)$, and $c_{2n}(r)$ that remain bounded as $r \to 0+$.

The ODE for $n = 0$ is integrated by first noting that, after multiplication by r, the left-hand side can be written as $(r c_0')'$; in the end, we find that the desired bounded solution is $c_0(r) = r^2 - 1$.

Multiplying the ODEs for $n \ge 1$ by r^2, we arrive at BVPs for Cauchy–Euler equations. Thus, for $n = 1$,

$$r^2 c_{11}''(r) + r c_{11}'(r) - c_{11}(r) = 0, \quad 0 < r < 1,$$
$$c_{11}(1) = 2,$$

$$r^2 c_{21}''(r) + r c_{21}'(r) - c_{21}(r) = 0, \quad 0 < r < 1,$$
$$c_{21}(1) = 0,$$

with bounded solutions $c_{11}(r) = 2r$ and $c_{21}(r) = 0$; for $n = 2$,

$$r^2 c_{12}''(r) + r c_{12}'(r) - 4c_{12}(r) = 0, \quad 0 < r < 1,$$
$$c_{12}(1) = 0,$$

$$r^2 c_{22}''(r) + r c_{22}'(r) - 4c_{22}(r) = 0, \quad 0 < r < 1,$$
$$c_{22}(1) = -1,$$

with bounded solutions $c_{12}(r) = 0$ and $c_{22}(r) = -r^2$; and for $n = 3, 4, \ldots,$

$$r^2 c_{1n}''(r) + r c_{1n}'(r) - n^2 c_{1n}(r) = 0, \quad 0 < r < 1,$$
$$c_{1n}(1) = 0,$$

$$r^2 c_{2n}''(r) + r c_{2n}'(r) - n^2 c_{2n}(r) = 0, \quad 0 < r < 1,$$
$$c_{2n}(1) = 0,$$

with bounded solutions $c_{1n}(r) = 0$ and $c_{2n}(r) = 0$. Therefore, the solution of the BVP is

$$u(r, \theta) = c_0(r) + c_{1n}(r)\Theta_{1n}(\theta) + c_{2n}(r)\Theta_{2n}(\theta)$$
$$= r^2 - 1 + 2r \cos\theta - r^2 \sin(2\theta). \quad \blacksquare$$

7.4. Other Equations

The method of eigenfunction expansion can also be applied to IBVPs or BVPs for more general partial differential equations, such as those mentioned in Section 4.4.

7.9. Example. Consider the IBVP

$$u_t(x,t) = u_{xx}(x,t) - 2u_x(x,t) + u(x,t) + 2te^x \sin(2\pi x),$$
$$0 < x < 1, \ t > 0,$$
$$u(0,t) = 0, \quad u(1,t) = 0, \quad t > 0,$$
$$u(x,0) = 0, \quad 0 < x < 1.$$

Separating the variables in the associated homogenous IBVP in a convenient way, we see that the eigenfunctions satisfy the regular Sturm–Liouville problem

$$X''(x) - 2X'(x) + \lambda X(x) = 0, \quad 0 < x < 1,$$
$$X(0) = 0, \quad X(1) = 0,$$

so, by the formulas in Remark 3.21(iii) with $a = -2$, $b = 0$, $c = 1$, and $L = 1$,

$$\lambda_n = n^2\pi^2 + 1, \quad X_n(x) = e^x \sin(n\pi x), \quad n = 1, 2, \ldots.$$

Consequently, we seek the solution of our given nonhomogeneous IBVP in the form

$$u(x,t) = \sum_{n=1}^{\infty} c_n(t) e^x \sin(n\pi x).$$

Following the procedure applied in the preceding examples, we conclude that the coefficients $c_n(t)$ are the solutions of the IVPs

$$c_n'(t) + (\lambda_n - 1)c_n(t) = \begin{cases} 2t, & n = 2, \\ 0, & n \neq 2, \end{cases} \quad t > 0,$$

$$c_n(0) = 0.$$

For $n = 2$ we have

$$c_2'(t) + 4\pi^2 c_2(t) = 2t, \quad t > 0,$$
$$c_2(0) = 0,$$

from which

$$c_2(t) = \frac{1}{8\pi^4}\left(e^{-4\pi^2 t} + 4\pi^2 t - 1\right).$$

For $n \neq 2$, we find that $c_n(t) = 0$. Therefore, the solution of the given IBVP is

$$u(x,t) = \frac{1}{8\pi^4}\left(e^{-4\pi^2 t} + 4\pi^2 t - 1\right)e^x \sin(2\pi x). \quad \blacksquare$$

7.10. Example. The same procedure applied to the IBVP

$$u_{tt}(x,t) + 2u_t(x,t) = u_{xx}(x,t) - u_x(x,t)$$
$$+ \left[2 + \left(9\pi^2 + \tfrac{1}{4}\right)t\right]e^{x/2}\sin(3\pi x), \quad 0 < x < 1, \ t > 0,$$
$$u(0,t) = 0, \quad u(1,t) = 0, \quad t > 0,$$
$$u(x,0) = 0, \quad u_t(x,0) = e^{x/2}\sin(3\pi x), \quad 0 < x < 1,$$

yields

$$\lambda_n = n^2\pi^2 + \tfrac{1}{4}, \quad X_n(x) = e^{x/2}\sin(n\pi x), \quad n = 1,2,\ldots,$$

which suggests that we seek the solution in the form

$$u(x,t) = \sum_{n=1}^{\infty} c_n(t)e^{x/2}\sin(n\pi x),$$

with the coefficients $c_n(t)$ satisfying

$$c_n''(t) + 2c_n'(t) + \lambda_n c_n(t) = \begin{cases} 2 + (9\pi^2 + \tfrac{1}{4})t, & n = 3, \\ 0, & n \neq 3, \end{cases} \quad t > 0,$$

$$c_n(0) = 0, \quad c_n'(0) = \begin{cases} 1, & n = 3, \\ 0, & n \neq 3. \end{cases}$$

Solving the two cases $n = 3$ and $n \neq 3$ separately, we arrive at

$$c_n(t) = \begin{cases} t, & n = 3, \\ 0, & n \neq 3. \end{cases}$$

Hence, the solution of the given IBVP is

$$u(x,t) = te^{x/2}\sin(3\pi x). \quad \blacksquare$$

7.11. Example. The method of eigenfunction expansion applied to the BVP

$$u_{xx}(x,y) + u_{yy}(x,y) - 4u_x(x,y) = (y^2 - y)e^{2x}\sin(\pi x),$$

$$0 < x < 1, \ 0 < y < 2,$$

$$u(0,y) = 0, \quad u(1,y) = 0, \quad 0 < y < 2,$$

$$u(x,0) = 2e^{2x}\sin(2\pi x), \quad u(x,2) = 0, \quad 0 < x < 1,$$

leads us to the eigenvalues and eigenfunctions

$$\lambda_n = n^2\pi^2 + 4, \quad X_n(x) = e^{2x}\sin(n\pi x), \quad n = 1,2,\ldots,$$

so we seek the solution in the form

$$u(x,y) = \sum_{n=1}^{\infty} c_n(y)e^{2x}\sin(n\pi x),$$

where the coefficients $c_n(y)$ are the solutions of the ODE boundary value problems

$$c_n''(y) - \lambda_n c_n(y) = \begin{cases} y^2 - y, & n = 1, \\ 0, & n \neq 1, \end{cases} \quad 0 < y < 2,$$

$$c_n(0) = \begin{cases} 2, & n = 2, \\ 0, & n \neq 2, \end{cases} \quad c_n(2) = 0.$$

The individual coefficients are now easily computed. For $n = 1$,

$$c_1''(y) - (\pi^2 + 4)c_1(y) = y^2 - y, \quad 0 < y < 2,$$
$$c_1(0) = 0, \quad c_1(2) = 0,$$

with solution

$$c_1(y) = \frac{1}{(\pi^2 + 4)^2} \left\{ 2\operatorname{csch}\left(2\sqrt{\pi^2 + 4}\right) \left[(\pi^2 + 5)\sinh\left(\sqrt{\pi^2 + 4}\,y\right) \right. \right.$$
$$\left. \left. - \sinh\left(\sqrt{\pi^2 + 4}\,(y - 2)\right) \right] + (\pi^2 + 4)(y - y^2) - 2 \right\};$$

for $n = 2$,

$$c_2''(y) - 4(\pi^2 + 1)c_2(y) = 0, \quad 0 < y < 2,$$
$$c_2(0) = 2, \quad c_2(2) = 0,$$

with solution

$$c_2(y) = -2\operatorname{csch}\left(4\sqrt{\pi^2 + 1}\right)\sinh\left(2\sqrt{\pi^2 + 1}\,(y - 2)\right);$$

and for $n \neq 1, 2$,

$$c_n''(y) + (n^2\pi^2 + 4)c_n(y) = 0, \quad 0 < y < 2,$$
$$c_n(0) = 0, \quad c_n(2) = 0,$$

with solution $c_n(y) = 0$. Consequently, the solution of the given BVP is written as

$$u(x,y) = c_1(y)e^{2x}\sin(\pi x) + c_2(y)e^{2x}\sin(2\pi x),$$

with $c_1(y)$ and $c_2(y)$ as determined above. ∎

Exercises

In (1)–(8) use the method of eigenfunction expansion to find the solution of
the IBVP

$$u_t(x,t) = u_{xx}(x,t) + q(x,t), \quad 0 < x < 1, \ t > 0,$$

$$u(0,t) = 0, \quad u(1,t) = 0, \quad t > 0,$$

$$u(x,0) = f(x), \quad 0 < x < 1,$$

with the functions q and f as indicated.

(1) $q(x,t) = 2t\sin(2\pi x), \quad f(x) = \sin(2\pi x) - 5\sin(4\pi x).$

(2) $q(x,t) = [3 + \pi^2(3t - 2)]\sin(\pi x) + (9\pi^2 t^2 + 2t)\sin(3\pi x),$
$\quad f(x) = -2\sin(\pi x).$

(3) $q(x,t) = e^{-t}\sin(3\pi x) - \sin(5\pi x), \quad f(x) = \sin(\pi x) + 2\sin(3\pi x).$

(4) $q(x,t) = [(\pi^2 - 1)e^{-t} - \pi^2]\sin(\pi x) + (4\pi^2 t + 4\pi^2 + 1)\sin(2\pi x)$
$\quad\quad\quad\quad + 48\pi^2\sin(4\pi x), \quad f(x) = \sin(2\pi x) + 3\sin(4\pi x).$

(5) $q(x,t) = (t - 1)\sin(\pi x), \quad f(x) = \sin(\pi x) + 2\sin(2\pi x).$

(6) $q(x,t) = \pi^2 e^{-4\pi^2 t}\sin(2\pi x) + (2t + 3)\sin(3\pi x),$
$\quad f(x) = 3\sin(\pi x) - 2\sin(2\pi x).$

(7) $q(x,t) = xt, \quad f(x) = \begin{cases} 1, & 0 < x \le 1/2, \\ 0, & 1/2 < x < 1. \end{cases}$

(8) $q(x,t) = \frac{1}{2}(x - 1)t, \quad f(x) = x.$

In (9)–(16) use the method of eigenfunction expansion to find the solution
of the IBVP

$$u_t(x,t) = u_{xx}(x,t) + q(x,t), \quad 0 < x < 1, \ t > 0,$$

$$u_x(0,t) = 0, \quad u_x(1,t) = 0, \quad t > 0,$$

$$u(x,0) = f(x), \quad 0 < x < 1,$$

with the functions q and f as indicated.

(9) $q(x,t) = 2 + \cos(2\pi x), \quad f(x) = 2\cos(\pi x) - \cos(2\pi x).$

(10) $q(x,t) = 2 + 16\pi^2\cos(2\pi x) + [(9\pi^2 - 1)e^{-t} + 9\pi^2]\cos(3\pi x),$
$\quad f(x) = 4\cos(2\pi x) + 2\cos(3\pi x).$

(11) $q(x,t) = t - t\cos(\pi x), \quad f(x) = 1 + 3\cos(4\pi x).$

(12) $q(x,t) = \pi^2 e^{-\pi^2 t}\cos(\pi x) + (1 - 2t)\cos(2\pi x)$, $f(x) = 2\cos(2\pi x)$.

(13) $q(x,t) = e^{-t}\cos(\pi x)$, $f(x) = 2 - \cos(3\pi x)$.

(14) $q(x,t) = 3t - 4$, $f(x) = \begin{cases} -1, & 0 < x \le 1/2, \\ 1, & 1/2 < x < 1. \end{cases}$

(15) $q(x,t) = -2xt$, $f(x) = 1 - 3\cos(2x)$.

(16) $q(x,t) = (1 - x)t$, $f(x) = x$.

In (17)–(20) use the method of eigenfunction expansion to find the solution of the IBVP

$$u_t(x,t) = u_{xx}(x,t) + q(x,t), \quad 0 < x < 1, \ t > 0,$$
$$u(0,t) = 0, \quad u_x(1,t) = 0, \quad t > 0,$$
$$u(x,0) = f(x), \quad 0 < x < 1,$$

with the functions q and f as indicated.

(17) $q(x,t) = \sin\left(\frac{3}{2}\pi x\right) - 2\sin\left(\frac{5}{2}\pi x\right)$, $f(x) = \sin\left(\frac{3}{2}\pi x\right)$.

(18) $q(x,t) = t\sin\left(\frac{1}{2}\pi x\right)$, $f(x) = \sin\left(\frac{1}{2}\pi x\right) + 2\sin\left(\frac{5}{2}\pi x\right)$.

(19) $q(x,t) = \frac{1}{4}\left[9\pi^2 t^2 + 2(4 - 9\pi^2)t - 8\right]\sin\left(\frac{3}{2}\pi x\right)$
$\qquad\qquad + \frac{1}{2}(25\pi^2 - 4)e^{-t}\sin\left(\frac{5}{2}\pi x\right)$, $f(x) = 2\sin\left(\frac{5}{2}\pi x\right)$.

(20) $q(x,t) = t\sin\left(\frac{1}{2}\pi x\right)$, $f(x) = \pi x$.

In (21)–(26) use the method of eigenfunction expansion to find the solution of the IBVP

$$u_{tt}(x,t) = u_{xx}(x,t) + q(x,t), \quad 0 < x < 1, \ t > 0,$$
$$u(0,t) = 0, \quad u(1,t) = 0, \quad t > 0,$$
$$u(x,0) = f(x), \quad u_t(x,0) = g(x), \quad 0 < x < 1,$$

with the functions q, f, and g as indicated.

(21) $q(x,t) = 2\sin(2\pi x)$, $f(x) = \sin(\pi x)$, $g(x) = -3\sin(2\pi x)$.

(22) $q(x,t) = (3\pi^2 - 4 - 2\pi^2 t^2)\sin(\pi x) + 9\pi^2(1 + t)\sin(3\pi x)$,
$\qquad f(x) = 3\sin(\pi x) + \sin(3\pi x)$, $g(x) = \sin(3\pi x)$.

(23) $q(x,t) = t\sin(\pi x)$, $f(x) = \sin(\pi x)$, $g(x) = 2\sin(\pi x) + 4\sin(3\pi x)$.

(24) $q(x,t) = 4\left[\pi^2 - (\pi^2 + 1)e^{-2t}\right]\sin(2\pi x) + 2(16\pi^2 - 1)\sin t\sin(4\pi x)$,
$\qquad f(x) = 0$, $g(x) = 2\sin(2\pi x) + 2\sin(4\pi x)$.

(25) $q(x,t) = (t+1)\sin(2\pi x)$, $f(x) = 0$, $g(x) = 2x - 1$.

(26) $q(x,t) = 2xt$, $f(x) = \begin{cases} 0, & 0 < x \le 1/2, \\ 2 & 1/2 < x < 1, \end{cases}$ $g(x) = -1$.

In (27)–(32) use the method of eigenfunction expansion to find the solution of the IBVP

$$u_{tt}(x,t) = u_{xx}(x,t) + q(x,t), \quad 0 < x < 1, \ t > 0,$$
$$u_x(0,t) = 0, \quad u_x(1,t) = 0, \quad t > 0,$$
$$u(x,0) = f(x), \quad u_t(x,0) = g(x), \quad 0 < x < 1,$$

with the functions q, f, and g as indicated.

(27) $q(x,t) = 3$, $f(x) = 1 + 2\cos(2\pi x)$, $g(x) = \cos(3\pi x)$.

(28) $q(x,t) = \pi^2(3-t)\cos(\pi x)$,
      $f(x) = 3\cos(\pi x)$, $g(x) = -\cos(\pi x) + 6\pi\cos(3\pi x)$.

(29) $q(x,t) = 1 + t\cos(2\pi x)$,
      $f(x) = 2\cos(2\pi x)$, $g(x) = 1 + \cos(\pi x) - \cos(2\pi x)$.

(30) $q(x,t) = (t-2)e^{-t} + 4\pi^2 t\cos(2\pi x)$,
      $f(x) = -\cos(3\pi x)$, $g(x) = 1 + \cos(2\pi x)$.

(31) $q(x,t) = -x$, $f(x) = x - 1$, $g(x) = 0$.

(32) $q(x,t) = 2xt$, $f(x) = x$, $g(x) = \begin{cases} 0, & 0 < x \le 1/2, \\ 1, & 1/2 < x < 1. \end{cases}$

In (33)–(48) use the method of eigenfunction expansion to find the solution of the BVP

$$u_{xx}(x,y) + u_{yy}(x,y) = q(x,y), \quad 0 < x < 1, \ 0 < y < 2,$$

with the function q and BCs (for $0 < x < 1, 0 < y < 2$) as indicated.

(33) $q(x,y) = \sin(2\pi x)$, $u(0,y) = 0$, $u(1,y) = 0$,
      $u(x,0) = \sin(\pi x) - 2\sin(3\pi x)$, $u(x,2) = -\sin(2\pi x)$.

(34) $q(x,y) = -\pi^2(2y+3)\sin(\pi x) + (1 - 4\pi^2)e^{-y}\sin(2\pi x)$,
      $u(0,y) = 0$, $u(1,y) = 0$,
      $u(x,0) = 3\sin(\pi x) + \sin(2\pi x)$, $u_y(x,2) = 2\sin(\pi x) - e^{-2}\sin(2\pi x)$.

(35) $q(x,y) = 2y\sin(\pi x)$, $u(0,y) = 0$, $u(1,y) = 0$,
$u_y(x,0) = x - 1$, $u_y(x,2) = 0$.

(36) $q(x,y) = \frac{1}{2}\pi^2$, $u(0,y) = 0$, $u(1,y) = 0$,
$u(x,0) = 0$, $u(x,2) = \frac{1}{2}x$.

(37) $q(x,y) = -\pi^2\sin(\pi y)$, $u(0,y) = 2\sin(\pi y)$, $u(1,y) = \sin\left(\frac{1}{2}\pi y\right)$,
$u(x,0) = 0$, $u(x,2) = 0$.

(38) $q(x,y) = \left[2(1 - \pi^2)e^x + \pi^2\right]\sin(\pi y) - \frac{9}{4}\pi^2(x + 3)\sin\left(\frac{3}{2}\pi y\right)$,
$u(0,y) = \sin(\pi y) + 3\sin\left(\frac{3}{2}\pi y\right)$,
$u_x(1,y) = 2e\sin(\pi y) + \sin\left(\frac{3}{2}\pi y\right)$, $u(x,0) = 0$, $u(x,2) = 0$.

(39) $q(x,y) = (x - 1)\sin(\pi y)$,
$u_x(0,y) = 0$, $u(1,y) = y - 2$, $u(x,0) = 0$, $u(x,2) = 0$.

(40) $q(x,y) = 3x\sin\left(\frac{1}{2}\pi y\right)$,
$u(0,y) = y$, $u(1,y) = 0$, $u(x,0) = 0$, $u(x,2) = 0$.

(41) $q(x,y) = -2 - 4\pi^2(2y - 1)\cos(2\pi x)$, $u_x(0,y) = 0$, $u_x(1,y) = 0$,
$u(x,0) = 1 - \cos(2\pi x)$, $u(x,2) = -3 + 3\cos(2\pi x)$.

(42) $q(x,y) = -2 + \pi^2(2 - y)\cos(\pi x) + (4 - 9\pi^2)e^{2y}\cos(3\pi x)$,
$u_x(0,y) = 0$, $u_x(1,y) = 0$,
$u_y(x,0) = \cos(\pi x) + 2\cos(3\pi x)$, $u(x,2) = -3 + e^4\cos(3\pi x)$.

(43) $q(x,y) = y^2\cos(2\pi x)$, $u_x(0,y) = 0$, $u_x(1,y) = 0$,
$u(x,0) = 2\cos(\pi x)$, $u_y(x,2) = -\cos(2\pi x)$.

(44) $q(x,y) = \sin y\cos(\pi x)$, $u_x(0,y) = 0$, $u_x(1,y) = 0$,
$u(x,0) = \cos(\pi x)$, $u(x,2) = 1$.

(45) $q(x,y) = (1 - x^2)\cos(\pi y)$,
$u(0,y) = -\cos(\pi y)$, $u(1,y) = 2\cos\left(\frac{1}{2}\pi y\right)$,
$u_y(x,0) = 0$, $u_y(x,2) = 0$.

(46) $q(x,y) = 2 - \frac{3}{4}\pi^2\cos\left(\frac{1}{2}\pi y\right) + 2(\pi^2 - 1)e^{-x}\cos(\pi y)$,
$u(0,y) = 3\cos\left(\frac{1}{2}\pi y\right) - 2\cos(\pi y)$, $u_x(1,y) = 2e^{-1}\cos(\pi y)$,
$u_y(x,0) = 0$, $u_y(x,2) = 0$.

(47) $q(x,y) = x\cos\left(\frac{3}{2}\pi y\right)$,
$u_x(0,y) = 0$, $u(1,y) = 2 - y$, $u_y(x,0) = 0$, $u_y(x,2) = 0$.

(48) $q(x,y) = 4y - 2$,
$u(0,y) = -\cos(\pi y)$, $u(1,y) = 0$, $u_y(x,0) = 0$, $u_y(x,2) = 0$.

In (49)–(54) use the method of eigenfunction expansion to find the solution of the IBVP

$$u_{rr}(r,\theta) + r^{-1}u_r(r,\theta) + r^{-2}u_{\theta\theta}(r,\theta) = q(r,\theta),$$

$$0 < x < 1, \quad -\pi < \theta < \pi,$$

$$u(r,\theta), \ u_r(r,\theta) \text{ bounded as } r \to 0+, \quad u(1,\theta) = f(\theta), \quad -\pi < \theta < \pi,$$

$$u(r,-\pi) = u(r,\pi), \quad u_\theta(r,-\pi) = u_\theta(r,\pi), \quad 0 < r < 1,$$

with the functions q and f as indicated.

(49) $q(r,\theta) = -8, \ \ f(\theta) = -1 + 2\sin\theta + 2\cos(3\theta)$.

(50) $q(r,\theta) = 9r + 6\cos\theta + 3\sin\theta, \ \ f(\theta) = 1 - 3\sin\theta$.

(51) $q(r,\theta) = 18r - 4 + 8r\cos\theta + (7r^2 - 5)\sin(3\theta),$
     $f(\theta) = 1 + \cos\theta + 2\sin(3\theta)$.

(52) $q(r,\theta) = 3\cos\theta + 10\cos(3\theta) - 2(2r^4 + 9r^3)e^{2r}\sin(4\theta),$
     $f(\theta) = \cos\theta - \cos(3\theta) - e^2\sin(4\theta)$.

(53) $q(r,\theta) = (1-r)\sin(2\theta), \ \ f(\theta) = \begin{cases} 0, & -\pi < \theta \le 0, \\ 1, & 0 < \theta < \pi. \end{cases}$

(54) $q(r,\theta) = r\cos\theta - 2\sin(3\theta), \ \ f(\theta) = \begin{cases} \theta, & -\pi < \theta \le 0, \\ -1, & 0 < \theta < \pi. \end{cases}$

In (55)–(58) use the method of eigenfunction expansion to find the solution of the IBVP

$$u_t(x,t) = u_{xx}(x,t) - u_x(x,t) + 2u(x,t) + q(x,t),$$

$$0 < x < 1, \ t > 0,$$

$$u(0,t) = 0, \quad u(1,t) = 0, \quad t > 0,$$

$$u(x,0) = f(x), \quad 0 < x < 1,$$

with the functions q and f as indicated.

(55) $q(x,t) = (2 - 3t)e^{x/4}\sin(\pi x), \ \ f(x) = -e^{x/4}\sin(\pi x)$.

(56) $q(x,t) = \frac{3}{4}(48\pi^2 - 5)e^{-t+x/4}\sin(3\pi x), \ \ f(x) = 2e^{x/4}\sin(3\pi x)$.

(57) $q(x,t) = e^{x/4}, \ \ f(x) = -e^{x/4}\sin(2\pi x)$.

(58) $q(x,t) = 2e^{x/4}\sin(\pi x), \ \ u(x,0) = x$.

In (59)–(62) use the method of eigenfunction expansion to find the solution of the IBVP

$$u_{tt}(x,t) + u_t(x,t) + u(x,t) = u_{xx}(x,t) - 2u_x(x,t) + q(x,t),$$
$$0 < x < 1, \ t > 0,$$
$$u(0,t) = 0, \quad u(1,t) = 0, \quad t > 0,$$
$$u(x,0) = f(x), \quad u_t(x,0) = g(x), \quad 0 < x < 1,$$

with the functions q, f, and g as indicated.

(59) $q(x,t) = \left[(\pi^2 + 2)t + 1\right]e^x \sin(\pi x), \quad f(x) = 0, \quad g(x) = e^x \sin(\pi x).$

(60) $q(x,t) = 4\left[(2\pi^2 + 1)t - \pi^2\right]e^x \sin(2\pi x),$
   $f(x) = -e^x \sin(2\pi x), \quad g(x) = 2e^x \sin(2\pi x).$

(61) $q(x,t) = e^x, \quad f(x) = -e^x \sin(\pi x), \quad g(x) = 0.$

(62) $q(x,t) = 2e^x \sin(2\pi x), \quad f(x) = 0, \quad g(x) = 1.$

In (63)–(66) use the method of eigenfunction expansion to find the solution of the BVP

$$u_{xx}(x,y) + u_{yy}(x,y) - 2u_x(x,y) + u(x,y) = q(x,t),$$
$$0 < x < 1, \ 0 < y < 2,$$
$$u(0,y) = 0, \quad u(1,y) = 0, \quad 0 < y < 2,$$
$$u(x,0) = f(x), \quad u(x,2) = g(x), \quad 0 < x < 1,$$

with the functions q, f, and g as indicated.

(63) $q(x,y) = -\pi^2(2y + 1)e^x \sin(\pi x),$
   $f(x) = e^x \sin(\pi x), \quad g(x) = 5e^x \sin(\pi x).$

(64) $q(x,y) = (2 - 3\pi^2 - \pi^2 y)e^x \sin(\pi x),$
   $f(x) = 3e^x \sin(\pi x), \quad g(x) = 7e^x \sin(\pi x).$

(65) $q(x,y) = 2e^{2x}, \quad f(x) = e^x \sin(2\pi x), \quad g(x) = 0.$

(66) $q(x,y) = 3e^x \sin(\pi x), \quad f(x) = 0, \quad g(x) = -1.$

Chapter 8
The Fourier Transformations

Some problems of practical importance are beyond the reach of the method of eigenfunction expansion. This is the case, for example, when the space variable is defined on the entire real line and where, as a consequence, there are no boundary points. This may lead to the problem in question having a continuum of eigenvalues instead of a countable set. In such situations we need to employ other techniques of solution. The Fourier transformations— developed, in fact, from the Fourier series representations of functions—are particularly useful tools when dealing with infinite or semi-infinite spatial regions because they are designed for exactly this type of setup and have the added advantage that they reduce by one the number of "active" variables in the given PDE problem.

8.1. The Full Fourier Transformation

Construction of the transformation. Consider, for simplicity, a function f that is continuous and periodic with period $2L$ on $\mathbb{R}$. Then we have the representation (see Chapter 2)

$$f(x) = \frac{1}{2}a_0 + \sum_{n=1}^{\infty}\left(a_n \cos\frac{n\pi x}{L} + b_n \sin\frac{n\pi x}{L}\right), \tag{8.1}$$

where

$$a_n = \frac{1}{L}\int_{-L}^{L} f(x)\cos\frac{n\pi x}{L}\,dx, \quad n = 0,1,2,\ldots,$$

$$b_n = \frac{1}{L}\int_{-L}^{L} f(x)\sin\frac{n\pi x}{L}\,dx, \quad n = 1,2,\ldots. \tag{8.2}$$

Using Euler's formula

$$e^{i\theta} = \cos\theta + i\sin\theta, \quad i^2 = -1,$$

and its alternative with θ replaced by $-\theta$, we get

$$\cos\theta = \tfrac{1}{2}(e^{i\theta} + e^{-i\theta}), \quad \sin\theta = -\tfrac{1}{2}i(e^{i\theta} - e^{-i\theta});$$

consequently, (8.1) with $\theta = n\pi x/L$ becomes

$$f(x) = \tfrac{1}{2}a_0 + \overbrace{\sum_{n=1}^{\infty} \tfrac{1}{2}(a_n - ib_n)e^{in\pi x/L}}^{\cos} + \overbrace{\sum_{n=1}^{\infty} \tfrac{1}{2}(a_n + ib_n)e^{-in\pi x/L}}^{\sin}.$$

Replacing n by $-n$ in the first sum above and noticing from (8.2) that $a_{-n} = a_n$ and $b_{-n} = -b_n$, from the last equality we obtain

$$f(x) = \tfrac{1}{2}a_0 + \sum_{n=-1}^{-\infty} \tfrac{1}{2}(a_n + ib_n)e^{-in\pi x/L} + \sum_{n=1}^{\infty} \tfrac{1}{2}(a_n + ib_n)e^{-in\pi x/L},$$

or

$$f(x) = \sum_{n=-\infty}^{\infty} c_n e^{-in\pi x/L},$$

where, by (8.2),

$$c_n = \tfrac{1}{2}(a_n + ib_n) = \frac{1}{2L} \int_{-L}^{L} f(x) \left(\cos \frac{n\pi x}{L} + i \sin \frac{n\pi x}{L} \right) dx$$

$$= \frac{1}{2L} \int_{-L}^{L} f(x) e^{in\pi x/L} dx \quad \text{(with } b_0 = 0\text{)};$$

hence,

$$f(x) = \sum_{n=-\infty}^{\infty} \left[\frac{1}{2L} \int_{-L}^{L} f(\xi) e^{in\pi \xi/L} d\xi \right] e^{-in\pi x/L}. \tag{8.3}$$

If f is not periodic in the proper sense of the word, then we may regard it as "periodic" with an "infinite period". Using some advanced calculus arguments, we find that, as $L \to \infty$, representation (8.3) for such a function takes the form

$$f(x) = \int_{-\infty}^{\infty} \left[\frac{1}{2\pi} \int_{-\infty}^{\infty} f(\xi) e^{i\omega \xi} d\xi \right] e^{-i\omega x} d\omega. \tag{8.4}$$

We define the (full) *Fourier transform* of f by

$$\mathcal{F}[f](\omega) = F(\omega) = \frac{1}{\sqrt{2\pi}} \int_{-\infty}^{\infty} f(x) e^{i\omega x} dx. \tag{8.5}$$

From (8.4) it is clear that the *inverse Fourier transform* of F is

$$\mathcal{F}^{-1}[F](x) = f(x) = \frac{1}{\sqrt{2\pi}} \int_{-\infty}^{\infty} F(\omega)e^{-i\omega x}d\omega. \tag{8.6}$$

The integral operators $\mathcal{F}$ and $\mathcal{F}^{-1}$ are called the *Fourier transformation* and *inverse Fourier transformation*, respectively. The variable ω is called the *transformation parameter.*

8.1. Remarks. (i) Formulas (8.5) and (8.6) are valid if

$$\int_{-\infty}^{\infty} |f(x)|\,dx < \infty;$$

that is, if f is absolutely integrable on $\mathbb{R}$. However, the Fourier transform may also be defined for some functions that do not have the above property. This is done in a generalized sense through a limiting process involving transforms of absolutely integrable functions.

(ii) The construction of the full Fourier transform can be extended to piecewise continuous functions f. In this case, $f(x)$ must be replaced by $\frac{1}{2}\left[f(x-) + f(x+)\right]$ in (8.4) and (8.6). ∎

8.2. Example. The Fourier transform of the function

$$f(x) = \begin{cases} 1, & -a \le x \le a, \\ 0 & \text{otherwise,} \end{cases}$$

where $a > 0$ is a constant, is

$$F(\omega) = \frac{1}{\sqrt{2\pi}} \int_{-\infty}^{\infty} f(x)e^{i\omega x}\,dx = \frac{1}{\sqrt{2\pi}} \int_{-a}^{a} e^{i\omega x}\,dx$$

$$= \frac{1}{\sqrt{2\pi}} \frac{1}{i\omega}(e^{i\omega a} - e^{-i\omega a}) = \sqrt{\frac{2}{\pi}} \frac{\sin(a\omega)}{\omega}. \quad \blacksquare$$

8.3. Definition. Let f and g be absolutely integrable on $\mathbb{R}$. By the *convolution* of f and g we understand the function $f * g$ defined by

$$(f * g)(x) = \frac{1}{\sqrt{2\pi}} \int_{-\infty}^{\infty} f(x - \xi)g(\xi)\,d\xi, \quad -\infty < x < \infty. \quad \blacksquare$$

8.4. Remark. Making the change of variable $x - \xi = \eta$ and then replacing η by ξ, we easily convince ourselves that

$$(f * g)(x) = \frac{1}{\sqrt{2\pi}} \int\limits_{-\infty}^{\infty} f(\xi) g(x - \xi) \, d\xi = (g * f)(x).$$

This means that the operation of convolution is commutative. ∎

8.5. Theorem. (i) $\mathcal{F}$ *is linear; that is,*

$$\mathcal{F}[c_1 f_1 + c_2 f_2] = c_1 \mathcal{F}[f_1] + c_2 \mathcal{F}[f_2]$$

for any functions f_1 and f_2 (to which $\mathcal{F}$ can be applied) and any numbers c_1 and c_2.

(ii) *If $u = u(x,t)$, $u(x,t) \to 0$ as $x \to \pm\infty$, and $\mathcal{F}[u](\omega,t) = U(\omega,t)$, then*

$$\mathcal{F}[u_x](\omega,t) = -i\omega U(\omega,t).$$

(iii) *If, in addition, $u_x(x,t) \to 0$ as $x \to \pm\infty$, then*

$$\mathcal{F}[u_{xx}](\omega,t) = -\omega^2 U(\omega,t).$$

(iv) *Time differentiation and the Fourier transformation with respect to x commute:*

$$\mathcal{F}[u_t](\omega,t) = \big(\mathcal{F}[u]\big)_t(\omega,t) = U'(\omega,t).$$

(v) *The Fourier transform of a convolution is given by the formula*

$$\mathcal{F}[f * g] = \mathcal{F}[f]\mathcal{F}[g].$$

8.6. Remarks. (i) In general, $\mathcal{F}[fg] \neq \mathcal{F}[f]\mathcal{F}[g]$.

(ii) It is obvious from its definition that $\mathcal{F}^{-1}$ is also linear.

(iii) The Fourier transforms of a few elementary functions are given in Table A2 in the Appendix. ∎

The properties of the Fourier transformation listed in Theorem 8.5 play an essential role in the solution of certain types of PDE problems. The solution strategy in such situations is best illustrated by examples.

The Cauchy problem for an infinite rod. Heat conduction in a very long uniform rod where the diffusion activity diminishes towards the endpoints is modeled mathematically by the IVP (called a *Cauchy problem*)

$$u_t(x,t) = ku_{xx}(x,t), \quad -\infty < x < \infty, \ t > 0,$$

$$u(x,t), \ u_x(x,t) \to 0 \quad \text{as } x \to \pm\infty, \ t > 0,$$

$$u(x,0) = f(x), \quad -\infty < x < \infty.$$

Adopting the notation

$$\mathcal{F}[u](\omega,t) = U(\omega,t), \quad \mathcal{F}[f](\omega) = F(\omega),$$

we apply $\mathcal{F}$ to the PDE and IC and use the properties of $\mathcal{F}$ in Theorem 8.5 to reduce the given IVP to an initial value problem for an ordinary differential equation in the so-called transform domain. This new IVP (in which ω is an "inert" parameter) is

$$U'(\omega,t) + k\omega^2 U(\omega,t) = 0, \quad t > 0,$$

$$U(\omega,0) = F(\omega),$$

with solution

$$U(\omega,t) = F(\omega)e^{-k\omega^2 t}. \tag{8.7}$$

To find the solution of the original IVP, we now need to compute the inverse Fourier transform of U. Let $P(\omega,t) = e^{-k\omega^2 t}$. Using formula 7 (with $a = (4kt)^{-1/2}$) in Table A2, we easily see that the inverse transform of P is

$$\mathcal{F}^{-1}[P](x,t) = p(x,t) = \frac{1}{\sqrt{2kt}} e^{-x^2/(4kt)}.$$

By Theorem 8.5(v), we can now write (8.7) as

$$\mathcal{F}[u](\omega,t) = \mathcal{F}[f](\omega)\mathcal{F}[p](\omega,t) = \mathcal{F}[f * p](\omega,t),$$

so $u = f * p$, or, by Definition 8.3,

$$u(x,t) = \frac{1}{\sqrt{2\pi}} \int_{-\infty}^{\infty} f(\xi)\frac{1}{\sqrt{2kt}} e^{-(x-\xi)^2/(4kt)} d\xi.$$

An alternative form for this is

$$u(x,t) = \int_{-\infty}^{\infty} G(x,t;\xi,0)u(\xi,0)\,d\xi, \qquad (8.8)$$

where

$$G(x,t;\xi,0) = \frac{1}{2\sqrt{\pi kt}}e^{-(x-\xi)^2/(4kt)}$$

is called the *Gauss–Weierstrass kernel* or *influence function*. Formula (8.8) shows how the initial temperature distribution $u(x,0)$ influences the subsequent evolution of the temperature in the rod. This type of representation formulas will be discussed in more detail in Chapter 10.

8.7. Example. In the IVP

$$u_t(x,t) = 2u_{xx}(x,t), \quad -\infty < x < \infty,\ t > 0,$$
$$u(x,t),\, u_x(x,t) \to 0 \quad \text{as } x \to \pm\infty,\ t > 0,$$
$$u(x,0) = f(x) = \begin{cases} -3, & |x| \le 1, \\ 0, & |x| > 1, \end{cases}$$

we have $k = 2$, so

$$G(x,t;\xi,0) = \frac{1}{2\sqrt{2\pi t}}e^{-(x-\xi)^2/(8t)},$$

which, replaced in (8.8), yields

$$u(x,t) = -\frac{3}{2\sqrt{2\pi t}}\int_{-1}^{1} e^{-(x-\xi)^2/(8t)}\,d\xi.$$

If instead of (8.8) we use formula 12 (with $a = 1$) in Table A2 to compute $F(\omega) = -3\sqrt{2/\pi}\,(\sin\omega)/\omega$ and then solve the transformed problem

$$U'(\omega,t) + 2\omega^2 U(\omega,t) = 0, \quad t > 0,$$
$$U(\omega,0) = -3\sqrt{\frac{2}{\pi}}\,\frac{\sin\omega}{\omega},$$

we find that

$$U(\omega,t) = -3\sqrt{\frac{2}{\pi}}\,\frac{\sin\omega}{\omega}e^{-2\omega^2 t},$$

which, through a combination of formulas 3 and 16, yields

$$u(x,t) = \mathcal{F}^{-1}[U](x,t) = \frac{3}{2}\left(\operatorname{erf}\frac{x-1}{2\sqrt{2t}} - \operatorname{erf}\frac{x+1}{2\sqrt{2t}}\right).$$

The function $y = \operatorname{erf}(x)$ is called the *error function*. The graphs of the error function and of the *complementary error function* defined on $(-\infty,\infty)$, respectively, by

$$y = \operatorname{erf}(x) = \frac{2}{\sqrt{\pi}}\int_0^x e^{-\xi^2}\,d\xi,$$

$$y = \operatorname{erfc}(x) = \frac{2}{\sqrt{\pi}}\int_x^\infty e^{-\xi^2}\,d\xi = 1 - \operatorname{erf}(x)$$

are shown in Fig. 8.1. ∎

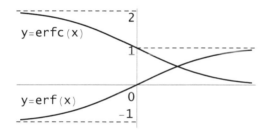

Fig. 8.1.

Vibration of an infinite string. The vibrations of a very long string initially at rest, with negligible body force and where the effects of mechanical activity at the endpoints are insignificant, are modeled by the IVP

$$u_{tt}(x,t) = c^2 u_{xx}(x,t), \quad -\infty < x < \infty,\ t > 0,$$
$$u(x,t),\ u_x(x,t) \to 0 \quad \text{as } x \to \pm\infty,\ t > 0,$$
$$u(x,0) = f(x), \quad u_t(x,0) = 0, \quad -\infty < x < \infty.$$

As above, let $\mathcal{F}[u](\omega,t) = U(\omega,t)$ and $\mathcal{F}[f](\omega) = F(\omega)$. Applying $\mathcal{F}$ to the PDE and ICs, we arrive at the ODE problem

$$U''(\omega,t) + c^2\omega^2 U(\omega,t) = 0, \quad t > 0,$$
$$U(\omega,0) = F(\omega), \quad U'(\omega,0) = 0,$$

with general solution $U(\omega,t) = C_1(\omega)\cos(c\omega t) + C_2(\omega)\sin(c\omega t)$, where $C_1(\omega)$ and $C_2(\omega)$ are arbitrary functions of the transformation parameter. Hence, using the ICs, we find that

$$U(\omega,t) = F(\omega)\cos(c\omega t).$$

By Euler's formula, we have $\cos(c\omega t) = \frac{1}{2}(e^{ic\omega t} + e^{-ic\omega t})$, so from (8.6) it follows that if f is continuous, then

$$u(x,t) = \mathcal{F}^{-1}[U](x,t) = \int_{-\infty}^{\infty} F(\omega)\cos(c\omega t)e^{-i\omega x}\,d\omega$$

$$= \frac{1}{2}\int_{-\infty}^{\infty} F(\omega)[e^{-i\omega(x-ct)} + e^{-i\omega(x+ct)}]\,d\omega$$

$$= \frac{1}{2}[f(x-ct) + f(x+ct)]. \tag{8.9}$$

This problem will be approached from a different angle in Chapter 12.

8.8. Example. In the IVP

$$u_{tt}(x,t) = u_{xx}(x,t), \quad -\infty < x < \infty, \; t > 0,$$

$$u(x,t),\, u_x(x,t) \to 0 \quad \text{as } x \to \pm\infty, \; t > 0,$$

$$u(x,0) = e^{-x^2}, \quad u_t(x,0) = 0, \quad -\infty < x < \infty,$$

we have $c = 1$ and $f(x) = e^{-x^2}$, so, by (8.9),

$$u(x,t) = \frac{1}{2}\left[e^{-(x-t)^2} + e^{-(x+t)^2}\right]. \quad \blacksquare$$

8.2. The Fourier Sine and Cosine Transformations

The Fourier sine and cosine transforms generalize the Fourier sine and cosine series, respectively, and are defined for functions f that are piecewise continuous on $0 < x < \infty$ by

sine $$\mathcal{F}_S[f](\omega) = F(\omega) = \sqrt{\frac{2}{\pi}}\int_0^{\infty} f(x)\sin(\omega x)\,dx,$$

cosine $$\mathcal{F}_C[f](\omega) = F(\omega) = \sqrt{\frac{2}{\pi}}\int_0^{\infty} f(x)\cos(\omega x)\,dx.$$

It can be shown that these transforms exist if f is absolutely integrable on $(0, \infty)$; that is,

$$\int_0^\infty |f(x)|\,dx < \infty.$$

The corresponding inverse transforms are

$$\mathcal{F}_S^{-1}[F](x) = f(x) = \sqrt{\frac{2}{\pi}} \int_0^\infty F(\omega)\sin(\omega x)\,d\omega,$$

derive the ω vs. the x in the regular transform

$$\mathcal{F}_C^{-1}[F](x) = f(x) = \sqrt{\frac{2}{\pi}} \int_0^\infty F(\omega)\cos(\omega x)\,d\omega.$$

The operators $\mathcal{F}_S$ and $\mathcal{F}_C$ are called the *Fourier sine transformation* and *Fourier cosine transformation*, respectively, while $\mathcal{F}_S^{-1}$ and $\mathcal{F}_C^{-1}$ are their inverse transformations. The comment made at the end of Remark 8.1(ii) applies here as well.

8.9. Example. The Fourier sine and cosine transforms of the function

$$f(x) = \begin{cases} 1, & 0 \le x \le a, \\ 0, & x > a, \end{cases} \qquad a = \text{const},$$

are

$$\mathcal{F}_S[f](\omega) = \sqrt{\frac{2}{\pi}} \int_0^\infty f(x)\sin(\omega x)\,dx = \sqrt{\frac{2}{\pi}} \int_0^a \sin(\omega x)\,dx$$
$$= \sqrt{\frac{2}{\pi}} \frac{1}{\omega}\left[1 - \cos(a\omega)\right],$$

$$\mathcal{F}_C[f](\omega) = \sqrt{\frac{2}{\pi}} \int_0^\infty f(x)\cos(\omega x)\,dx = \sqrt{\frac{2}{\pi}} \int_0^a \cos(\omega x)\,dx$$
$$= \sqrt{\frac{2}{\pi}} \frac{1}{\omega}\sin(a\omega). \quad \blacksquare$$

8.10. Remark. $\mathcal{F}_S$ ($\mathcal{F}_C$) can also be defined for functions on $\mathbb{R}$ if these functions are odd (even). $\blacksquare$

8.11. Theorem. (i) $\mathcal{F}_S$ *and* $\mathcal{F}_C$ *are linear operators.*

(ii) *If* $u = u(x,t)$ *and* $u(x,t) \to 0$ *as* $x \to \infty$, *then*

$$\mathcal{F}_S[u_x](\omega,t) = -\omega\mathcal{F}_C[u](\omega,t),$$
$$\mathcal{F}_C[u_x](\omega,t) = -\sqrt{\frac{2}{\pi}} u(0,t) + \omega\mathcal{F}_S[u](\omega,t).$$

(iii) *If, in addition,* $u_x(x,t) \to 0$ *as* $x \to \infty$, *then*

$u(0,t) \checkmark$ use sine

$$\mathcal{F}_S[u_{xx}](\omega,t) = \sqrt{\frac{2}{\pi}}\,\omega u(0,t) - \omega^2 \mathcal{F}_S[u](\omega,t),$$

$$\mathcal{F}_C[u_{xx}](\omega,t) = -\sqrt{\frac{2}{\pi}}\,u_x(0,t) - \omega^2 \mathcal{F}_C[u](\omega,t).$$

derivative → use cosine

(iv) *Time differentiation commutes with both the Fourier sine and cosine transformations:*

$$\mathcal{F}_S[u_t](\omega,t) = \big(\mathcal{F}_S[u]\big)_t(\omega,t),$$

$$\mathcal{F}_C[u_t](\omega,t) = \big(\mathcal{F}_C[u]\big)_t(\omega,t).$$

As Theorem 8.11(iii) indicates, the choice between using the sine transformation or the cosine transformation in the solution of a given IBVP depends on the type of BC prescribed at $x = 0$.

Brief lists of Fourier sine and cosine transforms are given in Tables A3 and A4 in the Appendix.

Heat conduction in a semi-infinite rod. The process of heat conduction in a long rod for which the temperature at the near endpoint is prescribed while the effects of the conditions at the far endpoint are negligible is modeled by the IBVP

$$u_t(x,t) = ku_{xx}(x,t), \quad x > 0,\ t > 0,$$
$$u(0,t) = g(t), \quad t > 0,$$
$$u(x,t),\, u_x(x,t) \to 0 \quad \text{as } x \to \infty,\ t > 0,$$
$$u(x,0) = f(x), \quad x > 0.$$

In view of the given BC, we use the sine transformation; thus, let

$$\mathcal{F}_S[u](\omega,t) = U(\omega,t), \quad \mathcal{F}_S[f](\omega) = F(\omega).$$

Applying $\mathcal{F}_S$ to the PDE and IC and using the properties in Theorem 8.11, we arrive at the ODE problem

$$U'(\omega,t) + k\omega^2 U(\omega,t) = \sqrt{\frac{2}{\pi}}\,k\omega g(t), \quad t > 0,$$
$$U(\omega,0) = F(\omega).$$

After finding U in the transform domain, we obtain the solution of the original IBVP as $u(x,t) = \mathcal{F}_S^{-1}[U](x,t)$.

8.12. Example. The IBVP

$$u_t(x,t) = u_{xx}(x,t), \quad x > 0,\ t > 0,$$

$$u(0,t) = 1, \quad t > 0,$$

$$u(x,t),\ u_x(x,t) \to 0 \quad \text{as } x \to \infty,\ t > 0,$$

$$u(x,0) = 0, \quad x > 0,$$

requires the use of the Fourier sine transformation because it is the temperature that is prescribed in the BC.

By Theorem 8.11, the transform $U(\omega,t) = \mathcal{F}_S[u](\omega,t)$ of u satisfies

$$U'(\omega,t) + \omega^2 U(\omega,t) = \sqrt{\frac{2}{\pi}}\,\omega, \quad t > 0,$$

$$U(\omega,0) = 0.$$

The solution of the transformed problem is

$$U(\omega,t) = \sqrt{\frac{2}{\pi}}\,\frac{1}{\omega}\left(1 - e^{-\omega^2 t}\right);$$

so, by formula 12 (with $a = 1/(2\sqrt{t})$) in Table A3,

$$u(x,t) = \mathcal{F}_S^{-1}[U](x,t) = \operatorname{erfc}\frac{x}{2\sqrt{t}}.$$

Vibrations of a semi-infinite string. The method applied in this case is similar to that used above.

8.13. Example. The solution of the IBVP

$$u_{tt}(x,t) = 4u_{xx}(x,t), \quad x > 0,\ t > 0,$$

$$u_x(0,t) = 0, \quad t > 0,$$

$$u(x,t),\ u_x(x,t) \to 0 \quad \text{as } x \to \infty,\ t > 0,$$

$$u(x,0) = e^{-x}, \quad u_t(x,0) = 0, \quad x > 0,$$

is obtained by means of the Fourier cosine transformation because the BC prescribes the x-derivative of u at $x = 0$. According to formula 5 (with

$a = 1$) in Table A4, the Fourier cosine transform of the function e^{-x} is $\sqrt{2/\pi}\,(1+\omega^2)^{-1}$; hence, by Theorem 8.11, $U(\omega,t) = \mathcal{F}_C[u](\omega,t)$ satisfies

$$U''(\omega,t) + 4\omega^2 U(\omega,t) = 0, \quad t > 0,$$

$$U(\omega,0) = \sqrt{\frac{2}{\pi}}\,\frac{1}{1+\omega^2},$$

with solution

$$U(\omega,t) = \sqrt{\frac{2}{\pi}}\,\frac{\cos(2\omega t)}{1+\omega^2}.$$

Then, by formula 14 (with $a = 2t$ and $b = 1$), we find that the solution of the IBVP is

$$u(x,t) = \mathcal{F}_C^{-1}[U](x,t) = \begin{cases} e^{-x}\cosh(2t), & t \le x/2, \\ e^{-2t}\cosh x, & t > x/2. \end{cases} \blacksquare$$

Equilibrium temperature in a semi-infinite strip. The steady-state temperature distribution in a semi-infinite strip $0 \le x \le L$, $y \ge 0$ is modeled by the BVP

$$u_{xx}(x,y) + u_{yy}(x,y) = 0, \quad 0 < x < L,\ y > 0,$$
$$u(0,y) = g_1(y), \quad u(L,y) = g_2(y), \quad y > 0,$$
$$g_1(y), g_2(y) \to 0 \quad \text{as } y \to \infty,$$
$$u(x,0) = f(x), \quad 0 < x < L,$$
$$u(x,y), u_y(x,y) \to 0 \quad \text{as } y \to \infty, \quad 0 < x < L.$$

Using the principle of superposition, we write the solution in the form

$$u(x,y) = u_1(x,y) + u_2(x,y),$$

where u_1 satisfies the given BVP with $g_1 = 0$ and $g_2 = 0$, and u_2 satisfies it with $f = 0$.

Thus, the first BVP is

$$(u_1)_{xx}(x,y) + (u_1)_{yy}(x,y) = 0, \quad 0 < x < L,\ y > 0,$$
$$u_1(0,y) = 0, \quad u_1(L,y) = 0, \quad y > 0,$$
$$u_1(x,0) = f(x), \quad 0 < x < L,$$
$$u_1(x,y), (u_1)_y(x,y) \to 0 \quad \text{as } y \to \infty,\ 0 < x < L,$$

which is solved by the method of separation of variables. Proceeding as in Section 5.3, from the PDE and BCs at $x = 0$ and $x = L$ we obtain

$$u_1(x,y) = \sum_{n=1}^{\infty} \sin \frac{n\pi x}{L} \left(A_n \cosh \frac{n\pi y}{L} + B_n \sinh \frac{n\pi y}{L} \right)$$

$$= \sum_{n=1}^{\infty} \sin \frac{n\pi x}{L} \left(a_n e^{n\pi y/L} + b_n e^{-n\pi y/L} \right).$$

Since $u_1(x,y) \to 0$ as $y \to \infty$, we must have $a_n = 0$, $n = 1,2,\dots$; hence,

$$u_1(x,y) = \sum_{n=1}^{\infty} b_n \sin \frac{n\pi x}{L} e^{-n\pi y/L}. \tag{8.10}$$

The BC at $y = 0$ now yields

$$u_1(x,0) = f(x) = \sum_{n=1}^{\infty} b_n \sin \frac{n\pi x}{L},$$

where, by (2.10), the Fourier sine series coefficients b_n are

$$b_n = \frac{2}{L} \int_0^L f(x) \sin \frac{n\pi x}{L} \, dx, \quad n = 1,2,\dots.$$

Consequently, the solution of the first BVP is given by (8.10) with the b_n computed by means of the above formula.

The second BVP is

$$(u_2)_{xx}(x,y) + (u_2)_{yy}(x,y) = 0, \quad 0 < x < L, \ y > 0,$$
$$u_2(0,y) = g_1(y), \quad u_2(L,y) = g_2(y), \ y > 0,$$
$$g_1(y), g_2(y) \to 0 \quad \text{as } y \to \infty,$$
$$u_2(x,0) = 0, \quad 0 < x < L,$$
$$u_2(x,y) \to 0, (u_2)_y(x,y) \to 0 \quad \text{as } y \to \infty, \ 0 < x < L.$$

In view of the BC at $y = 0$, we use $\mathcal{F}_S$ with respect to y and make the notation

$$\mathcal{F}_S[u_2](x,\omega) = U(x,\omega), \quad \mathcal{F}_S[g_1](\omega) = G_1(\omega), \quad \mathcal{F}_S[g_2](\omega) = G_2(\omega).$$

Then the above BVP reduces to

$$U''(x,\omega) - \omega^2 U(x,\omega) = 0, \quad 0 < x < L,$$
$$U(0,\omega) = G_1(\omega), \quad U(L,\omega) = G_2(\omega).$$

The general solution of the equation can be written as (see Remark 1.4)

$$U(x,\omega) = C_1(\omega) \sinh(\omega x) + C_2(\omega) \sinh\big(\omega(L - x)\big),$$

where C_1 and C_2 are arbitrary functions of ω. Using the BCs at $x = 0$ and $x = L$, we find that

$$U(x,\omega) = \operatorname{csch}(\omega L)\big[G_2(\omega)\sinh(\omega x) + G_1(\omega)\sinh\big(\omega(L - x)\big)\big]. \qquad (8.11)$$

The solution u_2 is now obtained by applying $\mathcal{F}_S^{-1}$ to the above equality.

8.14. Example. For the BVP

$$u_{xx}(x,y) + u_{yy}(x,y) = 0, \quad 0 < x < 1, \ y > 0,$$
$$u(0,y) = 0, \quad u(1,y) = g(y), \quad y > 0,$$
$$u(x,0) = 0, \quad 0 < x < 1,$$
$$u(x,y) \to 0, u_y(x,y) \to 0 \quad \text{as } y \to \infty, \ 0 < x < 1,$$

where

$$g(y) = \begin{cases} 1, & 0 < y \le 2, \\ 0, & y > 2, \end{cases}$$

we have $L = 1$, $g_1(y) = 0$, and $g_2(y) = g(y)$, so $G_1(\omega) = 0$ and, by formula 10 (with $a = 2$) in Table A3,

$$G_2(\omega) = \sqrt{\frac{2}{\pi}}\frac{1}{\omega}\big[1 - \cos(2\omega)\big] = \sqrt{\frac{2}{\pi}}\frac{2}{\omega}\sin^2\omega;$$

consequently, by (8.11),

$$U(x,\omega) = \sqrt{\frac{2}{\pi}}\frac{2}{\omega}\sin^2\omega \operatorname{csch}\omega \sinh(\omega x),$$

which yields

$$u(x,y) = \mathcal{F}_S^{-1}[U](x,y)$$
$$= \frac{4}{\pi}\int_0^\infty \frac{1}{\omega}\sin^2\omega \operatorname{csch}\omega \sinh(\omega x)\sin(\omega y)\,d\omega. \quad \blacksquare$$

8.3. Other Applications

The Fourier transformation method may also be applied to other suitable problems, such as those mentioned in Section 4.4, including some with nonhomogeneous PDEs.

8.15. Example. The IVP (Cauchy problem)

$$u_t(x,t) = u_{xx}(x,t) + 2u(x,t) + (1 - 4x^2t)e^{-x^2},$$

$$-\infty < x < \infty, \ t > 0,$$

$$u(x,t), \ u_x(x,t) \to 0 \quad \text{as } x \to \pm\infty, \ t > 0,$$

$$u(x,0) = 0, \quad -\infty < x < \infty,$$

models a diffusion process with a chain reaction and a source in a one-dimensional medium.

By formulas 8 and 10 (with $a = 1$) in Table A2,

$$\mathcal{F}[(1 - 4x^2t)e^{-x^2}] = \frac{1}{\sqrt{2}}\left[(\omega^2 - 2)t + 1\right]e^{-\omega^2/4};$$

hence, setting $\mathcal{F}[u](\omega,t) = U(\omega,t)$ and using formula 2, we see that U is the solution of the transformed problem

$$U'(\omega,t) + (\omega^2 - 2)U(\omega,t) = \frac{1}{\sqrt{2}}\left[(\omega^2 - 2)t + 1\right]e^{-\omega^2/4}, \quad t > 0,$$

$$U(\omega,0) = 0.$$

Direct calculation shows that

$$U(\omega,t) = \frac{1}{\sqrt{2}}te^{-\omega^2/4},$$

so, by formula 8, the solution of the given IVP is

$$u(x,t) = \mathcal{F}^{-1}[U](x,t) = te^{-x^2}. \quad \blacksquare$$

8.16. Example. The IBVP

$$u_{tt}(x,t) + 2u_t(x,t) = u_{xx}(x,t) + (2 + 4t - 4t^2)e^{-2x},$$

$$x > 0, \ t > 0,$$

$$u(0,t) = t^2, \quad t > 0,$$

$$u(x,t), \ u_x(x,t) \to 0 \quad \text{as } x \to \infty, \ t > 0,$$

$$u(x,0) = 0, \quad u_t(x,0) = 0, \quad x > 0,$$

describes the propagation of a dissipative wave along a semi-infinite string initially at rest, under the action of a prescribed displacement of the near endpoint and an external force. In view of the BC, we use the Fourier sine transformation to find the solution.

First, by formula 6 (with $a = 2$) in Table A3,

$$\mathcal{F}_S[(2 + 4t - 4t^2)e^{-2x}] = 2\sqrt{\frac{2}{\pi}} \frac{\omega}{4 + \omega^2} (1 + 2t - 2t^2);$$

hence, by Theorem 8.11 and the ICs, the transform $U(\omega,t) = \mathcal{F}_S[u](\omega,t)$ of u satisfies

$$U''(\omega,t) + 2U'(\omega,t) + \omega^2 U(\omega,t)$$
$$= \sqrt{\frac{2}{\pi}} \omega t^2 + 2\sqrt{\frac{2}{\pi}} \frac{\omega}{4 + \omega^2} (1 + 2t - 2t^2), \quad t > 0,$$

$$U(\omega,0) = 0, \quad U'(\omega,0) = 0.$$

Solving this ODE problem in the usual way, we find that

$$U(\omega,t) = \sqrt{\frac{2}{\pi}} \frac{\omega}{4 + \omega^2} t^2,$$

so, again by formula 6,

$$u(x,t) = \mathcal{F}_S^{-1}[U](x,t) = t^2 e^{-2x}. \quad \blacksquare$$

8.17. Example. The BVP

$$u_{xx}(x,y) + u_{yy}(x,y) + u(x,y) = \frac{x(y^4 + 8y^2 - 1)}{(1 + y^2)^3},$$
$$0 < x < 1, \ y > 0,$$

$$u(0,y) = 0, \quad u(1,y) = \frac{1}{1 + y^2}, \quad y > 0,$$

$$u_y(x,0) = 0, \quad 0 < x < 1,$$

$$u(x,y) \to 0, u_y(x,y) \to 0 \quad \text{as } y \to \infty, \ 0 < x < 1,$$

models the steady-state distribution of heat in a semi-infinite strip with a time-dependent source and insulated base. Since the BC at $y = 0$ prescribes the derivative u_y, we use the Fourier cosine transformation with respect to y to find the solution.

By formulas 9–12 (with $a = 1$) in Table A4, it is easy to see that

$$\mathcal{F}_C\left[\frac{x(y^4 + 8y^2 - 1)}{(1 + y^2)^3}\right] = \sqrt{\frac{\pi}{2}}(1 - \omega^2)e^{-\omega}x,$$

$$\mathcal{F}_C\left[\frac{1}{1 + y^2}\right] = \sqrt{\frac{\pi}{2}}e^{-\omega};$$

consequently, by Theorem 8.11, the transform $U(x,\omega) = \mathcal{F}_C[u](x,\omega)$ of u satisfies

$$U''(x,\omega) - (\omega^2 - 1)U(x,\omega) = \sqrt{\frac{\pi}{2}}(1 - \omega^2)e^{-\omega}x, \quad 0 < x < 1,$$

$$U(0,\omega) = 0, \quad U(1,\omega) = \sqrt{\frac{\pi}{2}}e^{-\omega}.$$

with solution

$$U(x,\omega) = \sqrt{\frac{\pi}{2}}xe^{-\omega};$$

so, by formula 9,

$$u(x,y) = \mathcal{F}_C^{-1}[U](x,y) = \frac{x}{1 + y^2}. \quad \blacksquare$$

Exercises

In (1)–(18) use a suitable Fourier transformation to find the solution of the IBVP for the heat equation

$$u_t(x,t) = ku_{xx}(x,t) + q(x,t), \quad t > 0,$$

with the constant k, interval for x, function q, BCs, and IC as indicated. In the exercises with underlined numerical labels express the answer as an integral.

(1) $k = 1$, $-\infty < x < \infty$, $q(x,t) = 0$,
    $u(x,t), u_x(x,t) \to 0$ as $x \to \pm\infty$, $u(x,0) = -3e^{-x^2}$.

(2) $k = 1$, $-\infty < x < \infty$, $q(x,t) = 0$,
    $u(x,t), u_x(x,t) \to 0$ as $x \to \pm\infty$, $u(x,0) = (1 - 2x^2)e^{-4x^2}$.

(3) $k = 2$, $-\infty < x < \infty$, $q(x,t) = \begin{cases} 1, & |x| \le 1, \\ 0, & |x| > 1, \end{cases}$
    $u(x,t), u_x(x,t) \to 0$ as $x \to \pm\infty$, $u(x,0) = 0$.

(4) $k = 2,$ $-\infty < x < \infty,$ $q(x,t) = 0,$

$u(x,t), u_x(x,t) \to 0$ as $x \to \pm\infty,$ $u(x,0) = \begin{cases} 1, & |x| \leq 1, \\ 0, & |x| > 1. \end{cases}$

(5) $k = 2,$ $-\infty < x < \infty,$ $q(x,t) = 2(8x^2t - 4x^2 - 4t + 1),$

$u(x,t), u_x(x,t) \to 0$ as $x \to \pm\infty,$ $u(x,0) = e^{-x^2}.$

(6) $k = 1,$ $-\infty < x < \infty,$ $q(x,t) = (5 - 16x^2)e^{t-2x^2},$

$u(x,t), u_x(x,t) \to 0$ as $x \to \pm\infty,$ $u(x,0) = e^{-2x^2}.$

(7) $k = 1,$ $x > 0,$ $q(x,t) = 0,$ $u(0,t) = 0,$

$u(x,t), u_x(x,t) \to 0$ as $x \to \infty,$ $u(x,0) = xe^{-x}.$

(8) $k = 1,$ $x > 0,$ $q(x,t) = 0,$ $u(0,t) = 1,$

$u(x,t), u_x(x,t) \to 0$ as $x \to \infty,$ $u(x,0) = e^{-x/2}.$

(9) $k = 2,$ $x > 0,$ $q(x,t) = \begin{cases} 1, & 0 < x \leq 1, \\ 0, & x > 1, \end{cases}$ $u(0,t) = t,$

$u(x,t), u_x(x,t) \to 0$ as $x \to \infty,$ $u(x,0) = 0.$

(10) $k = 2,$ $x > 0,$ $q(x,t) = 0,$ $u(0,t) = 1 - t,$

$u(x,t), u_x(x,t) \to 0$ as $x \to \infty,$ $u(x,0) = \begin{cases} -1, & 0 < x \leq 1, \\ 0, & x > 1. \end{cases}$

(11) $k = 2,$ $x > 0,$ $q(x,t) = 2(x - 2t - 1)e^{-x},$ $u(0,t) = 2t,$

$u(x,t), u_x(x,t) \to 0$ as $x \to \infty,$ $u(x,0) = -xe^{-x}.$

(12) $k = 1,$ $x > 0,$ $q(x,t) = -2(4xt - x - 4t + 2)e^{-2x},$ $u(0,t) = 1,$

$u(x,t), u_x(x,t) \to 0$ as $x \to \infty,$ $u(x,0) = e^{-2x}.$

(13) $k = 1,$ $x > 0,$ $q(x,t) = 0,$ $u_x(0,t) = -t,$

$u(x,t), u_x(x,t) \to 0$ as $x \to \infty,$ $u(x,0) = -2e^{-x}.$

(14) $k = 1,$ $x > 0,$ $q(x,t) = 0,$ $u_x(0,t) = 1,$

$u(x,t), u_x(x,t) \to 0$ as $x \to \infty,$ $u(x,0) = -xe^{-2x}.$

(15) $k = 2,$ $x > 0,$ $q(x,t) = \begin{cases} 1, & 0 < x \leq 1, \\ 0, & x > 1, \end{cases}$ $u_x(0,t) = t - 1,$

$u(x,t), u_x(x,t) \to 0$ as $x \to \infty,$ $u(x,0) = 0.$

(16) $k = 2,$ $x > 0,$ $q(x,t) = 0,$ $u_x(0,t) = t,$

$u(x,t), u_x(x,t) \to 0$ as $x \to \infty,$ $u(x,0) = \begin{cases} 1, & 0 < x \leq 2, \\ 0, & x > 2. \end{cases}$

(17) $k = 2,$ $x > 0,$ $q(x,t) = 2(8xt - x - 8t - 4)e^{-2x},$ $u_x(0,t) = -2t - 2,$

$u(x,t), u_x(x,t) \to 0$ as $x \to \infty,$ $u(x,0) = e^{-2x}.$

(18) $k = 1$, $x > 0$, $q(x,t) = -2e^{2t-x+1}$, $u_x(0,t) = -e^{2t+1}$,

  $u(x,t), u_x(x,t) \to 0$ as $x \to \infty$, $u(x,0) = e^{1-x}$.

In (19)–(36) use a suitable Fourier transformation to find the solution of the IBVP for the wave equation

$$u_{tt}(x,t) = c^2 u_{xx}(x,t) + q(x,t), \quad t > 0,$$

with the constant c, interval for x, function q, BCs, and ICs as indicated. In the exercises with underlined numerical labels express the answer as an integral.

(19) $c = 1$, $-\infty < x < \infty$, $q(x,t) = 0$,

  $u(x,t), u_x(x,t) \to 0$ as $x \to \pm\infty$,

  $u(x,0) = (2x - 1)e^{-x^2}$, $u_t(x,0) = 0$.

(20) $c = 1$, $-\infty < x < \infty$, $q(x,t) = 0$,

  $u(x,t), u_x(x,t) \to 0$ as $x \to \pm\infty$, $u(x,0) = 0$, $u_t(x,0) = x^2 e^{-4x^2}$.

(21) $c = 2$, $-\infty < x < \infty$, $q(x,t) = \begin{cases} 1, & |x| \le 1, \\ 0, & |x| > 1, \end{cases}$

  $u(x,t), u_x(x,t) \to 0$ as $x \to \pm\infty$, $u(x,0) = 0$, $u_t(x,0) = 0$.

(22) $c = 2$, $-\infty < x < \infty$, $q(x,t) = 0$,

  $u(x,t), u_x(x,t) \to 0$ as $x \to \pm\infty$,

  $u(x,0) = \begin{cases} 1, & |x| \le 2, \\ 0, & |x| > 2, \end{cases}$ $u_t(x,0) = 0$.

(23) $c = 2$, $-\infty < x < \infty$, $q(x,t) = 2(9 - 16x^2)e^{-t-x^2}$,

  $u(x,t), u_x(x,t) \to 0$ as $x \to \pm\infty$,

  $u(x,0) = 2e^{-x^2}$, $u_t(x,0) = -2e^{-2x^2}$.

(24) $c = 1$, $-\infty < x < \infty$, $q(x,t) = 2(1 - t)(2x^2 - 1)e^{-x^2}$,

  $u(x,t), u_x(x,t) \to 0$ as $x \to \pm\infty$, $u(x,0) = -e^{-x^2}$, $u_t(x,0) = e^{-x^2}$.

(25) $c = 1$, $x > 0$, $q(x,t) = 0$, $u(0,t) = 1$,

  $u(x,t), u_x(x,t) \to 0$ as $x \to \infty$, $u(x,0) = 3e^{-2x}$, $u_t(x,0) = 0$.

(26) $c = 1$, $x > 0$, $q(x,t) = 0$, $u(0,t) = t$,

  $u(x,t), u_x(x,t) \to 0$ as $x \to \infty$, $u(x,0) = 0$, $u_t(x,0) = xe^{-x}$.

(27) $c = 2$, $x > 0$, $q(x,t) = \begin{cases} 1, & 0 < x \le 1, \\ 0, & x > 1, \end{cases}$ $u(0,t) = 0$,

$u(x,t)$, $u_x(x,t) \to 0$ as $x \to \infty$,

$u(x,0) = 0$, $u_t(x,0) = \begin{cases} -1, & 0 < x \le 1, \\ 0, & x > 1. \end{cases}$

(28) $c = 2$, $x > 0$, $q(x,t) = 0$, $u(0,t) = t - 1$,

$u(x,t)$, $u_x(x,t) \to 0$ as $x \to \infty$,

$u(x,0) = \begin{cases} 1, & 0 < x \le 2, \\ 0, & x > 2, \end{cases}$ $u_t(x,0) = 0$.

(29) $c = 1$, $x > 0$, $q(x,t) = -3e^{2t-x-1}$, $u(0,t) = -e^{2t-1}$,

$u(x,t)$, $u_x(x,t) \to 0$ as $x \to \infty$,

$u(x,0) = -e^{-x-1}$, $u_t(x,0) = -2e^{-x-1}$.

(30) $c = 1$, $x > 0$, $q(x,t) = 2(-2x^2 - 2xt + 4x + 2t - 1)e^{-2x}$, $u(0,t) = 0$,

$u(x,t)$, $u_x(x,t) \to 0$ as $x \to \infty$, $u(x,0) = x^2 e^{-2x}$, $u_t(x,0) = xe^{-2x}$.

(31) $c = 1$, $x > 0$, $q(x,t) = 0$, $u_x(0,t) = -1$,

$u(x,t)$, $u_x(x,t) \to 0$ as $x \to \infty$, $u(x,0) = 3e^{-x}$, $u_t(x,0) = 0$.

(32) $c = 1$, $x > 0$, $q(x,t) = 0$, $u_x(0,t) = -t$,

$u(x,t)$, $u_x(x,t) \to 0$ as $x \to \infty$, $u(x,0) = 0$, $u_t(x,0) = -xe^{-2x}$.

(33) $c = 2$, $x > 0$, $q(x,t) = \begin{cases} t, & 0 < x \le 2, \\ 0, & x > 2, \end{cases}$ $u_x(0,t) = 0$,

$u(x,t)$, $u_x(x,t) \to 0$ as $x \to \infty$,

$u(x,0) = 0$, $u_t(x,0) = \begin{cases} -1, & 0 < x \le 2, \\ 0, & x > 2. \end{cases}$

(34) $c = 2$, $x > 0$, $q(x,t) = 0$, $u_x(0,t) = 2t$,

$u(x,t)$, $u_x(x,t) \to 0$ as $x \to \infty$,

$u(x,0) = \begin{cases} 1, & 0 < x \le 1, \\ 0, & x > 1, \end{cases}$ $u_t(x,0) = 0$.

(35) $c = 1$, $x > 0$, $q(x,t) = (xt - 2t - 2)e^{-x}$, $u_x(0,t) = -t - 2$,

$u(x,t)$, $u_x(x,t) \to 0$ as $x \to \infty$,

$u(x,0) = 2e^{-x}$, $u_t(x,0) = -xe^{-x}$.

(36) $c = 1$, $x > 0$, $q(x,t) = 4(3x - xt + t - 4)e^{-2x}$, $u_x(0,t) = t - 5$,

$u(x,t)$, $u_x(x,t) \to 0$ as $x \to \infty$,

$u(x,0) = (1 - 3x)e^{-2x}$, $u_t(x,0) = xe^{-2x}$.

In (37)–(58) use a suitable Fourier transformation to find the solution of the BVP for the Laplace (Poisson) equation

$$u_{xx}(x,y) + u_{yy}(x,y) = q(x,y), \quad 0 < x < 1,$$

with the interval for y, function q, and BCs as indicated. In the exercises with underlined numerical labels express the answer as an integral.

(37) $-\infty < y < \infty$, $q(x,y) = 0$, $u(0,y) = 0$, $u(1,y) = 2e^{-3y^2}$,
$u(x,y)$, $u_y(x,y) \to 0$ as $y \to \pm\infty$.

(38) $-\infty < y < \infty$, $q(x,y) = 0$, $u(0,y) = y^2 e^{-y^2}$, $u(1,y) = 0$,
$u(x,y)$, $u_y(x,y) \to 0$ as $y \to \pm\infty$.

(39) $-\infty < y < \infty$, $q(x,y) = \begin{cases} x, & |y| \le 1, \\ 0 & |y| > 1, \end{cases}$ $u(0,y) = 0$, $u(1,y) = 0$,
$u(x,y)$, $u_y(x,y) \to 0$ as $y \to \pm\infty$.

(40) $-\infty < y < \infty$, $q(x,y) = 0$, $u(0,y) = \begin{cases} 2, & |y| \le 1, \\ 0, & |y| > 1, \end{cases}$ $u(1,y) = 0$,
$u(x,y)$, $u_y(x,y) \to 0$ as $y \to \pm\infty$.

(41) $-\infty < y < \infty$, $q(x,y) = 2(8x^2y^2 - 2x^2 + 16y^2 - 3)e^{-2y^2}$,
$u(0,y) = 2e^{-2y^2}$, $u(1,y) = 3e^{-2y^2}$,
$u(x,y)$, $u_y(x,y) \to 0$ as $y \to \pm\infty$.

(42) $-\infty < y < \infty$, $q(x,y) = -2(2y^2 + 1)e^{2x-y^2}$,
$u(0,y) = -e^{-y^2}$, $u(1,y) = -e^{2-y^2}$,
$u(x,y)$, $u_y(x,y) \to 0$ as $y \to \pm\infty$.

(43) $y > 0$, $q(x,y) = 0$, $u(0,y) = -e^{-2y}$, $u(1,y) = 0$,
$u(x,0) = x$, $u(x,y)$, $u_y(x,y) \to 0$ as $y \to \infty$.

(44) $y > 0$, $q(x,y) = 0$, $u(0,y) = 0$, $u(1,y) = 2e^{-y}$,
$u(x,0) = x - 1$, $u(x,y)$, $u_y(x,y) \to 0$ as $y \to \infty$.

(45) $y > 0$, $q(x,y) = \begin{cases} 1, & 0 < y \le 1, \\ 0, & y > 1, \end{cases}$
$u(0,y) = \begin{cases} -2, & 0 < y \le 1, \\ 0, & y > 1, \end{cases}$ $u(1,y) = 0$,
$u(x,0) = 0$, $u(x,y)$, $u_y(x,y) \to 0$ as $y \to \infty$.

(46) $y > 0$, $q(x,y) = 0$, $u(0,y) = 0$, $u(1,y) = \begin{cases} 1, & 0 < y \le 2, \\ 0, & y > 2, \end{cases}$
$u(x,0) = 1 - 2x$, $u(x,y)$, $u_y(x,y) \to 0$ as $y \to \infty$.

(47) $y > 0$, $q(x,y) = (x^2 - y + 4)e^{-y}$,
$u(0,y) = -ye^{-y}$, $u(1,y) = (1-y)e^{-y}$,
$u(x,0) = x^2$, $u(x,y), u_y(x,y) \to 0$ as $y \to \infty$.

(48) $y > 0$, $q(x,y) = (x^2 + y^2 - 2y + 1)e^{-2y}$,
$u(0,y) = y^2 e^{-2y}$, $u(1,y) = (y^2 + 1)e^{-2y}$,
$u(x,0) = x^2$, $u(x,y), u_y(x,y) \to 0$ as $y \to \infty$.

(49) $y > 0$, $q(x,y) = (5y + 1)e^{x-2y}$,
$u(0,y) = (y + 1)e^{-2y}$, $u_x(1,y) = (y+1)e^{1-2y}$,
$u(x,0) = e^x$, $u(x,y), u_y(x,y) \to 0$ as $y \to \infty$.

(50) $y > 0$, $q(x,y) = 4(1 + x - xy)e^{-2y}$,
$u_x(0,y) = -ye^{-2y}$, $u(1,y) = (1 - y)e^{-2y}$,
$u(x,0) = 1$, $u(x,y), u_y(x,y) \to 0$ as $y \to \infty$.

(51) $y > 0$, $q(x,y) = 0$, $u(0,y) = 2e^{-y}$, $u(1,y) = 0$,
$u_y(x,0) = x - 1$, $u(x,y), u_y(x,y) \to 0$ as $y \to \infty$.

(52) $y > 0$, $q(x,y) = 0$, $u(0,y) = 0$, $u(1,y) = e^{-2y}$,
$u_y(x,0) = 2x$, $u(x,y), u_y(x,y) \to 0$ as $y \to \infty$.

(53) $y > 0$, $q(x,y) = \begin{cases} -1, & 0 < y \le 1, \\ 0, & y > 1, \end{cases}$ $u(0,y) = 0$, $u(1,y) = 0$,
$u_y(x,0) = 1$, $u(x,y), u_y(x,y) \to 0$ as $y \to \infty$.

(54) $y > 0$, $q(x,y) = 0$, $u(0,y) = \begin{cases} 1, & 0 < y \le 2, \\ 0, & y > 2, \end{cases}$ $u(1,y) = 0$,
$u_y(x,0) = x^2 - 1$, $u(x,y), u_y(x,y) \to 0$ as $y \to \infty$.

(55) $y > 0$, $q(x,y) = 2(4xy + 2y^2 - 4x - 4y + 1)e^{-2y}$,
$u(0,y) = y^2 e^{-2y}$, $u(1,y) = (y^2 + 2y)e^{-2y}$,
$u_y(x,0) = 2x$, $u(x,y), u_y(x,y) \to 0$ as $y \to \infty$.

(56) $y > 0$, $q(x,y) = (5y - 2)e^{2x-y}$, $u(0,y) = ye^{-y}$, $u(1,y) = ye^{2-y}$,
$u_y(x,0) = e^{2x}$, $u(x,y), u_y(x,y) \to 0$ as $y \to \infty$.

(57) $y > 0$, $q(x,y) = (x^2 + x - 1)e^{-y}$, $u_x(0,y) = e^{-y}$, $u(1,y) = e^{-y}$,
$u_y(x,0) = 1 - x - x^2$, $u(x,y), u_y(x,y) \to 0$ as $y \to \infty$.

(58) $y > 0$, $q(x,y) = 4(3xy - 3x - 2)e^{-2y}$,
$u(0,y) = -2e^{-2y}$, $u_x(1,y) = 3ye^{-2y}$,
$u_y(x,0) = 3x + 4$, $u(x,y), u_y(x,y) \to 0$ as $y \to \infty$.

Chapter 9
The Laplace Transformation

Usually works on linear problems (handwritten)

The Fourier transformations are used mainly with respect to the space variables. In certain circumstances, however, for reasons of expedience or necessity, it is desirable to eliminate time as an active variable. This is achieved by means of the Laplace transformation. Problems where the spatial part of the domain is unbounded but the solution is not expected to decay fast enough away from the origin are particularly suited to this method.

9.1. Definition and Properties

First, we introduce a couple of useful mathematical entities.

9.1. Definition. The function H defined by

$$H(t) = \begin{cases} 0, & t < 0, \quad - \\ 1, & t \geq 0, \quad + \end{cases}$$

is called the *Heaviside (unit step) function*. It is clear that, more generally, for any real number t_0,

$$H(t - t_0) = \begin{cases} 0, & t < t_0, \quad - \\ 1, & t \geq t_0. \quad \blacksquare + \end{cases}$$

9.2. Remark. H is piecewise continuous. As mentioned in Remark 2.4(ii), the type of analysis we are performing is not affected by the value of a piecewise continuous function at its points of discontinuity. For our purposes, therefore, the value $H(0) = 1$ is chosen purely for convenience, to have H correctly defined as a function on the entire real line, but it is otherwise unimportant. ∎

In mathematical modeling it is often necessary to cater for a special type of physical data such as unit impulses and point sources. Let us assume, for example, that a unit impulse is produced by a constant force of magnitude $1/\varepsilon$ acting over a very short time interval $(t_0 - \varepsilon/2, t_0 + \varepsilon/2)$, $\varepsilon > 0$. We may express this mathematically by taking the force to be   *(very small + #)* (handwritten)

$$g_{t_0,\varepsilon}(t) = \begin{cases} 1/\varepsilon, & t_0 - \varepsilon/2 < t < t_0 + \varepsilon/2, \\ 0 & \text{otherwise,} \end{cases}$$

and computing the total impulse as

$$\int_{-\infty}^{\infty} g_{t_0,\varepsilon}(t)\, dt = \int_{t_0-\varepsilon/2}^{t_0+\varepsilon/2} \frac{1}{\varepsilon}\, dt = \frac{1}{\varepsilon}\cdot \varepsilon = 1.$$

We note that the above integral is equal to 1 irrespective of the value of ε. If we now want to regard the impulse as being produced at the single moment $t = t_0$, we need to consider a limiting process and introduce some sort of limit of $g_{t_0,\varepsilon}$ as $\varepsilon \to 0$, which we denote by $\delta(t - t_0)$.

9.3. Definition. The mathematical object δ defined by

(i) $\delta(t - t_0) = 0$ for all $t \neq t_0$,

(ii) $\int_{-\infty}^{\infty} \delta(t - t_0)\, dt = 1$

is called the *Dirac delta*. ■

9.4. Remarks. (i) From Definition 9.3 it is obvious that δ cannot be ascribed a finite value at $t = t_0$, because then its integral over $\mathbb{R}$ would be 0, not 1. Consequently, δ is not a function. Strictly speaking, δ is a so-called *distribution* (generalized function) and its proper handling requires a special formalism that goes beyond the scope of this book.

(ii) If $t < t_0$, we have

$$\int_{-\infty}^{t} \delta(\tau - t_0)\, d\tau = \int_{-\infty}^{t} 0\, d\tau = 0;$$

if $t > t_0$, we can find $\varepsilon > 0$ sufficiently small so that $t_0 + \varepsilon/2 < t$; hence, using the function $g_{t_0,\varepsilon}$ introduced above, we see that

$$\int_{-\infty}^{t} \delta(\tau - t_0)\, d\tau = \lim_{\varepsilon \to 0} \int_{-\infty}^{t} g_{t_0,\varepsilon}(\tau)\, d\tau = \lim_{\varepsilon \to 0} \int_{t_0-\varepsilon/2}^{t_0+\varepsilon/2} \frac{1}{\varepsilon}\, d\tau = 1.$$

Therefore, combining these results, we may write

$$\int_{-\infty}^{t} \delta(\tau - t_0)\, d\tau = H(t - t_0),$$

which means that, in a certain generalized sense,

$$H'(t - t_0) = \delta(t - t_0).$$

(iii) If f is continuous, then, by the mean value theorem, there is t', $t - \varepsilon/2 < t' < t + \varepsilon/2$, such that

$$\int_{-\infty}^{\infty} f(\tau)\delta(t-\tau)\,d\tau = \lim_{\varepsilon \to 0} \int_{-\infty}^{\infty} f(\tau)g_{t,\varepsilon}(\tau)\,d\tau = \lim_{\varepsilon \to 0} \int_{t-\varepsilon/2}^{t+\varepsilon/2} f(\tau)\frac{1}{\varepsilon}\,d\tau$$

$$= \lim_{\varepsilon \to 0} \frac{1}{\varepsilon}\left[\left(t + \frac{\varepsilon}{2}\right) - \left(t - \frac{\varepsilon}{2}\right)\right] f(t') = f(t). \quad (9.1)$$

In distribution theory, δ is, in fact, defined rigorously by a formula of this type and not as in Definition 9.3. Like differentiation, integration on the left-hand side above is understood in a generalized, distributional sense.

(iv) The Dirac delta may also be used in problems formulated on semi-infinite or finite intervals. In such cases its symbol stands for the "restriction" of this distribution to the corresponding interval. ∎

9.5. Definition. The *Laplace transform* of a function $f(t)$, $0 < t < \infty$, is defined by

$$\mathcal{L}[f](s) = F(s) = \int_0^\infty f(t)e^{-st}\,dt.$$

Here s is the *transformation parameter*. The corresponding *inverse Laplace transform*, computed by means of complex variable techniques, is

$$\mathcal{L}^{-1}[F](t) = f(t) = \frac{1}{2\pi i}\int_{c-i\infty}^{c+i\infty} F(s)e^{st}\,ds.$$

The operators $\mathcal{L}$ and $\mathcal{L}^{-1}$ are called the *Laplace transformation* and *inverse Laplace transformation*, respectively. ∎

9.6. Remark. $\mathcal{L}$ is applicable to a wider class of functions than $\mathcal{F}$. ∎

The following assertion gives sufficient conditions for the existence of the Laplace transform of a function.

9.7. Theorem. *If*

(i) f *is piecewise continuous on* $[0,\infty)$;

(ii) *there are constants C and α such that* $|f(t)| \leq Ce^{\alpha t}$, $0 < t < \infty$,

then $\mathcal{L}[f](s) = F(s)$ *exists for all $s > \alpha$.*

9.8. Examples. (i) The function $f(t) = 1$, $t > 0$, satisfies the conditions in Theorem 9.7 with $C = 1$ and $\alpha = 0$, and we have

$$F(s) = \int_0^\infty e^{-st}\,dt = -\frac{1}{s}\left[e^{-st}\right]_0^\infty = \frac{1}{s}, \quad s > 0.$$

(ii) For $f(t) = e^{2t}$, $t > 0$, Theorem 9.7 holds with $C = 1$ and $\alpha = 2$. In this case we have

$$F(s) = \int_0^\infty e^{2t}e^{-st}\,dt = \int_0^\infty e^{(2-s)t}\,dt = \frac{1}{s-2}, \quad s > 2.$$

(iii) The rate of growth of the function $f(t) = e^{t^2}$, $t > 0$, as $t \to \infty$ exceeds the exponential growth prescribed in Theorem 9.7. It turns out that this function does not have a Laplace transform. ∎

9.9. Theorem. (i) $\mathcal{L}$ *is linear; that is,*

$$\mathcal{L}[c_1 f_1 + c_2 f_2] = c_1\mathcal{L}[f_1] + c_2\mathcal{L}[f_2]$$

for any functions f_1, f_2 *(to which* $\mathcal{L}$ *can be applied) and any numbers* c_1, c_2.

(ii) *If* $u = u(x,t)$ *and* $\mathcal{L}[u](x,s) = U(x,s)$, *then*

$$\mathcal{L}[u_t](x,s) = sU(x,s) - u(x,0),$$
$$\mathcal{L}[u_{tt}](x,s) = s^2U(x,s) - su(x,0) - u_t(x,0).$$

(iii) *For the same type of function* u, *differentiation with respect to* x *and the Laplace transformation commute:*

$$\mathcal{L}[u_x](x,s) = (\mathcal{L}[u])_x(x,s) = U'(x,s).$$

(iv) *If we adopt a definition of the convolution* $f * g$ *of two functions* f *and* g *that is slightly different from Definition 8.3, namely,*

$$(f * g)(t) = \int_0^t f(\tau)g(t-\tau)\,d\tau = \int_0^t f(t-\tau)g(\tau)\,d\tau = (g * f)(t), \quad (9.2)$$

then

$$\mathcal{L}[f * g] = \mathcal{L}[f]\mathcal{L}[g].$$

9.10. Remarks. (i) As in the case of the Fourier transformations, in general $\mathcal{L}[fg] \neq \mathcal{L}[f]\mathcal{L}[g]$.

(ii) Clearly, $\mathcal{L}^{-1}$ is also linear. ∎

The Laplace transforms of some frequently used functions are listed in Table A5 in the Appendix.

9.11. Example. To find the inverse Laplace transform of the function $1/(s(s^2 + 1))$, we first see from Table A5 that $1/s$ is the transform of the constant function 1 and $1/(s^2 + 1)$ is the transform of the function $\sin t$. Therefore, by Theorem 9.9(iv), we can write symbolically

$$\frac{1}{s(s^2+1)} = \frac{1}{s}\frac{1}{s^2+1} = \mathcal{L}[1]\mathcal{L}[\sin t] = \mathcal{L}\big[1 * (\sin t)\big],$$

so

$$\mathcal{L}^{-1}\left[\frac{1}{s(s^2+1)}\right] = 1 * (\sin t) = \int_0^t \sin\tau \, d\tau = 1 - \cos t.$$

Alternatively, we can split the given function into partial fractions as

$$\frac{1}{s(s^2+1)} = \frac{1}{s} - \frac{s}{s^2+1}$$

and then use the linearity of $\mathcal{L}^{-1}$ to obtain the above result. ∎

9.12. Example. The simplest way to find the inverse Laplace transform of $(3s^2 + 2s + 12)/(s(s^2 + 4))$ by direct calculation is to establish the partial fraction decomposition

$$\frac{3s^2+2s+12}{s(s^2+4)} = \frac{3}{s} + \frac{2}{s^2+2^2}$$

and then to apply formulas 5 and 8 in Table A5 to arrive at

$$\mathcal{L}^{-1}\left[\frac{3s^2+2s+12}{s(s^2+4)}\right] = 3 + \sin(2t). \ \blacksquare$$

The next assertion lists two other helpful properties of the Laplace transformation.

9.13. Theorem. *If $\mathcal{L}[f](s) = F(s)$, then*

(i) $\mathcal{L}[e^{at}f](s) = F(s - a), \ s > a = $ const;

(ii) $\mathcal{L}[H(t - b)f(t - b)](s) = e^{-bs}F(s), \ b = $ const > 0.

9.14. Example. Since

$$\mathcal{L}[\sin(2t)] = \frac{2}{s^2 + 4}, \quad \mathcal{L}[\cos(3t)] = \frac{s}{s^2 + 9}, \quad \mathcal{L}[t^2] = \frac{2}{s^3},$$

from Theorem 9.13 it follows that

$$\mathcal{L}\big[e^{-t}\sin(2t) + e^{5t}\cos(3t) - 2(t-2)^2 H(t-2)\big]$$

$$= \frac{2}{(s+1)^2 + 4} + \frac{s-5}{(s-5)^2 + 9} - \frac{4}{s^3}e^{-2s}. \quad \blacksquare$$

9.15. Example. Similarly, we have

$$\mathcal{L}^{-1}\left[\frac{s-1}{s^2 - 2s + 10} - \frac{1}{s^2 + 4}e^{-s}\right]$$

$$= \mathcal{L}^{-1}\left[\frac{s-1}{(s-1)^2 + 3^2} - \frac{1}{s^2 + 2^2}e^{-s}\right]$$

$$= e^t\cos(3t) - \tfrac{1}{2}H(t-1)\sin\big(2(t-1)\big). \quad \blacksquare$$

9.2. Applications

The signal problem for the wave equation. Consider a very long elastic string of negligible weight, initially at rest, where the vertical displacement (signal) is prescribed at the near endpoint and where the mechanical activity diminishes considerably towards the far endpoint. Such a problem is modeled mathematically by an IBVP of the form

$$u_{tt}(x,t) = c^2 u_{xx}(x,t), \quad x > 0, \ t > 0,$$

$$u(0,t) = f(t), \quad t > 0,$$

$$u(x,t) \text{ bounded as } x \to \infty, \ t > 0,$$

$$u(x,0) = 0, \quad u_t(x,0) = 0, \quad x > 0.$$

Introducing the notation

$$\mathcal{L}[u](x,s) = U(x,s), \quad \mathcal{L}[f](s) = F(s),$$

applying $\mathcal{L}$ to the PDE and BC, and using the properties of $\mathcal{L}$ in Theorem 9.9, we arrive at the transformed problem

$$s^2 U(x,s) = c^2 U''(x,s), \quad x > 0,$$

$$U(0,s) = F(s), \quad U(x,s) \text{ bounded as } x \to \infty.$$

The ODE in the above problem can be rewritten in the form

$$U''(x,s) - (s/c)^2 U(x,s) = 0,$$

and its general solution is

$$U(x,s) = C_1(s)e^{(s/c)x} + C_2(s)e^{-(s/c)x},$$

where $C_1(s)$ and $C_2(s)$ are arbitrary functions of the transformation parameter. Since $U(x,s)$ needs to be bounded as $x \to \infty$, we must have $C_1(s) = 0$. Then the BC yields $C_2(s) = F(s)$, so

$$U(x,s) = F(s)e^{-(s/c)x} = F(s)e^{-(x/c)s}.$$

Consequently, by Theorem 9.13(ii), the solution of the original IBVP is

$$u(x,t) = \mathcal{L}^{-1}[F(s)e^{-(x/c)s}] = H(t - x/c)f(t - x/c)$$
$$= \begin{cases} 0, & 0 < t < x/c, \\ f(t - x/c), & t \geq x/c. \end{cases}$$

This solution can also be expressed as

$$u(x,t) = \begin{cases} f(t - x/c), & x \leq ct, \\ 0, & x > ct. \end{cases} \tag{9.3}$$

We see that $u(x,t)$ is constant when $x - ct = \text{const}$. Physically, this means that the solution is a wave of fixed shape (determined by the BC function f) with velocity $dx/dt = c$. Formula (9.3) indicates that at time t the signal originating from $x = 0$ has not reached the points $x > ct$, which are still in the initial state of rest.

9.16. Remark. The same inversion result can also be obtained by means of convolution. Since, by formula 12 in Table A5, $e^{-(x/c)s} = \mathcal{L}[\delta(t - x/c)]$, we can write

$$U(x,s) = F(s)e^{-(x/c)s} = \mathcal{L}[f]\mathcal{L}[\delta(t - x/c)] = \mathcal{L}[f * \delta(t - x/c)],$$

from which we conclude that

$$u(x,t) = \mathcal{L}^{-1}[U](x,t) = f * \delta(t - x/c) = \int_0^t f(\tau)\delta(t - x/c - \tau)\,d\tau.$$

If $t < x/c$, then $t - x/c - \tau < 0$ for $0 \leq \tau \leq t$, so $\delta(t - x/c - \tau) = 0$; hence, $u(x,t) = 0$ for $t < x/c$, or, equivalently, for $x > ct$.

If $t > x/c$, then $t - x/c - \tau = 0$ at $0 < \tau = t - x/c < t$; therefore, by (9.1), for $x < ct$,

$$u(x,t) = \int_0^t f(\tau)\delta(t - x/c - \tau)\,d\tau = \int_0^\infty f(\tau)\delta(t - x/c - \tau)\,d\tau = f(t - x/c).$$

This result is the same as (9.3). ∎

Heat conduction in a semi-infinite rod. A very long rod without sources, with the near endpoint kept in open air of zero temperature, with negligible thermal activity at the far endpoint, and with a constant initial temperature distribution, is modeled by the IBVP

$$u_t(x,t) = u_{xx}(x,t), \quad x > 0, \ t > 0,$$
$$u_x(0,t) - u(0,t) = 0, \quad t > 0,$$
$$u(x,t) \text{ bounded as } x \to \infty, \ t > 0,$$
$$u(x,0) = u_0 = \text{const}, \quad x > 0.$$

Let $\mathcal{L}[u](x,s) = U(x,s)$. Applying $\mathcal{L}$ to the PDE and BCs, we arrive at the transformed problem

$$U''(x,s) - sU(x,s) + u_0 = 0, \quad x > 0,$$
$$U'(0,s) - U(0,s) = 0,$$
$$U(x,s) \text{ bounded as } x \to \infty.$$

The general solution of the equation is

$$U(x,s) = C_1(s)e^{\sqrt{s}\,x} + C_2(s)e^{-\sqrt{s}\,x} + \frac{1}{s}u_0,$$

where $C_1(s)$ and $C_2(s)$ are arbitrary functions of the transformation parameter. Since U needs to be bounded as $x \to \infty$, we must have $C_1(s) = 0$. Then, differentiating U and using the BC at $x = 0$, we see that

$$-C_2(s)\sqrt{s} - C_2(s) - \frac{1}{s}u_0 = 0,$$

from which $C_2(s) = -u_0/(s(\sqrt{s}+1))$; hence,

$$U(x,s) = u_0\left[- \frac{1}{s(\sqrt{s}+1)}e^{-\sqrt{s}\,x} + \frac{1}{s}\right].$$

After some manipulation (coupled with the use of a more comprehensive Laplace transform table than A5), it can be shown that the solution of the original IBVP is

$$u(x,t) = \mathcal{L}^{-1}[U](x,t) = u_0\left[1 - \operatorname{erfc}\left(\frac{x}{2\sqrt{t}}\right) - \operatorname{erfc}\left(\sqrt{t}+\frac{x}{2\sqrt{t}}\right)e^{x+t}\right].$$

Other IBVPs for a semi-infinite rod can be solved by the same method.

9.17. Example. Consider the IBVP

$$u_t(x,t) = u_{xx}(x,t) + \sin x, \quad x > 0,\ t > 0,$$
$$u(0,t) = 2t - 1, \quad t > 0,$$
$$u(x,t) \text{ bounded as } x \to \infty,\ t > 0,$$
$$u(x,0) = 1, \quad x > 0.$$

Setting $\mathcal{L}[u](x,s) = U(x,s)$, as above, and applying the Laplace transformation to the PDE and BC, we arrived at the ODE problem

$$U''(x,s) - sU(x,s) = -1 - \frac{1}{s}\sin x, \quad x > 0,$$
$$U(0,s) = \frac{2}{s^2} - \frac{1}{s}, \quad x > 0,$$
$$U(x,s) \text{ bounded as } x \to \infty.$$

The general solution of the equation, written as the sum of the complementary function and a particular integral, is

$$U(x,s) = C_1(s)e^{\sqrt{s}\,x} + C_2(s)e^{-\sqrt{s}\,x} + \frac{1}{s} + \frac{1}{s(s+1)}\sin x.$$

Since U has to remain bounded as $x \to \infty$, it follows that $C_1(s) = 0$. Then, applying the BC, we find that $C_2(s) = 2/s^2 - 2/s$; hence, given the partial fraction decomposition

$$\frac{1}{s(s+1)} = \frac{1}{s} - \frac{1}{s+1},$$

we arrive at

$$U(x,s) = \left(\frac{2}{s^2} - \frac{2}{s}\right)e^{-\sqrt{s}\,x} + \frac{1}{s} + \left(\frac{1}{s} - \frac{1}{s+1}\right)\sin x.$$

By formulas 5–7, 15, and 16 in Table A5, the solution of the original IBVP is

$$u(x,t) = \mathcal{L}^{-1}[U](x,t)$$

$$= -2x\sqrt{\frac{t}{\pi}}\,e^{-x^2/(4t)} + (x^2 + 2t - 2)\operatorname{erfc}\left(\frac{x}{2\sqrt{t}}\right)$$

$$+ 1 + (1 - e^{-t})\sin x. \quad\blacksquare$$

Finite rod with temperature prescribed on the boundary. The IBVP

$$w_t(x,t) = w_{xx}(x,t), \quad 0 < x < 1,\ t > 0,$$

$$w(0,t) = 0, \quad w(1,t) = 1, \quad t > 0,$$

$$w(x,0) = 0, \quad 0 < x < 1,$$

is of a type that we have already encountered. The equilibrium solution in this case, computed as in Section 6.1, is $w_\infty(x) = x$. Using this solution, we reduce the problem to a similar one where both BCs are homogeneous and which can thus be solved by the method of separation of variables. Putting all the results together, we obtain

$$w(x,t) = x + \sum_{n=1}^{\infty} (-1)^n \frac{2}{n\pi} \sin(n\pi x) e^{-n^2\pi^2 t}. \tag{9.4}$$

The same IBVP can also be solved by using the Laplace transformation with respect to t. If we write $\mathcal{L}[w](x,s) = W(x,s)$, then from the PDE and BCs we find that W is the solution of the BVP

$$W''(x,s) - sW(x,s) = 0, \quad 0 < x < 1,$$

$$W(0,s) = 0, \quad W(1,s) = \frac{1}{s}.$$

The general solution of the transformed equation can be written in the form (see Remark 1.4)

$$W(x,s) = C_1(s)\cosh(\sqrt{s}\,x) + C_2(s)\sinh(\sqrt{s}\,x),$$

with $C_1(s)$ and $C_2(s)$ determined from the BCs. Applying these conditions leads to

$$C_1(s) = 0, \quad C_2(s)\sinh\sqrt{s} = \frac{1}{s},$$

from which

$$C_2(s) = \frac{1}{s \sinh \sqrt{s}};$$

hence,

$$W(x,s) = \frac{\sinh\left(\sqrt{s}\,x\right)}{s \sinh \sqrt{s}}.$$

Then, using the inverse transformation and comparing with (9.4), we see that

$$w(x.t) = \mathcal{L}^{-1}\left[\frac{1}{s}\frac{\sinh\left(\sqrt{s}\,x\right)}{\sinh \sqrt{s}}\right] = x + \sum_{n=1}^{\infty}(-1)^n\frac{2}{n\pi}\sin(n\pi x)e^{-n^2\pi^2 t}.$$

Now consider the more general IBVP

$$u_t(x,t) = u_{xx}(x,t), \quad 0 < x < 1,\ t > 0,$$
$$u(0,t) = 0, \quad u(1.t) = f(t), \quad t > 0,$$
$$u(x,0) = 0, \quad 0 < x < 1,$$

and let $\mathcal{L}[u](x,s) = U(x,s)$ and $\mathcal{L}[f](s) = F(s)$. Applying $\mathcal{L}$ to the PDE and BCs, we arrive at the BVP

$$U''(x,s) - sU(x,s) = 0, \quad 0 < x < 1,$$
$$U(0,s) = 0, \quad U(1,s) = F(s).$$

Proceeding exactly as above, we find that

$$U(x,s) = F(s)\frac{\sinh\left(\sqrt{s}\,x\right)}{\sinh \sqrt{s}} = F(s)\left\{s\left[\frac{1}{s}\frac{\sinh\left(\sqrt{s}\,x\right)}{\sinh \sqrt{s}}\right]\right\}$$

$$= F(s)\left[sW(x,s)\right]. \tag{9.5}$$

Since $w(x,0) = 0$ in the IBVP for w, it follows that

$$\mathcal{L}[w_t](x,s) = sW(x,s) - w(x,0) = sW(x,s);$$

therefore, by (9.5) and Theorem 9.9(iv),

$$\mathcal{L}[u] = U = F(sW) = \mathcal{L}[f]\mathcal{L}[w_t] = \mathcal{L}[f * w_t].$$

Using (9.2) and integration by parts, we now obtain

$$u(x,t) = (f * w_t)(x,t) = \int_0^t f(t-\tau)w_\tau(x,\tau)\,d\tau$$

$$= f(t-\tau)w(x,\tau)\big|_{\tau=0}^{\tau=t} + \int_0^t w(x,\tau)f'(t-\tau)\,d\tau$$

$$= f(0)w(x,t) - f(t)w(x,0) + \int_0^t w(x,t-\tau)f'(\tau)\,d\tau$$

$$= \int_0^t w(x,t-\tau)f'(\tau)\,d\tau + f(0)w(x,t),$$

where we have used the condition $w(x,0) = 0$ and the commutativity of the convolution operation.

This result shows how the solution of a problem with more general BCs can sometimes be obtained from that of a problem with simpler ones.

9.18. Remark. If we replace the BC $w(1,t) = 1$ by

$$w(1,t) = \delta(t),$$

then the above formula becomes

$$u(x,t) = \int_0^t w(x,t-\tau)f(\tau)\,d\tau. \quad \blacksquare$$

Diffusion–convection problems. Suppose that a chemical substance is being poured at a constant rate into a straight, narrow, clean river that flows with a constant velocity. The concentration $u(x,t)$ of the substance at a distance x downstream at time t is the solution of the IBVP

$$u_t(x,t) = \sigma u_{xx}(x,t) - v u_x(x,t), \quad x > 0,\ t > 0,$$
$$u(0,t) = \alpha = \text{const}, \quad t > 0,$$
$$u(x,0) = 0, \quad x > 0,$$

where σ is the diffusion coefficient, $v = \text{const} > 0$ is the velocity of the river, $\alpha = \text{const} > 0$ is related to the substance discharge rate, and the second term on the right-hand side in the PDE accounts for the convection effect of the water flow on the substance.

If the river is slow, then the convection term is much smaller than the diffusion term and the PDE assumes the approximate form

$$u_t(x,t) = \sigma u_{xx}(x,t), \quad x > 0, \ t > 0,$$

which is the diffusion equation. If the river is fast, then the approximation is given by the convection equation

$$u_t(x,t) = -v u_x(x,t), \quad x > 0, \ t > 0.$$

Since we have already studied the diffusion (heat) equation, we now turn our attention to the convection and combined cases.

(i) The IBVP for pure convection is

$$u_t(x,t) = -v u_x(x,t), \quad x > 0, \ t > 0,$$
$$u(0,t) = \alpha, \quad t > 0,$$
$$u(x,0) = 0, \quad x > 0.$$

Let $\mathcal{L}[u](x,s) = U(x,s)$. Applying the Laplace transformation to the PDE and BC, we arrive at the problem

$$vU'(x,s) + sU(x,s) = 0, \quad x > 0,$$
$$U(0,s) = \frac{1}{s}\alpha,$$

with solution

$$U(x,s) = \frac{1}{s}\alpha e^{-(s/v)x} = \frac{1}{s}\alpha e^{-(x/v)s}.$$

Since $\mathcal{L}^{-1}\big[e^{-as}/s\big] = H(t-a)$, we set $a = x/v$ to find that the solution of the original IBVP is

$$u(x,t) = \mathcal{L}^{-1}[U](x,t) = \mathcal{L}^{-1}\left[\frac{1}{s}\alpha e^{-(x/v)s}\right]$$
$$= \alpha H(t - x/v) = \begin{cases} 0, & 0 < t < x/v, \\ \alpha, & t \geq x/v. \end{cases}$$

Thus, the substance reaches a fixed position x at time $t = x/v$; after that, the concentration of the substance at x remains constant (equal to the concentration of the substance at the point where it is poured into the

river). The line $t = x/v$ in the (x,t)-plane is the advancing wave front of
the substance (see Fig. 9.1).

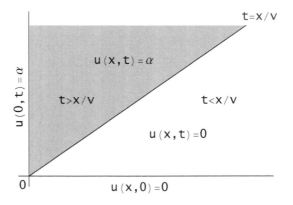

Fig. 9.1. The front wave and the regions behind it and ahead of it.

(ii) We now consider a very long river with the substance already uniformly distributed in it from the source of the river up to the observation
point $x = 0$, and assume that both diffusion and convection effects are
significant. This mixed diffusion–convection problem in an infinite one-
dimensional medium is modeled by the IVP

$$u_t(x,t) = \sigma u_{xx}(x,t) - v u_x(x,t), \quad -\infty < x < \infty, \ t > 0,$$
$$u(x,t), u_x(x,t) \to 0 \quad \text{as } x \to \pm\infty, \ t > 0,$$
$$u(x,0) = 1 - H(x), \quad -\infty < x < \infty.$$

We already know two possible methods for solving this problem: we can
apply the Laplace transformation with respect to t or the (full) Fourier
transformation with respect to x. In light of the discussion in (i) above, however, we indicate a third one, which consists in changing the x-coordinate
by connecting it to the wave front through the combination

$$\xi = x - vt. \tag{9.6}$$

Clearly, $\xi = 0$ means that the point (x,t) is on the wave front, $\xi > 0$ means
that (x,t) is ahead of the wave front, and $\xi < 0$ means that (x,t) is behind

the wave front. We also write

$$u(x,t) = u(\xi + vt,t) = w(\xi,t).$$

Hence, by the chain rule,

$$u_t = w_\xi \xi_t + w_t = -vw_\xi + w_t,$$
$$u_x = w_\xi \xi_x = w_\xi,$$
$$u_{xx} = (w_\xi)_\xi \xi_x = w_{\xi\xi}.$$

Since $t = 0$ yields $x = \xi$, the above IVP becomes

$$w_t(\xi,t) = \sigma w_{\xi\xi}(\xi,t), \quad -\infty < \xi < \infty, \ t > 0,$$
$$w(\xi,t), w_\xi(\xi,t) \to 0 \quad \text{as } \xi \to \pm\infty, \ t > 0,$$
$$w(\xi,0) = 1 - H(\xi), \quad -\infty < \xi < \infty.$$

This problem was solved earlier by means of the Fourier transformation (see Section 8.1), and its solution is

$$w(\xi,t) = \frac{1}{2\sqrt{\pi\sigma t}} \int_{-\infty}^{\infty} [1 - H(y)] e^{-(\xi-y)^2/(4\sigma t)} \, dy$$

$$= \frac{1}{2\sqrt{\pi\sigma t}} \int_{-\infty}^{0} e^{-(\xi-y)^2/(4\sigma t)} \, dy.$$

Then, by (9.6), we find that the solution of our IVP in terms of the original variables x and t is

$$u(x,t) = \frac{1}{2\sqrt{\pi\sigma t}} \int_{-\infty}^{0} e^{-(x-vt-y)^2/(4\sigma t)} \, dy.$$

Loss transmission line. Problems of this type can also be solved by the Laplace transformation method.

9.19. Example. Consider the IBVP

$$u_{tt}(x,t) + 4u_t(x,t) + 4u(x,t) = u_{xx}(x,t) - 1, \quad x > 0, \ t > 0,$$
$$u(0,t) = 0, \quad u(x,t) \text{ bounded as } x \to \infty, \quad t > 0,$$
$$u(x,0) = 1, \quad u_t(x,0) = 0, \quad x > 0.$$

If we write, as usual, $\mathcal{L}[u](x,s) = U(x,s)$ and apply the Laplace transformation to the PDE and BC, we arrive at the ODE problem

$$U''(x,s) - (s+2)^2 U(x,s) = \frac{1 - 4s - s^2}{s}, \quad x > 0,$$

$$U(0,s) = 0, \quad U(x,s) \text{ bounded as } x \to \infty,$$

with general solution

$$U(x,s) = C_1(s)e^{(s+2)x} + C_2(s)e^{-(s+2)x} + \frac{s^2 + 4s - 1}{s(s+2)^2}.$$

The boundedness requirement implies that $C_1(s) = 0$, and the BC yields

$$U(x,s) = \frac{s^2 + 4s - 1}{s(s+2)^2} \left[1 - e^{-(s+2)x}\right].$$

This can also be written in the form

$$U(x,s) = F(s) - F(s)e^{-sx}e^{-2x}, \tag{9.7}$$

where, using partial fractions, we have

$$F(s) = \frac{s^2 + 4s - 1}{s(s+2)^2} = -\frac{1}{4s} + \frac{5}{4(s+2)} + \frac{5}{2(s+2)^2}.$$

By formulas 3 and 5–7 in Table A5,

$$f(t) = \mathcal{L}^{-1}[F](t) = -\tfrac{1}{4} + \tfrac{5}{4}e^{-2t} + \tfrac{5}{2}te^{-2t} = \tfrac{1}{4}\left[5(2t+1)e^{-2t} - 1\right].$$

Applying formula 2 in (9.7), we now find that

$$u(x,t) = \mathcal{L}^{-1}[U](x,t) = f(t) - f(t-x)H(t-x)e^{-2x}$$
$$= \tfrac{1}{4}\left[5(2t+1)e^{-2t} - 1\right] - \tfrac{1}{4}\left[5(2t-2x+1)e^{-2t} - e^{-2x}\right]H(t-x). \quad \blacksquare$$

Exercises

In (1)–(4) use Table A5 to compute the Laplace transform of the given function f.

(1) $f(t) = e^{-t}\sin(3t) - 3t^4$.

(2) $f(t) = e^{4t}\cos(2t) + 4(t-3)^3 H(t-3)$.

(3) $f(t) = e^{-2t}(\cos t - 3\sin t) - 2t^2 H(t-1)$.

(4) $f(t) = t^3 e^{t-2} - 3\sin^2\left(\tfrac{1}{2}t\right)$.

In (5)–(8) use Table A5 to compute the inverse Laplace transform of the given function F.

(5) $F(s) = \dfrac{2s+1}{s^2-2s+26}$.

(6) $F(s) = \dfrac{3s+2}{s^2+6s+25} - \dfrac{2s}{(s-1)^2}e^{-s}$.

(7) $F(s) = \dfrac{3s+1}{s}e^{-\sqrt{s/2}}$.

(8) $F(s) = \dfrac{2-s}{s^2}e^{-3\sqrt{s}}$.

In (9)–(18) use the Laplace transformation to find the solution of the PDE

$$u_t(x,t) = ku_{xx}(x,t) + q(x,t), \quad x>0,\ t>0,$$

for the coefficient k, function q, and BC and IC as indicated, under the condition that $u(x,t)$ be bounded as $x \to \infty$, $t>0$.

(9) $k=1$, $q(x,t)=1$, $u(0,t)=t+1$, $u(x,0)=\sin(2x)$.

(10) $k=4$, $q(x,t)=1$, $u(0,t)=2-t$, $u(x,0)=-2$.

(11) $k=1$, $q(x,t) = \begin{cases} -3e^{-3t}+2(x-1)^2-4t, & 0<x\le 1, \\ -3e^{-3t}, & x>1, \end{cases}$
$u_x(0,t)=-4t$, $u(x,0)=1$.

(12) $k=1$, $q(x,t) = \begin{cases} (1-t)e^{-t}+6, & 0<x\le 1, \\ (1-t)e^{-t}, & x>1, \end{cases}$
$u_x(0,t)=6$, $u(x,0) = \begin{cases} 3(2x-x^2), & 0<x\le 1, \\ 3, & x>1. \end{cases}$

(13) $k=1$, $q(x,t)=(2t+1)\cos x$, $u_x(0,t)=0$, $u(x,0)=-\cos x$.

(14) $k=2$, $q(x,t)=-(2t+3)e^{-x}$, $u(0,t)=t+2$, $u(x,0)=2e^{-x}$.

(15) $k=1$, $q(x,t)=5e^{-x}\big[\cos(2t)+\sin(2t)\big]$,
$u(0,t)=-3\cos(2t)+\sin(2t)$, $u(x,0)=-3e^{-x}$.

(16) $k=1$, $q(x,t)=-e^{-2t}[2\cos t + (5x^2+4)\sin t]$,
$u_x(0,t)=0$, $u(x,0)=x^2$.

(17) $k=4$, $q(x,t)=3e^{-3t}$, $u(0,t)=3-e^{-3t}$, $u(x,0)=1$.

(18) $k=4$, $q(x,t)=-2e^{2t-x}$, $u(0,t)=e^{2t}-2$, $u(x,0)=e^{-x}$.

In (19)–(28) use the Laplace transformation to find the solution of the PDE

$$u_{tt}(x,t) = c^2 u_{xx}(x,t) + q(x,t), \quad x>0,\ t>0,$$

for the coefficient c, function q, and BC and ICs as indicated, under the condition that $u(x,t)$ be bounded as $x \to \infty$, $t>0$.

(19) $c=1$, $q(x,t)=1$, $u(0,t)=t$, $u(x,0)=0$, $u_t(x,0)=-1$.

(20) $c = 1$, $q(x,t) = -1$, $u(0,t) = t + 1$, $u(x,0) = 2$, $u_t(x,0) = 0$.

(21) $c = 2$, $q(x,t) = e^{-3t}$, $u_x(0,t) = 2t$, $u(x,0) = 0$, $u_t(x,0) = -1$.

(22) $c = 2$, $q(x,t) = t$, $u_x(0,t) = 2e^{-t}$, $u(x,0) = 1$, $u_t(x,0) = 0$.

(23) $c = 1$, $q(x,t) = -(2 + 3e^{-2t})e^{-x}$, $u(0,t) = 2 - e^{-2t}$,
$\quad u(x,0) = e^{-x}$, $u_t(x,0) = 2e^{-x}$.

(24) $c = 2$, $q(x,t) = 2(2t + 5e^{-t})\sin x$, $u_x(0,t) = 2e^{-t} + t$,
$\quad u(x,0) = 2\sin x$, $u_t(x,0) = -\sin x$.

(25) $c = 1$, $q(x,t) = e^{-t/2} - 4\sin(2x)$, $u_x(0,t) = -2$,
$\quad u(x,0) = 4 - \sin(2x)$, $u_t(x,0) = -2$.

(26) $c = 1$, $q(x,t) = -8e^{-x}\sin(2t)$, $u(0,t) = \sin(2t)$,
$\quad u(x,0) = 0$, $u_t(x,0) = 2e^{-x}$.

(27) $c = 2$, $q(x,t) = 2$, $u_x(0,t) = \cos t$, $u(x,0) = 0$, $u_t(x,0) = 1$.

(28) $c = 1$, $q(x,t) = e^{-x}$, $u(0,t) = -e^{-t}$, $u(x,0) = -2$, $u_t(x,0) = 0$.

In (29)–(32) use the Laplace transformation to find the solution of the PDE

$$u_t(x,t) = u_{xx}(x,t) - 2u_x(x,t) + u(x,t) + q(x,t), \quad x > 0,\ t > 0,$$

for the function q and the BC and IC as indicated, under the condition that $u(x,t)$ be bounded as $x \to \infty$, $t > 0$.

(29) $q(x,t) = 0$, $u(0,t) = -2$, $u(x,0) = 0$.

(30) $q(x,t) = 0$, $u(0,t) = t$, $u(x,0) = 0$.

(31) $q(x,t) = 2(3t + 1)\cos x + 3\sin x$, $u_x(0,t) = 3t + 1$, $u(x,0) = \sin x$.

(32) $q(x,t) = \cos x \cos t - 2\sin x \sin t$, $u_x(0,t) = 0$, $u(x,0) = 0$.

In (33)–(36) use the Laplace transformation to find the solution of the PDE

$$u_{tt}(x,t) + 2u_t(x,t) + u(x,t) = u_{xx}(x,t) + q(x,t), \quad x > 0,\ t > 0,$$

for the function q and BC and ICs as indicated, under the condition that $u(x,t)$ be bounded as $x \to \infty$, $t > 0$.

(33) $q(x,t) = 1$, $u(0,t) = t + 2$, $u(x,0) = -1$, $u_t(x,0) = 0$.

(34) $q(x,t) = -e^{-2t}$, $u_x(0,t) = 3t$, $u(x,0) = 0$, $u_t(x,0) = 2$.

(35) $q(x,t) = 5e^{-2x}$, $u_x(0,t) = 2(2 - t)$,
$\quad u(x,0) = -2e^{-2x}$, $u_t(x,0) = e^{-2x}$.

(36) $q(x,t) = 4e^{-x}\big[\sin(2t) - \cos(2t)\big]$, $u(0,t) = -\sin(2t)$,
$\quad u(x,0) = 0$, $u_t(x,0) = -2e^{-x}$.

Chapter 10
The Method of Green's Functions

The types of problems we have considered for the heat, wave, and Laplace equations have solutions that are determined uniquely by the prescribed data (boundary conditions, initial conditions, and any nonhomogeneous term in the equation). It is natural, therefore, to seek a formula that gives the solution directly in terms of the data. Such closed-form solutions are constructed by means of the so-called Green's function of the given problem, and are of great importance in practical applications.

10.1. The Heat Equation

The equilibrium problem. The equilibrium temperature distribution in a finite rod with internal sources and zero temperature at the endpoints is modeled by a BVP of the form (see Section 6.1)

$$u''(x) = -\frac{1}{k} q(x), \quad 0 < x < L,$$
$$u(0) = 0, \quad u(L) = 0. \tag{10.1}$$

For convenience, we have omitted the subscript ∞ from the symbol of the steady-state solution, but have kept the factor $-1/k$ since we will later make a comparison between the solutions of the equilibrium and time-dependent problems.

If we have just one unit source concentrated at a point ξ, $0 < \xi < L$, then $q(x) = \delta(x - \xi)$ and the two-point solution $G(x,\xi)$ of the above BVP satisfies

$$G_{xx}(x,\xi) = -\frac{1}{k} \delta(x - \xi), \quad 0 < x < L,$$
$$G(0,\xi) = 0, \quad G(L,\xi) = 0. \tag{10.2}$$

The function $G(x,\xi)$ can be computed explicitly. Since $H_x(x-\xi) = \delta(x-\xi)$ (see Remark 9.4(ii)), from (10.2) it follows that

$$G_x(x,\xi) = -\frac{1}{k} H(x - \xi) + C_1(\xi) = \begin{cases} C_1(\xi), & x < \xi, \\ -\dfrac{1}{k} + C_1(\xi), & x > \xi, \end{cases}$$

from which

$$G(x,\xi) = \begin{cases} xC_1(\xi) + C_2(\xi), & x < \xi, \\ x\left[C_1(\xi) - \dfrac{1}{k}\right] + C_3(\xi), & x > \xi, \end{cases} \tag{10.3}$$

where C_1, C_2, and C_3 are arbitrary functions of ξ. Using the BCs in (10.2), we find that

$$C_2(\xi) = 0, \quad L\left[C_1(\xi) - \frac{1}{k}\right] + C_3(\xi) = 0.$$

Hence, $C_3(\xi) = -L\left[C_1(\xi) - 1/k\right]$, and (10.3) becomes

$$G(x,\xi) = \begin{cases} xC_1(\xi), & x < \xi, \\ (x - L)\left[C_1(\xi) - \dfrac{1}{k}\right], & x > \xi. \end{cases} \tag{10.4}$$

If $G(x,\xi)$ had a jump (H-type) discontinuity at $x = \xi$, then G_x would have a δ-type singularity at $x = \xi$. Since this is not the case, we must conclude that $G(x,\xi)$ is continuous at $x = \xi$; in other words, $G(\xi-,\xi) = G(\xi+,\xi)$, which, in view of (10.4), leads to

$$\xi C_1(\xi) = (\xi - L)\left[C_1(\xi) - \frac{1}{k}\right].$$

Thus, $C_1(\xi) = (L - \xi)/(kL)$, and (10.4) yields

$$G(x,\xi) = \begin{cases} \dfrac{x}{kL}(L - \xi), & x \le \xi, \\ \dfrac{\xi}{kL}(L - x), & x > \xi. \end{cases} \tag{10.5}$$

Clearly, $G(x,\xi) = G(\xi,x)$.

Using integration by parts, we find that for two smooth functions u and v on $[0, L]$,

$$\int_0^L (u''v - v''u)\,dx = \left[u'v - v'u\right]_0^L - \int_0^L (u'v' - v'u')\,dx$$

$$= \left[u'(L)v(L) - v'(L)u(L)\right] - \left[u'(0)v(0) - v'(0)u(0)\right]. \tag{10.6}$$

This is known as *Green's formula*. If u is now the solution of (10.1) and $v = G$ is the solution of (10.2), then the right-hand side in (10.6) vanishes and we can write

$$\int_0^L \left[u(x)\delta(x - \xi) - G(x,\xi)q(x) \right] dx = 0.$$

By (9.1), interchanging x and ξ and recalling that $G(x,\xi) = G(\xi,x)$, we obtain

$$u(x) = \int_0^L G(x,\xi)q(\xi)\, d\xi. \tag{10.7}$$

$G(x,\xi)$, called the *Green's function* of the BVP (10.1), is the temperature at x due to a concentrated unit heat source at ξ. Formula (10.7) shows the aggregate influence of all the sources $q(\xi)$ in the rod on the temperature at x.

A representation formula similar to (10.7) can also be derived for nonhomogeneous BCs. Suppose that the BCs in (10.1) are replaced by $u(0) = a$ and $u(L) = b$. Then (10.6) with the same choice of u and v as above becomes

$$\int_0^L \left[u(x)\delta(x - \xi) - G(x,\xi)q(x) \right] dx = -k \left[u(x)G_x(x,\xi) \right]_{x=0}^{x=L},$$

so

$$u(x) = \int_0^L G(x,\xi)q(\xi)\, d\xi - k\left[bG_\xi(x,L) - aG_\xi(x,0) \right].$$

By (10.5),

$$G_\xi(x,\xi) = \begin{cases} -\dfrac{x}{kL}, & x \le \xi, \\[2mm] -\dfrac{x - L}{kL}, & x > \xi; \end{cases}$$

consequently, the desired representation formula is

$$u(x) = \int_0^L G(x,\xi)q(\xi)\, d\xi + b\frac{x}{L} + a\left(1 - \frac{x}{L} \right). \tag{10.8}$$

10.1. Example. To compute the steady-state solution for the IBVP

$$u_t(x,t) = u_{xx}(x,t) + x - 1, \quad 0 < x < 1,\ t > 0,$$
$$u(0,t) = 2, \quad u(1,t) = -1, \quad t > 0,$$
$$u(x,0) = f(x), \quad 0 < x < 1,$$

we use (10.8) with $k = 1$, $L = 1$, $a = 2$, $b = -1$, and $q(x) = x - 1$. First, by (10.5),

$$G(x,\xi) = \begin{cases} x(1-\xi), & x \le \xi, \\ \xi(1-x), & x > \xi, \end{cases}$$

$$G_\xi(x,\xi) = \begin{cases} -x, & x < \xi, \\ 1-x, & x > \xi; \end{cases}$$

hence,

$$\int_0^1 G(x,\xi)q(\xi)\,d\xi = \int_0^x \xi(1-x)(\xi-1)\,d\xi + \int_x^1 x(1-\xi)(\xi-1)\,d\xi$$

$$= (1-x)\int_0^x (\xi^2 - \xi)\,d\xi - x\int_x^1 (\xi-1)^2\,d\xi$$

$$= -\tfrac{1}{6}x^3 + \tfrac{1}{2}x^2 - \tfrac{1}{3}x.$$

Since $bx/L + a(1 - x/L) = -x + 2(1-x) = 2 - 3x$, the equilibrium solution (10.8) of the given IBVP is

$$u(x) = -\tfrac{1}{6}x^3 + \tfrac{1}{2}x^2 - \tfrac{10}{3}x + 2. \quad \blacksquare$$

10.2. Example. In the case of the IBVP

$$u_t(x,t) = u_{xx}(x,t) + q(x), \quad 0 < x < 1, \ t > 0,$$
$$u(0,t) = 1, \quad u(1,t) = 3, \quad t > 0,$$
$$u(x,0) = f(x), \quad 0 < x < 1,$$

with

$$q(x) = \begin{cases} 2, & 0 < x \le 1/2, \\ -1, & 1/2 < x < 1, \end{cases}$$

we notice that the function G is the same as in the preceding example, whereas $a = 1$ and $b = 3$. Given that the expressions of q and G change at $x = 1/2$ and $x = \xi$, respectively, we split the computation of the equilibrium solution (10.8) into two parts.

(i) If $0 < x < 1/2$, the first term on the right-hand side in (10.8) is written as a sum of three integrals, one for each of the intervals $0 < \xi \le x$, $x < \xi \le 1/2$, and $1/2 < \xi < 1$. Thus, (10.8) yields

$$u(x) = \int_0^x 2\xi(1-x)\,d\xi + \int_x^{1/2} 2x(1-\xi)\,d\xi + \int_{1/2}^1 -x(1-\xi)\,d\xi + 2x + 1$$

$$= -x^2 + \tfrac{21}{8}x + 1.$$

(ii) If $1/2 < x < 1$, the three integrals are over the intervals $0 < \xi \le 1/2$, $1/2 < \xi \le x$, and $x < \xi < 1$:

$$u(x) = \int_0^{1/2} 2\xi(1-x)\,d\xi + \int_{1/2}^x -\xi(1-x)\,d\xi + \int_x^1 -x(1-\xi)\,d\xi + 2x + 1$$

$$= \tfrac{1}{2}x^2 + \tfrac{9}{8}x + \tfrac{11}{8}.$$

It is easy to verify that

$$u\left(\tfrac{1}{2}-\right) = u\left(\tfrac{1}{2}+\right) = \tfrac{33}{16},$$
$$u'\left(\tfrac{1}{2}-\right) = u'\left(\tfrac{1}{2}+\right) = \tfrac{13}{8},$$

but that

$$u''\left(\tfrac{1}{2}-\right) = -2, \quad u''\left(\tfrac{1}{2}+\right) = 1,$$

which confirms that, as expected, the discontinuity of q at $x = 1/2$ has reduced the smoothness of the solution. ∎

10.3. Remark. The Green's function can be expanded in a double Fourier series. In view of the eigenfunctions of the Sturm–Liouville problem associated with (10.1) (see Section 5.1), the continuity of G, and the symmetry $G(x,\xi) = G(\xi,x)$, it seems reasonable to seek a series representation of the form

$$G(x,\xi) = \sum_{m=1}^\infty \left(\sum_{n=1}^\infty b_{mn} \sin \frac{n\pi x}{L} \right) \sin \frac{m\pi\xi}{L}. \tag{10.9}$$

Differentiating (10.9) term by term twice with respect to x and substituting in the ODE in (10.2), we obtain

$$\sum_{m=1}^\infty \left[\sum_{n=1}^\infty \left(\frac{n\pi}{L} \right)^2 b_{mn} \sin \frac{n\pi x}{L} \right] \sin \frac{m\pi\xi}{L} = \frac{1}{k}\delta(x-\xi).$$

Multiplying this equality by $\sin(p\pi x/L)$, $p = 1, 2, \ldots$, integrating over $[0, L]$, and using (2.5) and (9.1), we find that

$$\left(\frac{p\pi}{L}\right)^2 \frac{kL}{2} \sum_{m=1}^{\infty} b_{mp} \sin\frac{m\pi\xi}{L} = \sin\frac{p\pi\xi}{L}, \quad p = 1, 2, \ldots.$$

Theorem 3.20(ii) now implies that the only nonzero coefficients above are $b_{pp} = 2L/(kp^2\pi^2)$, so series (10.9) is

$$G(x,\xi) = \sum_{n=1}^{\infty} \frac{2L}{kn^2\pi^2} \sin\frac{n\pi x}{L} \sin\frac{n\pi\xi}{L}. \quad \blacksquare$$

The time-dependent problem. A finite rod with internal sources and zero temperature at the endpoints is modeled by the IBVP

$$u_t(x,t) = ku_{xx}(x,t) + q(x,t), \quad 0 < x < L, \ t > 0,$$
$$u(0,t) = 0, \quad u(L,t) = 0, \quad t > 0,$$
$$u(x,0) = f(x), \quad 0 < x < L.$$

(According to the arguments presented in Chapter 6, we may consider homogeneous BCs without loss of generality.) This problem can be solved by the method of eigenfunction expansion (see Section 7.1), so let

$$u(x,t) = \sum_{n=1}^{\infty} u_n(t) \sin\frac{n\pi x}{L},$$

$$q(x,t) = \sum_{n=1}^{\infty} q_n(t) \sin\frac{n\pi x}{L}, \quad f(x) = \sum_{n=1}^{\infty} f_n \sin\frac{n\pi x}{L},$$

where

$$q_n(t) = \frac{2}{L} \int_0^L q(x,t) \sin\frac{n\pi x}{L}\,dx, \quad f_n = \frac{2}{L} \int_0^L f(x) \sin\frac{n\pi x}{L}\,dx. \quad (10.10)$$

Replacing these series in the PDE, we find in the usual way that the u_n, $n = 1, 2, \ldots$, must satisfy

$$u_n'(t) + k\left(\frac{n\pi}{L}\right)^2 u_n(t) = q_n(t), \quad t > 0,$$

$$u_n(0) = f_n.$$

This problem is solved, for example, by means of the integrating factor

$$\exp\left\{\int k\left(\frac{n\pi}{L}\right)^2 dt\right\} = e^{k(n\pi/L)^2 t}.$$

Thus, taking the IC for u_n into account, we obtain

$$u_n(t) = e^{-k(n\pi/L)^2 t}\left[\int_0^t q_n(\tau)e^{k(n\pi/L)^2 \tau} d\tau + C\right]$$

$$= f_n e^{-k(n\pi/L)^2 t} + e^{-k(n\pi/L)^2 t}\int_0^t q_n(\tau)e^{k(n\pi/L)^2 \tau} d\tau;$$

so, by (10.10),

$$u(x,t) = \sum_{n=1}^{\infty}\left[f_n e^{-k(n\pi/L)^2 t} + e^{-k(n\pi/L)^2 t}\int_0^t q_n(\tau)e^{k(n\pi/L)^2 \tau} d\tau\right]\sin\frac{n\pi x}{L}$$

$$= \sum_{n=1}^{\infty}\left\{\left[\frac{2}{L}\int_0^L f(\xi)\sin\frac{n\pi\xi}{L} d\xi\right]e^{-k(n\pi/L)^2 t}\right.$$

$$\left. + e^{-k(n\pi/L)^2 t}\int_0^t\left[\frac{2}{L}\int_0^L q(\xi,\tau)\sin\frac{n\pi\xi}{L} d\xi\right]e^{k(n\pi/L)^2 \tau} d\tau\right\}\sin\frac{n\pi x}{L}$$

$$= \int_0^L f(\xi)\left[\sum_{n=1}^{\infty}\frac{2}{L}\sin\frac{n\pi x}{L}\sin\frac{n\pi\xi}{L}e^{-k(n\pi/L)^2 t}\right] d\xi$$

$$+ \int_0^L\int_0^t q(\xi,\tau)\left[\sum_{n=1}^{\infty}\frac{2}{L}\sin\frac{n\pi x}{L}\sin\frac{n\pi\xi}{L}e^{-k(n\pi/L)^2 (t-\tau)}\right] d\tau\, d\xi.$$

If we now define the Green's function of this problem by

$$G(x,t;\xi,\tau) = \sum_{n=1}^{\infty}\frac{2}{L}\sin\frac{n\pi x}{L}\sin\frac{n\pi\xi}{L}e^{-k(n\pi/L)^2 (t-\tau)}, \quad \tau < t, \qquad (10.11)$$

then the solution of the IBVP can be written in the form

$$u(x,t) = \int_0^L G(x,t;\xi,0)f(\xi) d\xi + \int_0^L\int_0^t G(x,t;\xi,\tau)q(\xi,\tau) d\tau\, d\xi. \qquad (10.12)$$

The first term in this formula represents the influence of the initial temperature in the rod on the subsequent temperature at any point x and any time t. The second term represents the influence of all the sources in the rod at all times $0 < \tau < t$ on the temperature at x and t. This expresses what is known as the *causality principle*.

10.4. Example. We use (10.11) and (10.12) to compute the solution of the IBVP

$$u_t(x,t) = u_{xx}(x,t) + t(x-1), \quad 0 < x < 1,\ t > 0,$$
$$u(0,t) = 0, \quad u(1,t) = 0, \quad t > 0,$$
$$u(x,0) = x, \quad 0 < x < 1.$$

Here $k = 1$, $L = 1$, $q(x,t) = t(x-1)$, and $f(x) = x$, so

$$G(x,t;\xi,\tau) = \sum_{n=1}^{\infty} 2\sin(n\pi x)\sin(n\pi\xi)e^{-n^2\pi^2(t-\tau)};$$

hence,

$$u(x,t) = \int_0^1 G(x,t;\xi,0)\xi\,d\xi + \int_0^1 \int_0^t G(x,t;\xi,\tau)\tau(\xi-1)\,d\tau\,d\xi$$

$$= \sum_{n=1}^{\infty} 2\sin(n\pi x)\left[\int_0^1 \xi\sin(n\pi\xi)\,d\xi\right]e^{-n^2\pi^2 t}$$

$$+ \sum_{n=1}^{\infty} 2\sin(n\pi x)\left[\int_0^1 (\xi-1)\sin(n\pi\xi)\,d\xi\right]\left[\int_0^t \tau e^{-n^2\pi^2(t-\tau)}\,d\tau\right].$$

Integrating by parts, we see that

$$\int_0^1 \xi\sin(n\pi\xi)\,d\xi = (-1)^{n+1}\frac{1}{n\pi},$$

$$\int_0^1 (\xi-1)\sin(n\pi\xi)\,d\xi = -\frac{1}{n\pi},$$

$$\int_0^t \tau e^{-n^2\pi^2(t-\tau)}\,d\tau = \frac{1}{n^2\pi^2}t - \frac{1}{n^4\pi^4}(1-e^{-n^2\pi^2 t}).$$

We now put these results together and obtain the solution of the IBVP in the form

$$u(x,t) = \sum_{n=1}^{\infty} \frac{2}{n\pi} \left[(-1)^{n+1} e^{-n^2\pi^2 t} \right.$$
$$\left. - \frac{1}{n^2\pi^2} t + \frac{1}{n^4\pi^4} \left(1 - e^{-n^2\pi^2 t} \right) \right] \sin(n\pi x). \ \blacksquare$$

10.5. Remark. Green's functions and representation formulas in terms of such functions can also be constructed for IBVPs with other types of BCs, and for IBVPs where the space variable takes values in a semi-infinite or infinite interval (see, for example, (8.8)). ∎

10.2. The Laplace Equation

The equilibrium temperature in a thin, uniform rectangular plate with time-independent sources and zero temperature on the boundary is the solution of the BVP

$$(\Delta u)(x,y) = q(x,y), \quad 0 < x < L, \ 0 < y < K,$$
$$u(x,0) = 0, \quad u(x,K) = 0, \quad 0 < x < L,$$
$$u(0,y) = 0, \quad u(L,y) = 0, \quad 0 < y < K.$$

As in Section 10.1, let $G(x,y;\xi,\eta)$ be the effect at (x,y) produced by just one unit source located at a point (ξ,η), $0 < \xi < L$, $0 < \eta < K$. Then G is the solution of the BVP

$$\Delta(x,y)G(x,y;\xi,\eta) = \delta(x-\xi, y-\eta), \quad 0 < x < L, \ 0 < y < K,$$
$$G(x,0;\xi,\eta) = 0, \quad G(x,K;\xi,\eta) = 0, \quad 0 < x < L,$$
$$G(0,y;\xi,\eta) = 0, \quad G(L,y;\xi,\eta) = 0, \quad 0 < y < K,$$

where $\delta(x-\xi, y-\eta) = \delta(x-\xi)\delta(y-\eta)$ and $\Delta(x,y)$ indicates that the Laplacian is applied with respect to the variables x, y. Also, let D be the rectangle where the problem is formulated, that is,

$$D = \{(x,y) : 0 < x < L, \ 0 < y < K\},$$

and let ∂D be the four-sided boundary of D.

Since ∂D is piecewise smooth, we use the divergence theorem to find that for a pair of smooth functions u and v,

$$\int_D (u\Delta v - v\Delta u)\,da = \int_D (u\,\mathrm{div}\,\mathrm{grad}\,v - v\,\mathrm{div}\,\mathrm{grad}\,u)\,da$$

$$= \int_D \big[\,\mathrm{div}(u\,\mathrm{grad}\,v) - (\mathrm{grad}\,u)\cdot(\mathrm{grad}\,v)$$

$$-\,\mathrm{div}(v\,\mathrm{grad}\,u) + (\mathrm{grad}\,v)\cdot(\mathrm{grad}\,u)\big]\,da$$

$$= \int_{\partial S} \big[(u\,\mathrm{grad}\,v)\cdot n - (v\,\mathrm{grad}\,u)\cdot n\big]\,ds$$

$$= \int_{\partial D} (uv_n - vu_n)\,ds, \qquad (10.13)$$

where da and ds are the elements of area and arc, respectively, and the subscript n denotes the derivative in the direction of the unit outward normal to the boundary. (The normal is not defined at the four corner points, but this does not influence the outcome.)

Equality (10.13) is Green's formula for functions of two space variables. If u is the solution of the given BVP and v is replaced by G, then the homogeneous BCs satisfied by both u and G make the right-hand side in (10.13) vanish and the formula reduces to

$$\int_D \big[u(x,y)\delta(x-\xi)\delta(y-\eta) - q(x,y)G(x,y;\xi,\eta)\big]\,da(x,y) = 0,$$

where $da(x,y)$ indicates that integration is performed with respect to x, y. By (9.1), this yields

$$u(\xi,\eta) = \int_D G(x,y;\xi,\eta)q(x,y)\,da(x,y). \qquad (10.14)$$

At the same time, applying (10.13) with $u(x,y)$ replaced by $G(x,y;\xi,\eta)$ and $v(x,y)$ replaced by $G(x,y;\rho,\sigma)$ and making use once more of (9.1), we obtain the symmetry

$$G(\xi,\eta;\rho,\sigma) = G(\rho,\sigma;\xi,\eta).$$

Then, as a simple interchange of variables shows, (10.14) becomes the representation formula

$$u(x,y) = \int_D G(x,y;\xi,\eta)q(\xi,\eta)\,da(\xi,\eta). \tag{10.15}$$

$G(x,y;\xi,\eta)$ is called the *Green's function* of the given BVP. Formula (10.15) shows the effect of all the sources in D on the temperature at the point (x,y).

10.6. Remark. To find a Fourier series representation for G we recall that the two-dimensional eigenvalue problem (5.54) associated with our BVP has the eigenvalue-eigenfunction pairs

$$\lambda_{nm} = \left(\frac{n\pi}{L}\right)^2 + \left(\frac{m\pi}{K}\right)^2,$$

$$S_{nm} = \sin\frac{n\pi x}{L}\sin\frac{m\pi y}{K}, \quad n,m = 1,2,\ldots.$$

Consequently, it seems reasonable to seek an expansion of the form

$$G(x,y;\xi,\eta) = \sum_{n=1}^{\infty}\sum_{m=1}^{\infty} c_{nm}(\xi,\eta)S_{nm}(x,y)$$

$$= \sum_{n=1}^{\infty}\sum_{m=1}^{\infty} c_{nm}(\xi,\eta)\sin\frac{n\pi x}{L}\sin\frac{m\pi y}{K}.$$

If we replace this series in the equation satisfied by G, then

$$\Delta(x,y)G(x,y;\xi,\eta) = \sum_{n=1}^{\infty}\sum_{m=1}^{\infty} c_{nm}(\xi,\eta)(\Delta S_{nm})(x,y)$$

$$= -\sum_{n=1}^{\infty}\sum_{m=1}^{\infty} \lambda_{nm}c_{nm}(\xi,\eta)S_{nm}(x,y)$$

$$= \delta(x-\xi)\delta(y-\eta).$$

Multiplying both sides above by $S_{pq}(x,y)$, integrating over D, and taking (9.1) into account, we arrive at

$$c_{pq}(\xi,\eta) = -\frac{4}{LK\lambda_{pq}}S_{pq}(\xi,\eta).$$

Hence, the double Fourier sine series for G is

$$G(x,y;\xi,\eta)$$
$$= -\frac{4}{LK} \sum_{n=1}^{\infty} \sum_{m=1}^{\infty} \frac{\sin(n\pi\xi/L)\sin(m\pi\eta/K)}{(n\pi/L)^2 + (m\pi/K)^2} \sin\frac{n\pi x}{L} \sin\frac{m\pi y}{K}. \quad \blacksquare$$

$$(10.16)$$

10.7. Example. To compute the solution of the BVP

$$u_{xx}(x,y) + u_{yy}(x,y) = -5\pi^2 \sin(\pi x)\sin(2\pi y),$$
$$0 < x < 1,\ 0 < y < 2,$$
$$u(x,0) = 0, \quad u(x,2) = 0, \quad 0 < x < 1,$$
$$u(0,y) = 0, \quad u(1,y) = 0, \quad 0 < y < 2,$$

we notice that here

$$L = 1, \quad K = 2, \quad q(x,t) = -5\pi^2 \sin(\pi x)\sin(2\pi y).$$

Consequently, by (10.16),

$$G(x,y;\xi,\eta) = -2 \sum_{n=1}^{\infty} \sum_{m=1}^{\infty} \frac{\sin(n\pi\xi)\sin(m\pi\eta/2)}{n^2\pi^2 + m^2\pi^2/4} \sin(n\pi x)\sin\frac{m\pi y}{2},$$

so from (10.15) and (2.5) it follows that

$$u(x,y) = \int_0^2 \int_0^1 (-2) \sum_{n=1}^{\infty} \sum_{m=1}^{\infty} \frac{4\sin(n\pi\xi)\sin(m\pi\eta/2)}{\pi^2(4n^2 + m^2)} \sin(n\pi x)\sin\frac{m\pi y}{2}$$
$$\times (-5\pi^2)\sin(\pi\xi)\sin(2\pi\eta)\,d\xi\,\delta\eta$$

$$= 40 \sum_{n=1}^{\infty} \sum_{m=1}^{\infty} \frac{1}{4n^2 + m^2} \left(\int_0^1 \sin(\pi\xi)\sin(n\pi\xi)\,d\xi \right)$$
$$\times \left(\int_0^2 \sin(2\pi\eta)\sin\frac{m\pi\eta}{2}\,d\eta \right) \sin(n\pi x)\sin\frac{m\pi y}{2}$$

$$= \left(\frac{40}{4\cdot 1^2 + 4^2} \cdot \frac{1}{2} \cdot 1 \right) \sin(\pi x)\sin(2\pi y) = \sin(\pi x)\sin(2\pi y). \quad \blacksquare$$

10.3. The Wave Equation

The vibrations of an infinite string are described by the IVP

$$u_{tt}(x,t) = c^2 u_{xx}(x,t) + q(x,t), \quad -\infty < x < \infty, \ t > 0,$$

$$u(x,t), \ u_x(x,t) \to 0 \quad \text{as } x \to \pm\infty, \ t > 0,$$

$$u(x,0) = f(x), \quad u_t(x,0) = g(x), \quad -\infty < x < \infty.$$

If we have a unit force acting at a point ξ at time $\tau > 0$, then its influence $G(x,t;\xi,\tau)$ on the vertical vibration of a point x at time t is the solution of the IVP

$$G_{tt}(x,t;\xi,\tau) = c^2 G_{xx}(x,t;\xi,\tau) + \delta(x - \xi, t - \tau),$$

$$-\infty < x < \infty, \ t > 0,$$

$$G(x,t;\xi,\tau), \ G_x(x,t;\xi,\tau) \to 0 \quad \text{as } x \to \pm\infty, \ t > 0,$$

$$G(x,t;\xi,\tau) = 0, \quad -\infty < x < \infty, \ t < \tau,$$

where $\delta(x - \xi, t - \tau) = \delta(x - \xi)\delta(t - \tau)$ and the IC reflects the physical reality that the displacement of the point x is not affected by the unit force at ξ until this force has acted at time τ.

As we did in Chapter 8 in the case of the Cauchy problem for the heat equation, we find the function G by means of the full Fourier transformation. First, we note that, by (9.1),

$$\mathcal{F}[\delta(x - \xi)] = \frac{1}{\sqrt{2\pi}} \int_{-\infty}^{\infty} \delta(x - \xi)e^{i\omega x}\, dx = \frac{1}{\sqrt{2\pi}} e^{i\omega\xi}.$$

Therefore, if we write $\mathcal{F}[G](\omega,t;\xi,\tau) = \tilde{G}(\omega,t;\xi,\tau)$ and apply $\mathcal{F}$ to the PDE and IC satisfied by G, we arrive at the transformed problem

$$\tilde{G}_{tt}(\omega,t;\xi,\tau) + c^2\omega^2\tilde{G}(\omega,t;\xi,\tau) = \frac{1}{\sqrt{2\pi}} e^{i\omega\xi}\delta(t - \tau), \quad t > 0,$$

$$\tilde{G}(\omega,t;\xi,\tau) = 0, \quad t < \tau. \tag{10.17}$$

Since $\delta(t - \tau) = 0$ for $t \neq \tau$, the solution of (10.17) is

$$\tilde{G}(\omega,t;\xi,\tau) = \begin{cases} 0, & t < \tau, \\ C_1\cos\left(c\omega(t - \tau)\right) + C_2\sin\left(c\omega(t - \tau)\right), & t > \tau, \end{cases} \tag{10.18}$$

where C_1 and C_2 are arbitrary functions of ω, ξ, and τ. Requiring $\tilde{G}$ to be continuous at $t = \tau$ yields $C_1 = 0$. To find C_2, we consider an interval $[\tau_1, \tau_2]$ such that $0 < \tau_1 < \tau < \tau_2$ and integrate (10.17) with respect to t over this interval:

$$\tilde{G}_t(\omega, \tau_2; \xi, \tau) - \tilde{G}_t(\omega, \tau_1; \xi, \tau) + c^2 \omega^2 \int_{\tau_1}^{\tau_2} \tilde{G}(\omega, t; \xi, \tau) \, dt$$

$$= \frac{1}{\sqrt{2\pi}} e^{i\omega\xi} \int_{\tau_1}^{\tau_2} \delta(t - \tau) \, dt = \frac{1}{\sqrt{2\pi}} e^{i\omega\xi} \int_{-\infty}^{\infty} \delta(t - \tau) \, dt = \frac{1}{\sqrt{2\pi}} e^{i\omega\xi}.$$

By (10.18),

$$\tilde{G}_t(\omega, \tau_1; \xi, \tau) = 0,$$

$$\tilde{G}_t(\omega, \tau_2; \xi, \tau) = c\omega C_2 \cos \big(c\omega(\tau_2 - \tau) \big).$$

If we now let $\tau_1, \tau_2 \to \tau$, from the continuity of G at $t = \tau$ it follows that $C_2 = e^{i\omega\xi}/(\sqrt{2\pi}\, c\omega)$; hence,

$$\tilde{G}(\omega, t; \xi, \tau) = \begin{cases} 0, & t < \tau, \\ \dfrac{1}{\sqrt{2\pi}\, c} e^{i\omega\xi} \dfrac{\sin \big(c\omega(t - \tau) \big)}{\omega}, & t > \tau. \end{cases}$$

According to formulas 12 and 3 in Table A2 in the Appendix,

$$\mathcal{F}^{-1}\left[\sqrt{\frac{2}{\pi}} \frac{\sin(a\omega)}{\omega} \right] = H(a - |x|),$$

$$\mathcal{F}^{-1}\big[e^{i\omega a} \mathcal{F}[f](\omega) \big] = f(x - a);$$

so, setting $a = c(t - \tau)$ and $a = \xi$, respectively, we obtain

$$G(x, t; \xi, \tau) = \frac{1}{2c} H\big(c(t - \tau) - |x - \xi| \big). \tag{10.19}$$

The diagram in Fig. 10.1 shows the values of G in the upper half $(t > 0)$ of the (x, t)-plane, computed from (10.19). Using a similar diagram, it is easy to see that (10.19) can also be written in the form

$$G(x, t; \xi, \tau) = \frac{1}{2c} \big[H\big((x - \xi) + c(t - \tau) \big) - H\big((x - \xi) - c(t - \tau) \big) \big]. \tag{10.20}$$

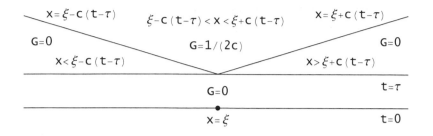

Fig. 10.1.

A procedure analogous to, but more involved than, that followed in the case of the Laplace equation can also be devised for the wave equation to obtain a symmetry relation for G and a representation formula for a solution u in terms of G. Thus, in its most general form the latter is

$$u(x,t) = \int_0^t \int_a^b G(x,t;\xi,\tau)q(\xi,\tau)\,d\xi\,d\tau$$

$$+ \int_a^b \left[G(x,t;\xi,0)u_\tau(\xi,0) - G_\tau(x,t;\xi,0)u(\xi,0)\right]d\xi$$

$$- c^2 \int_0^t \left[G_\xi(x,t;\xi,\tau)u(\xi,\tau) - G(x,t;\xi,\tau)u_\xi(\xi,\tau)\right]_{\xi=a}^{\xi=b}d\tau, \quad (10.21)$$

where q is the forcing term and a and b are the points where the BCs are prescribed. $G(x,t;\xi,\tau)$ is called the *Green's function* for the wave equation.

When $-\infty < x < \infty$, as in our problem, the corresponding formula is obtained from the one above by letting $a \to -\infty$ and $b \to \infty$ and taking into account that $G(x,t;\xi,\tau) = 0$ for $|x|$ sufficiently large.

10.8. Example. Consider the IBVP

$$u_{tt}(x,t) = u_{xx}(x,t) + q(x,t), \quad -\infty < x < \infty, \ t > 0,$$
$$u(x,t), u_x(x,t) \to 0 \quad \text{as } x \to \pm\infty, \ t > 0,$$
$$u(x,0) = 0, \quad u_t(x,0) = 0, \quad -\infty < x < \infty,$$

where

$$q(x,t) = \begin{cases} t, & -1 < x < 1, \ t > 0, \\ 0 & \text{otherwise.} \end{cases}$$

By formulas (10.21) and (10.20) with $c = 1$, the solution of this problem is computed as

$$u(x,t) = \int_0^t \int_{-\infty}^{\infty} G(x,t;\xi,\tau)q(\xi,\tau)\,d\xi\,d\tau$$

$$= \int_0^t \int_{-\infty}^{\infty} \tfrac{1}{2}[H(x-\xi+t-\tau) - H(x-\xi-t+\tau)]q(\xi,\tau)\,d\xi\,d\tau$$

$$= \frac{1}{2}\int_0^t \int_{-\infty}^{x+t-\tau} q(\xi,\tau)\,d\xi\,d\tau - \frac{1}{2}\int_0^t \int_{-\infty}^{x-t+\tau} q(\xi,\tau)\,d\xi\,d\tau;$$

that is,

$$u(x,t) = \frac{1}{2}\int_0^t \int_{x-t+\tau}^{x+t-\tau} q(\xi,\tau)\,d\xi\,d\tau.$$

The value of the solution at, say, $(x,t) = (3,2)$ is

$$u(3,2) = \frac{1}{2}\int_0^2 \int_{1+\tau}^{5-\tau} q(\xi,\tau)\,d\xi\,d\tau.$$

To calculate the above integral, we sketch the lines $\xi = 1+\tau$ and $\xi = 5-\tau$ in the (ξ,τ) system of axes and identify the domain of integration (see Fig. 10.2). Since, as the diagram shows, q is zero in this domain, it follows that $u(3,2) = 0$.

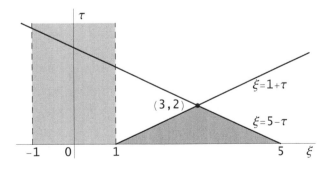

Fig. 10.2. The domain of integration for $(x,t) = (3,2)$.

Similarly,

$$u(2,2) = \frac{1}{2} \int_0^2 \int_\tau^{4-\tau} q(\xi,\tau)\, d\xi\, \delta\tau.$$

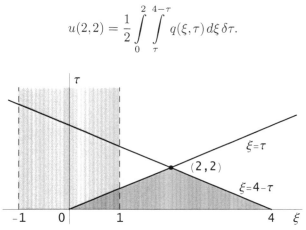

Fig. 10.3. The domain of integration for $(x,t) = (2,2)$.

Following the same procedure and taking into account the intersection of the domain of integration and the semi-infinite strip where q is nonzero (see Fig. 10.3), we see that

$$u(2,2) = \frac{1}{2} \int_0^1 \int_\tau^1 \tau\, d\xi\, d\tau = \frac{1}{2} \int_0^1 \tau\xi\Big|_{\xi=\tau}^{\xi=1} d\tau = \frac{1}{2} \int_0^1 \tau(1-\tau)\, d\tau = \frac{1}{12}.$$

Finally, to compute

$$u(1,3) = \frac{1}{2} \int_0^3 \int_{-2+\tau}^{4-\tau} q(\xi,\tau)\, d\xi\, d\tau,$$

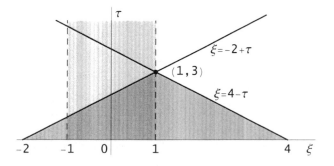

Fig. 10.4. The domain of integration for $(x,t) = (1,3)$.

we make use of the sketch in Fig. 10.4 and obtain

$$u(1,3) = \frac{1}{2} \int_0^1 \int_{-1}^1 \tau \, d\xi \, d\tau + \frac{1}{2} \int_1^3 \int_{-2+\tau}^1 \tau \, d\xi \, d\tau = \frac{13}{6}.$$

Alternatively, by changing the order of integration,

$$u(1,3) = \frac{1}{2} \int_{-1}^1 \int_0^{\xi+2} \tau \, d\tau \, d\xi = \frac{1}{4} \int_{-1}^1 (\xi+2)^2 \, d\xi = \frac{13}{6}. \quad \blacksquare$$

10.9. Remark. If there is no forcing term ($q = 0$) in the general IVP with $-\infty < x < \infty$, then the representation formula (10.21) reduces to

$$u(x,t) = \int_{-\infty}^{\infty} \left[G(x,t;\xi,0)u_\tau(\xi,0) - G_\tau(x,t;\xi,0)u(\xi,0) \right] d\xi. \qquad (10.22)$$

This formula can be further simplified by using the explicit form of G. By (10.20) and Remark 9.4(ii), according to which

$$H'(\tau - a) = \delta(\tau - a),$$

we have

$$G_\tau(x,t;\xi,\tau) = -\tfrac{1}{2} \left[\delta\big((x - \xi) + c(t - \tau)\big) + \delta\big((x - \xi) - c(t - \tau)\big) \right];$$

so, by the definition of δ and H (see Section 9.1), (10.22) becomes

$$u(x,t) = \frac{1}{2} \int_{-\infty}^{\infty} \left[\delta(x - \xi - ct) + \delta(x - \xi + ct) \right] f(\xi) \, d\xi$$

$$+ \frac{1}{2c} \int_{-\infty}^{\infty} \left[H(x - \xi + ct) - H(x - \xi - ct) \right] g(\xi) \, d\xi$$

$$= \frac{1}{2} \left[f(x + ct) + f(x - ct) \right] + \frac{1}{2c} \int_{x-ct}^{x+ct} g(\xi) \, d\xi.$$

This formula, called *d'Alembert's solution*, is revisited in Chapter 12, where it is be established by another method. $\blacksquare$

Exercises

In (1)–(10) construct the appropriate Green's function and solve the one-dimensional steady-state heat equation

$$ku''(x) + q(x) = 0, \quad 0 < x < L,$$

with the constants k and L, function q, and BCs as indicated.

(1) $k = 2$, $L = 1$, $q(x) = 2x - 1$, $u(0) = 1$, $u(1) = -2$.

(2) $k = 1$, $L = 2$, $q(x) = \begin{cases} 1, & 0 < x \leq 1, \\ x & 1 < x < 2, \end{cases}$ $u(0) = 3$, $u(2) = 1$.

(3) $k = 1$, $L = 2$, $q(x) = 4$, $u(0) = 2$, $u'(2) = -3$.

(4) $k = 2$, $L = 1$, $q(x) = 1 - x$, $u(0) = -1$, $u'(1) = 1$.

(5) $k = 2$, $L = 2$, $q(x) = \begin{cases} -1, & 0 < x \leq 1, \\ 2, & 1 < x < 2, \end{cases}$ $u(0) = -2$, $u'(2) = 3$.

(6) $k = 1$, $L = 1$, $q(x) = \begin{cases} 1, & 0 < x \leq 1/2, \\ x + 1, & 1/2 < x < 1, \end{cases}$ $u(0) = 4$, $u'(1) = 2$.

(7) $k = 3$, $L = 1$, $q(x) = -2$, $u'(0) = -3$, $u(1) = 4$.

(8) $k = 2$, $L = 2$, $q(x) = x - 2$, $u'(0) = 1$, $u(2) = -3$.

(9) $k = 1$, $L = 1$, $q(x) = \begin{cases} 2, & 0 < x \leq 1/2, \\ -3, & 1/2 < x < 1, \end{cases}$ $u'(0) = -2$, $u(1) = 5$.

(10) $k = 1$, $L = 2$, $q(x) = \begin{cases} x - 1, & 0 < x \leq 1, \\ 1, & 1 < x < 2, \end{cases}$ $u'(0) = 3$, $u(2) = -3$.

In (11)–(20) construct a series representation of the appropriate Green's function and solve the one-dimensional heat equation

$$u_t(x,t) = ku_{xx}(x,t) + q(x,t), \quad 0 < x < L, \ t > 0,$$

with the constants k and L, function q, and BCs and IC as indicated.

(11) $k = 2$, $L = 1$, $q(x,t) = t\sin(\pi x)$,
$\quad$ $u(0,t) = 0$, $u(1,t) = 0$, $u(x,0) = -\sin(2\pi x)$.

(12) $k = 1$, $L = 2$, $q(x,t) = x(t - 2)$,
$\quad$ $u(0,t) = 0$, $u(2,t) = 0$, $u(x,0) = 1$.

(13) $k = 1$, $L = 2$, $q(x,t) = \begin{cases} 1, & 0 < x \le 1, \\ -1, & 1 < x < 2, \end{cases}$

$u(0,t) = 0$, $u(2,t) = 0$, $u(x,0) = 1 - x$.

(14) $k = 3$, $L = 1$, $q(x,t) = \begin{cases} x, & 0 < x \le 1/2, \\ 0, & 1/2 < x < 1, \end{cases}$

$u(0,t) = 0$, $u(1,t) = 0$, $u(x,0) = \begin{cases} 0, & 0 < x \le 1/2, \\ 2, & 1/2 < x < 1. \end{cases}$

(15) $k = 1$, $L = 1$, $q(x,t) = 2xt$,

$u_x(0,t) = 0$, $u_x(1,t) = 0$, $u(x,0) = x$.

(16) $k = 2$, $L = 1$, $q(x,t) = \begin{cases} -1, & 0 < x \le 1/2, \\ 2, & 1/2 < x < 1, \end{cases}$

$u_x(0,t) = 0$, $u_x(1,t) = 0$, $u(x,0) = x - 1$.

(17) $k = 2$, $L = 2$, $q(x,t) = t$,

$u(0,t) = 0$, $u_x(2,t) = 0$, $u(x,0) = -1$.

(18) $k = 1$, $L = 2$, $q(x,t) = \begin{cases} 0, & 0 < x \le 1, \\ 1, & 1 < x < 2, \end{cases}$

$u(0,t) = 0$, $u_x(2,t) = 0$, $u(x,0) = x$.

(19) $k = 1$, $L = 1$, $q(x,t) = \begin{cases} -1, & 0 < x \le 1/2, \\ 0, & 1/2 < x < 1, \end{cases}$

$u_x(0,t) = 0$, $u(1,t) = 0$, $u(x,0) = \begin{cases} 0, & 0 < x \le 1/2, \\ 2, & 1/2 < x < 1. \end{cases}$

(20) $k = 1$, $L = 2$, $q(x,t) = 1$,

$u_x(0,t) = 0$, $u(2,t) = 0$, $u(x,0) = x - 2$.

In (21)–(30) construct a series representation of the appropriate Green's function and solve the nonhomogeneous Laplace equation

$$u_{xx}(x,y) + u_{yy}(x,y) = q(x,y), \quad 0 < x < L, \ 0 < y < K,$$

with the constants L and K, function q, and BCs as indicated.

(21) $L = 1$, $K = 2$, $q(x,y) = \sin(\pi y)$,

$u(0,y) = 0$, $u(1,y) = 0$, $u(x,0) = 0$, $u(x,2) = 0$.

(22) $L = 1$, $K = 1$, $q(x,y) = (y - 1)\cos(\pi x)$,

$u_x(0,y) = 0$, $u_x(1,y) = 0$, $u(x,0) = 0$, $u(x,1) = 0$.

(23) $L = 1$, $K = 1$, $q(x,y) = x\sin\left(\frac{3}{2}\pi y\right)$,
$u(0,y) = 0$, $u(1,y) = 0$, $u(x,0) = 0$, $u_y(x,1) = 0$.

(24) $L = 2$, $K = 1$, $q(x,y) = \cos\left(\frac{1}{4}\pi x\right)$,
$u_x(0,y) = 0$, $u(2,y) = 0$, $u(x,0) = 0$, $u(x,1) = 0$.

(25) $L = 2$, $K = 1$, $q(x,y) = 2\cos(3\pi y)$,
$u(0,y) = 0$, $u_x(2,y) = 0$, $u_y(x,0) = 0$, $u_y(x,1) = 0$.

(26) $L = 1$, $K = 2$, $q(x,y) = -\sin\left(\frac{3}{2}\pi x\right)$,
$u(0,y) = 0$, $u_x(1,y) = 0$, $u_y(x,0) = 0$, $u(x,2) = 0$.

(27) $L = 1$, $K = 1$, $q(x,y) = 1$,
$u(0,y) = 0$, $u(1,y) = 0$, $u(x,0) = 0$, $u(x,1) = 0$.

(28) $L = 1$, $K = 2$, $q(x,y) = -2$,
$u(0,y) = 0$, $u(1,y) = 0$, $u(x,0) = 0$, $u_y(x,2) = 0$.

(29) $L = 1$, $K = 1$, $q(x,y) = x$,
$u_x(0,y) = 0$, $u(1,y) = 0$, $u(x,0) = 0$, $u_y(x,1) = 0$.

(30) $L = 2$, $K = 1$, $q(x,y) = y$,
$u(0,y) = 0$, $u(2,y) = 0$, $u_y(x,0) = 0$, $u_y(x,1) = 0$.

In (31)–(40) construct an integral representation of the solution in terms of the appropriate Green's function for the IVP

$$u_{tt}(x,t) = c^2 u_{xx}(x,t) + q(x,t), \quad -\infty < x < \infty, \ t > 0,$$
$$u(x,t), u_x(x,t) \to 0 \text{ as } x \to \pm\infty,$$
$$u(x,0) = 0, \quad u_t(x,0) = 0, \quad -\infty < x < \infty,$$

with the constant c and function q as indicated, then compute the solution at the given point x and time t.

(31) $c = 1$, $q(x,t) = \begin{cases} x, & -1 < x < 2, \ t > 0, \\ 0 & \text{otherwise}, \end{cases}$ $(x,t) = (-1,2)$.

(32) $c = 1$, $q(x,t) = \begin{cases} xt, & -1 < x < 3, \ t > 0, \\ 0 & \text{otherwise}, \end{cases}$ $(x,t) = (2,3)$.

(33) $c = 2$, $q(x,t) = \begin{cases} x - t, & -2 < x < 1, \ t > 0, \\ 0 & \text{otherwise}, \end{cases}$ $(x,t) = (1,2)$.

(34) $c = 2$, $q(x,t) = \begin{cases} 2x + t, & -3 < x < -1, \ t > 0, \\ 0 & \text{otherwise}, \end{cases}$ $(x,t) = (-2,1)$.

(35) $c = 1$, $q(x,t) = \begin{cases} 1 - x + 2t, & 0 < x < 4, \ t > 1, \\ 0 & \text{otherwise,} \end{cases}$ $(x,t) = (3,2)$.

(36) $c = 1$, $q(x,t) = \begin{cases} xt + 2, & -5 < x < 0, \ 0 < t < 1, \\ 0 & \text{otherwise,} \end{cases}$ $(x,t) = (-4,2)$.

(37) $c = 2$, $q(x,t) = \begin{cases} 2 - t, & -8 < x < 0, \ t > 0, \\ 0 & \text{otherwise,} \end{cases}$ $(x,t) = (-3,4)$.

(38) $c = 2$, $q(x,t) = \begin{cases} x - 1, & 3 < x < 10, \ t > 0, \\ 0 & \text{otherwise,} \end{cases}$ $(x,t) = (5,3)$.

(39) $c = 1$, $q(x,t) = \begin{cases} 2x - t - 1, & 0 < x < 5, \ 1/2 < t < 3/2, \\ 0 & \text{otherwise,} \end{cases}$ $(x,t) = (4,2)$.

(40) $c = 1$, $q(x,t) = \begin{cases} x(t - 1), & -3 < x < 2, \ t > 0, \\ 0 & \text{otherwise,} \end{cases}$ $(x,t) = (-1,4)$.

Chapter 11
General Second-Order Linear Partial Differential Equations with Two Independent Variables

Having studied several solution procedures for the heat, wave, and Laplace equations, we need to explain why we have chosen these particular models in preference to others. In Chapter 4 we mentioned that these were typical examples of what we called parabolic, hyperbolic, and elliptic equations, respectively. Below we present a systematic discussion of the general second-order linear PDE in two independent variables and show how such an equation can be reduced to its simplest form. It will be seen that if the equation has constant coefficients, then its dominant part—that is, the sum of the terms containing the highest-order derivatives with respect to each of the variables—consists of the same terms as one of the above three equations. This gives us a good indication of what solution technique we should use, and what kind of behavior to expect from the solution.

11.1. The Canonical Form

Classification. The general form of a second-order linear PDE in two independent variables is

$$A(x,y)u_{xx} + B(x,y)u_{xy} + C(x,y)u_{yy} + D(x,y)u_x + E(x,y)u_y$$
$$+ F(x,y)u = G(x,y), \qquad (11.1)$$

where $u = u(x,y)$ is the unknown function and $A, \ldots, G$ are given coefficients. (In particular, some or all of these coefficients may be constant.)

11.1. Definition. (i) If $B^2 - 4AC > 0$, (11.1) is called a *hyperbolic* equation.

(ii) If $B^2 - 4AC = 0$, (11.1) is called a *parabolic* equation.

(iii) If $B^2 - 4AC < 0$, (11.1) is called an *elliptic* equation. ∎

11.2. Example. For the one-dimensional wave equation

$$u_{tt} - c^2 u_{xx} = 0$$

we have (considering t in place of y)

$$A = -c^2, \ C = 1, \ B = D = E = F = G = 0;$$

hence, $B^2 - 4AC = 4c^2 > 0$, which means that the equation is hyperbolic at all points in the (x,t)-plane. ■

11.3. Example. In the case of the one-dimensional heat equation

$$u_t - k u_{xx} = 0$$

we have

$$A = -k, \quad E = 1, \quad B = C = D = F = G = 0,$$

so $B^2 - 4AC = 0$: the equation is parabolic in the entire (x,t)-plane. ■

11.4. Example. The (two-dimensional) Laplace equation

$$\Delta u = u_{xx} + u_{yy} = 0$$

is obtained for

$$A = 1, \ C = 1, \ B = D = E = F = G = 0;$$

therefore, $B^2 - 4AC = -4 < 0$, which means that this equation is elliptic throughout the (x,y)-plane. ■

11.5. Example. The equation

$$u_{xx} - \sqrt{y}\, u_{xy} + x u_{yy} + (2x + y)u_x - 3y u_y + 4u = \sin(x^2 - 2y), \quad y > 0,$$

fits the general form with

$$A = 1, \quad B = -\sqrt{y}, \quad C = x,$$
$$D = 2x + y, \quad E = -3y, \quad F = 4, \quad G = \sin(x^2 - 2y),$$

so $B^2 - 4AC = y - 4x$. Consequently,

(i) if $y > 4x$, the equation is hyperbolic;

(ii) if $y = 4x$, the equation is parabolic;

(iii) if $y < 4x$, the equation is elliptic.

In other words, the type of this equation at (x,y) depends on where the point lies in the half plane $y > 0$ (see Fig. 11.1). ∎

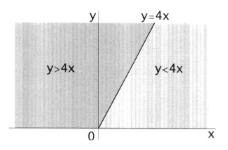

Fig. 11.1. The regions corresponding to the different equation types.

Reduction to the canonical form. We introduce new coordinates

$$r = r(x,y), \quad s = s(x,y),$$

or, conversely,

$$x = x(r,s), \quad y = y(r,s),$$

and write $u\big(x(r,s), y(r,s)\big) = v(r,s)$. By the chain rule,

$$u_x = v_r r_x + v_s s_x, \quad u_y = v_r r_y + v_s s_y,$$

and

$$
\begin{aligned}
u_{xx} &= (u_x)_x = (v_r r_x + v_s s_x)_x = (v_r)_x r_x + v_r (r_x)_x + (v_s)_x s_x + v_s (s_x)_x \\
&= \big[(v_r)_r r_x + (v_r)_s s_x\big] r_x + v_r r_{xx} + \big[(v_s)_r r_x + (v_s)_s s_x\big] s_x + v_s s_{xx} \\
&= v_{rr} r_x^2 + 2 v_{rs} r_x s_x + v_{ss} s_x^2 + v_r r_{xx} + v_s s_{xx},
\end{aligned}
$$

$$u_{yy} = v_{rr} r_y^2 + 2 v_{rs} r_y s_y + v_{ss} s_y^2 + v_r r_{yy} + v_s s_{yy},$$

$$u_{xy} = v_{rr} r_x r_y + v_{rs}(r_x s_y + r_y s_x) + v_{ss} s_x s_y + v_r r_{xy} + v_s s_{xy},$$

where the last two expressions have been calculated in the same way as the first one. Replacing all the derivatives in (11.1) and gathering the like terms, we arrive at the new equality

$$
\bar{A}(r,s)v_{rr} + \bar{B}(r,s)v_{rs} + \bar{C}(r,s)v_{ss} + \bar{D}(r,s)v_r + \bar{E}(r,s)v_s \\
+ \bar{F}(r,s)v = \bar{G}(r,s), \quad (11.2)
$$

where the new coefficients $\bar{A}, \ldots, \bar{G}$ are related to the old ones $A, \ldots, G$ through the formulas

$$\bar{A} = Ar_x^2 + Br_x r_y + Cr_y^2,$$

$$\bar{B} = 2Ar_x s_x + B(r_x s_y + r_y s_x) + 2Cr_y s_y,$$

$$\bar{C} = As_x^2 + Bs_x s_y + Cs_y^2,$$

$$\bar{D} = Ar_{xx} + Br_{xy} + Cr_{yy} + Dr_x + Er_y, \qquad (11.3)$$

$$\bar{E} = As_{xx} + Bs_{xy} + Cs_{yy} + Ds_x + Es_y,$$

$$\bar{F} = F,$$

$$\bar{G} = G.$$

We now choose r and s so that $\bar{A} = \bar{C} = 0$. Since neither $r = r(x,y)$ nor $s = s(x,y)$ can be a constant, it follows that at least one of r_x, r_y and at least one of s_x, s_y must be nonzero. Suppose, for definiteness, that r_y and s_y are nonzero. Setting the expression of $\bar{A}$ in (11.3) equal to zero and dividing through by r_y^2, we arrive at the equation

$$A\left(\frac{r_x}{r_y}\right)^2 + B\left(\frac{r_x}{r_y}\right) + C = 0; \qquad (11.4)$$

a similar procedure applied to the expression of $\bar{C}$ in (11.3) leads to the equality

$$A\left(\frac{s_x}{s_y}\right)^2 + B\left(\frac{s_x}{s_y}\right) + C = 0. \qquad (11.5)$$

From (11.4) and (11.5) it follows that

$$\frac{r_x}{r_y} = \frac{-B + \sqrt{B^2 - 4AC}}{2A},$$

$$\frac{s_x}{s_y} = \frac{-B - \sqrt{B^2 - 4AC}}{2A}. \qquad (11.6)$$

It is now obvious that the nature of the solutions of (11.6) depends on whether the given equation is hyperbolic, parabolic, or elliptic.

11.2. Hyperbolic Equations

Since here $B^2 - 4AC > 0$, there are two distinct equations (11.6). Consider the differential equations

$$\frac{dy}{dx} = \frac{B - \sqrt{B^2 - 4AC}}{2A},$$
$$\frac{dy}{dx} = \frac{B + \sqrt{B^2 - 4AC}}{2A}, \tag{11.7}$$

called the *characteristic equations* for (11.1), and let

$$\varphi(x,y) = c_1, \quad \psi(x,y) = c_2, \quad c_1, c_2 \text{ arbitrary constants,}$$

be the families of their solution curves, called *characteristics*. Along these curves we have, respectively,

$$d\varphi = \varphi_x dx + \varphi_y dy = 0, \quad d\psi = \psi_x dx + \psi_y dy = 0,$$

which yields

$$\frac{\varphi_x}{\varphi_y} = -\frac{dy}{dx} = \frac{-B + \sqrt{B^2 - 4AC}}{2A},$$
$$\frac{\psi_x}{\psi_y} = -\frac{dy}{dx} = \frac{-B - \sqrt{B^2 - 4AC}}{2A}.$$

Consequently, the functions

$$r = \varphi(x,y), \quad s = \psi(x,y) \tag{11.8}$$

are solutions of (11.6); that is, they define the change of variables that produces $\bar{A} = \bar{C} = 0$. Using (11.3) and (11.8), we now compute the new coefficients $\bar{A}, \ldots, \bar{G}$ and write out the canonical form (11.2) as

$$\bar{B}(r,s)v_{rs} + \bar{D}(r,s)v_r + \bar{E}(r,s)v_s + \bar{F}(r,s)v = \bar{G}(r,s). \tag{11.9}$$

11.6. Remark. The hyperbolic equation has an alternative canonical form. If we introduce new variables α, β by means of the formulas

$$\alpha = \tfrac{1}{2}(r + s), \quad \beta = \tfrac{1}{2}(r - s)$$

and write $v\big(r(\alpha,\beta), s(\alpha,\beta)\big) = w(\alpha,\beta)$, then

$$v_r = w_\alpha \alpha_r + w_\beta \beta_r = \tfrac{1}{2} w_\alpha + \tfrac{1}{2} w_\beta,$$

$$v_s = w_\alpha \alpha_s + w_\beta \beta_s = \tfrac{1}{2} w_\alpha - \tfrac{1}{2} w_\beta,$$

$$v_{rs} = (v_r)_s = \tfrac{1}{2}[(w_\alpha + w_\beta)_\alpha \alpha_s + (w_\alpha + w_\beta)_\beta \beta_s]$$

$$= \tfrac{1}{2}[\tfrac{1}{2}(w_{\alpha\alpha} + w_{\beta\alpha}) - \tfrac{1}{2}(w_{\alpha\beta} + w_{\beta\beta})] = \tfrac{1}{4} w_{\alpha\alpha} - \tfrac{1}{4} w_{\beta\beta},$$

and (11.9) becomes

$$\bar{B}(w_{\alpha\alpha} - w_{\beta\beta}) + 2(\bar{D} + \bar{E})w_\alpha + 2(\bar{D} - \bar{E})w_\beta + 4\bar{F}w = 4\bar{G}.$$

The one-dimensional wave equation $u_{tt} - c^2 u_{xx} = 0$ can be brought to this form with new coefficients $\bar{D} = \bar{E} = \bar{F} = \bar{G} = 0$ by means of the substitution $\tau = ct$. ■

11.7. Example. The coefficients of the PDE

$$y u_{xx} + 3y u_{xy} + 3u_x = 0, \quad y \neq 0,$$

are $A = y$, $B = 3y$, $C = 0$, $D = 3$, and $E = F = G = 0$. They yield $B^2 - 4AC = 9y^2 > 0$, so the equation is hyperbolic at all points (x, y) with $y \neq 0$. Here the characteristic equations (11.7) are

$$y'(x) = 0, \quad y'(x) = 3,$$

with general solutions $y = c_1$ and $y = 3x + c_2$, respectively. Hence, we can take our coordinate transformation to be

$$r = y, \quad s = y - 3x,$$

which, by (11.3), leads to $\bar{B} = -9y = -9r$, $\bar{E} = -9$, and $\bar{D} = \bar{F} = \bar{G} = 0$ (we already know that $\bar{A} = \bar{C} = 0$). Replacing in (11.9), we obtain the canonical form

$$r v_{rs} + v_s = 0, \quad v(r, s) = u\big(x(r,s), y(r,s)\big).$$

Writing this equation in the form $r(v_s)_r + v_s = 0$ and using, for example, the integrating factor method, we find that $v_s(r, s) = (1/r)C(s)$, where $C(s)$ is an arbitrary function. A second integration, this time with respect to s, produces the solution $v(r, s) = (1/r)\varphi(s) + \psi(r)$, or, in terms of the original variables x and y,

$$u(x, y) = v(r(x, y), s(x, y)) = \frac{1}{y}\varphi(y - 3x) + \psi(y),$$

where φ and ψ are arbitrary one-variable functions. ■

11.8. Example. For the equation

$$u_{xx} + u_{xy} - 2u_{yy} - 3u_x - 6u_y = 18x - 9y$$

we have

$$A = 1, \quad B = 1, \quad C = -2, \quad D = -3, \quad E = -6, \quad F = 0, \quad G = 18x - 9y.$$

Since $B^2 - 4AC = 9 > 0$, this PDE is hyperbolic at all points in the (x, y)-plane. By (11.7), the characteristic equations are

$$y'(x) = 2, \quad y'(x) = -1,$$

with solutions $y = 2x + c_1$ and $y = -x + c_2$, respectively, $c_1, c_2 = \text{const}$; hence, the coordinate transformation is

$$r = y - 2x, \quad s = y + x.$$

From (11.3) it follows that $\bar{B} = -9$, $\bar{E} = -9$, $\bar{G} = -9r$, and $\bar{D} = \bar{F} = 0$; therefore, the canonical form (11.9) of the given PDE is

$$v_{rs} + v_s = r, \quad v(r, s) = u(x(r, s), y(r, s)),$$

or $(v_s)_r + v_s = r$, which yields $v_s(r, s) = r - 1 + C(s)e^{-r}$. Then

$$v(r, s) = s(r - 1) + \varphi(s)e^{-r} + \psi(r),$$

so the general solution of the PDE is

$$u(x, y) = v\big(r(x, y), s(x, y)\big)$$
$$= (x + y)(y - 2x - 1) + \varphi(x + y)e^{2x - y} + \psi(y - 2x),$$

where φ and ψ are arbitrary one-variable functions. ∎

11.9. Example. Consider the IVP (with the variable t replaced by y to facilitate the use of the general formulas already derived in this chapter)

$$2u_{xx} - 5u_{xy} + 2u_{yy} = -36x - 18y, \quad -\infty < x < \infty, \ y > 0,$$
$$u(x, 0) = 4x^3 + 3x, \quad u_y(x, 0) = 12x^2 + 4, \quad -\infty < x < \infty.$$

Here

$$A = 2, \quad B = -5, \quad C = 2, \quad D = E = F = 0, \quad G = -36x - 18y,$$

so $B^2 - 4AC = 9 > 0$, which means that the equation is hyperbolic. From (11.7) it follows that the characteristic equations are

$$y'(x) = -2, \quad y'(x) = -\tfrac{1}{2},$$

with solutions $y = -2x + c_1$ and $y = -\tfrac{1}{2}x + c_2$, where $c_1, c_2 = \text{const}$. Therefore, we operate the coordinate transformation

$$r = 2x + y, \quad s = x + 2y$$

and, using (11.3), find that $\bar{B} = -9$, $\bar{D} = \bar{E} = \bar{F} = 0$, and $\bar{G} = -18r$. This yields the canonical form (see (11.9))

$$v_{rs} = 2r, \quad v(r,s) = u(x(r,s), y(r,s)),$$

with general solution

$$v(r,s) = r^2 s + \varphi(r) + \psi(s).$$

Hence, the general solution of the given PDE is

$$u(x,y) = (2x + y)^2 (x + 2y) + \varphi(2x + y) + \psi(x + 2y),$$

where, as before, φ and ψ are arbitrary one-variable functions.

To apply the ICs, we first differentiate u with respect to y to obtain

$$u_y(x,y) = 2(2x + y)(x + 2y) + 2(2x + y)^2 + \varphi'(2x + y) + 2\psi'(x + 2y)$$

and then set $y = 0$ in u and u_y. Canceling out the like terms, we arrive at the system of equations

$$\varphi(2x) + \psi(x) = 3x,$$
$$\varphi'(2x) + 2\psi'(x) = 4.$$

Differentiating the first equation with respect to x, we find that

$$2\varphi'(2x) + \psi'(x) = 3,$$

which, combined with the second equation, leads to $\psi'(x) = 5/3$; consequently, $\psi(x) = \tfrac{5}{3}x + c$, where c is an arbitrary constant.

Next,

$$\varphi(2x) = 3x - \psi(x) = 3x - \tfrac{5}{3}x - c = \tfrac{2}{3}(2x) - c,$$

so $\varphi(x) = \tfrac{2}{3}x - c$. Then the solution of the IVP is

$$u(x,y) = (2x + y)^2(x + 2y) + \tfrac{2}{3}(2x + y) + \tfrac{5}{3}(x + 2y)$$
$$= (2x + y)^2(x + 2y) + 3x + 4y. \quad \blacksquare$$

11.3. Parabolic Equations

Since here $B^2 - 4AC = 0$, from (11.6) we see that r and s satisfy the same ODE, which means that we can make only one of $\bar{A}$ and $\bar{C}$ zero. Let $\bar{A} = 0$. Then (11.6) reduces to

$$\frac{r_x}{r_y} = -\frac{B}{2A},$$

from which

$$\frac{dy}{dx} = -\frac{r_x}{r_y} = \frac{B}{2A}. \tag{11.10}$$

The general solution of this equation provides us with the function $r(x,y)$.

Now $B^2 - 4AC = 0$ implies that $AC \geq 0$ and $B = 2\sqrt{AC}$. Without loss of generality, we may also assume that $A, C \geq 0$. Then, by (11.3),

$$\bar{B} = 2Ar_x s_x + B(r_x s_y + r_y s_x) + 2Cr_y s_y$$
$$= 2\left[Ar_x s_x + \sqrt{A}\sqrt{C}(r_x s_y + r_y s_x) + Cr_y s_y\right]$$
$$= 2\left[\sqrt{A}r_x(\sqrt{A}s_x + \sqrt{C}s_y) + \sqrt{C}r_y(\sqrt{A}s_x + \sqrt{C}s_y)\right]$$
$$= 2(\sqrt{A}r_x + \sqrt{C}r_y)(\sqrt{A}s_x + \sqrt{C}s_y).$$

But, according to (11.10),

$$\frac{r_x}{r_y} = -\frac{B}{2A} = -\frac{2\sqrt{A}\sqrt{C}}{2A} = -\frac{\sqrt{C}}{\sqrt{A}},$$

so $\bar{B} = 0$. Consequently, s can be chosen arbitrarily, in any manner that does not "clash" with r given by (11.10) (more precisely, so that the Jacobian of the transformation is nonzero). The canonical form in this case is

$$\bar{C}(r,s)v_{ss} + \bar{D}(r,s)v_r + \bar{E}(r,s)v_s + \bar{F}(r,s)v = \bar{G}(r,s). \tag{11.11}$$

11.10. Remark. The one-dimensional heat equation $u_t - k u_{xx} = 0$ is of the form (11.11) with $\bar{C} = -k$, $\bar{D} = 1$, and $\bar{E} = \bar{F} = \bar{G} = 0$. ∎

11.11. Example. The coefficients of the PDE

$$u_{xx} + 2u_{xy} + u_{yy} = 0$$

are

$$A = 1, \quad B = 2, \quad C = 1, \quad D = E = F = G = 0.$$

Since $B^2 - 4AC = 0$, the equation is parabolic, and its characteristic equation (11.10) is $y'(x) = 1$, with general solution $y = x + c$, $c = $ const. Hence, we take $r = y - x$; for the function s we can make any suitable choice, for example, $s = y$. Then

$$\bar{C} = 1, \quad \bar{D} = \bar{E} = \bar{F} = \bar{G} = 0$$

(we already know that $\bar{A} = \bar{B} = 0$), which, by (11.11), leads to the canonical form

$$v_{ss} = 0.$$

Integrating twice with respect to s, we obtain the general solution

$$v(r, s) = s\varphi(r) + \psi(r),$$

or, in terms of x and y,

$$u(x, y) = v\big(r(x,y), s(x,y)\big) = y\varphi(y - x) + \psi(y - x),$$

where φ and ψ are arbitrary one-variable functions. ∎

11.12. Example. For the equation

$$4u_{xx} + 12u_{xy} + 9u_{yy} - 9u = 9$$

we have

$$A = 4, \quad B = 12, \quad C = 9, \quad D = E = 0, \quad F = -9, \quad G = 9,$$

so $B^2 - 4AC = 0$; that is, the PDE is parabolic. By (11.10), the characteristic equation is $y'(x) = 3/2$, with general solution $y = (3/2)x + c$, or

$2y - 3x = c'$, $c' = $ const; consequently, we can take $r = 2y - 3x$ and, say, $s = y$, as above. Then

$$\bar{C} = 9, \quad \bar{F} = -9, \quad \bar{G} = 9,$$

which, replaced in (11.11), yields the canonical form

$$v_{ss} - v = 1.$$

The general solution of this equation is

$$v(r,s) = \varphi(r)\cosh s + \psi(r)\sinh s - 1,$$

or, for the given PDE,

$$\begin{aligned}
u(x,y) &= v(r(x,y),s(x,y)) \\
&= \varphi(2y - 3x)\cosh y + \psi(2y - 3x)\sinh y - 1,
\end{aligned}$$

where φ and ψ are arbitrary one-variable functions. ∎

11.13. Example. The PDE

$$x^2 u_{xx} + 2xy u_{xy} + y^2 u_{yy} + (x + y)u_x = x^2 - 2y,$$

considered at all points (x,y) except the origin, has coefficients

$$A = x^2, \quad B = 2xy, \quad C = y^2,$$
$$D = x + y, \quad E = F = 0, \quad G = x^2 - 2y.$$

Since $B^2 - 4AC = 4x^2y^2 - 4x^2y^2 = 0$, the equation is parabolic and its characteristic equation is

$$y'(x) = \frac{y}{x},$$

with solution $y = cx$, $c = $ const. Hence, we may use the transformation

$$r = \frac{y}{x}, \quad s = y,$$

under which the new coefficients, given by (11.3), are

$$\bar{C} = s^2, \quad \bar{D} = -r - r^2, \quad \bar{E} = \bar{F} = 0, \quad \bar{G} = \frac{s^2}{r^2} - 2s.$$

In view of (11.11), we arrive at the canonical form

$$r^2 s^2 v_{ss} - (r^4 + r^3)v_r = s^2 - 2r^2 s, \quad v(r,s) = u(x(r,s),y(r,s)). \quad ∎$$

11.4. Elliptic Equations

The procedure in this case is the same as for hyperbolic equations, but, since this time $B^2 - 4AC < 0$, the characteristic curves are complex. However, a real canonical form can still be obtained.

11.14. Example. The coefficients of the PDE

$$u_{xx} + 2u_{xy} + 5u_{yy} + u_x = 0$$

are

$$A = 1, \quad B = 2, \quad C = 5, \quad D = 1, \quad E = F = G = 0.$$

Thus, $B^2 - 4AC = -16 < 0$, so the equation is elliptic. By (11.7), the characteristic equations are

$$y'(x) = 1 - 2i, \quad y'(x) = 1 + 2i,$$

with general solutions

$$y = (1 - 2i)x + c_1, \quad y = (1 + 2i)x + c_2,$$

respectively. Therefore, the coordinate transformation is

$$r = y - (1 - 2i)x, \quad s = y - (1 + 2i)x.$$

Then $\bar{B} = 16$, $\bar{D} = -(1 - 2i)$, $\bar{E} = -(1 + 2i)$, and $\bar{F} = \bar{G} = 0$ ($\bar{A} = \bar{C} = 0$ because of the transformation), which yields the complex canonical form

$$16v_{rs} - (1 - 2i)v_r - (1 + 2i)v_s = 0, \quad v(r, s) = u(x(r, s), y(r, s)).$$

Performing the second transformation

$$\alpha = \frac{1}{2}(r + s), \quad \beta = \frac{1}{2i}(r - s),$$

we easily arrive at the new (real) canonical form

$$4(w_{\alpha\alpha} + w_{\beta\beta}) - w_\alpha + 2w_\beta = 0, \quad w(\alpha, \beta) = v(r(\alpha, \beta), s(\alpha, \beta)). \quad \blacksquare$$

11.15. Remark. The Laplace equation $u_{xx} + u_{yy} = 0$ is elliptic, with $A = C = 1$ and $B = D = E = F = G = 0$. $\blacksquare$

Exercises

In (1)–(6) discuss the type (hyperbolic, parabolic, or elliptic) of the given PDE and sketch a graph in the (x, y)-plane to illustrate your findings.

(1) $u_{xx} + (x - 1)u_{xy} + u_{yy} - 2x^2 u_x + 3xy u_y + 2u = \sin x.$

(2) $u_{xx} + \sqrt{y}\, u_{xy} - (x - 2)u_{yy} + 2u_y - (x - y)u = e^x \sin y, \quad y > 0.$

(3) $yu_{xx} - xu_{xy} + yu_{yy} - 3x^2 u_y = xe^{-xy}.$

(4) $2xu_{xx} - u_{xy} + (y + 1)u_{yy} - xu_x + (x - 2y)u = x + 2y^2.$

(5) $(x + 2)u_{xx} + 2(x + y)u_{xy} + 2(y - 1)u_{yy} - 3x^2 u_x = x^3 y^3.$

(6) $4xu_{xx} + 4yu_{xy} + (4 - x)u_{yy} - 2xy u_y + 2u = x(2y + 1).$

In (7)–(16) verify that the given PDE is hyperbolic everywhere in the (x, y)-plane, reduce it to its canonical form, and find its general solution.

(7) $2u_{xx} - 7u_{xy} + 3u_{yy} = -150x - 50y.$

(8) $u_{xx} + u_{xy} - 2u_{yy} = 72(2x^2 + xy - y^2) - 9.$

(9) $3u_{xx} + 2u_{xy} - u_{yy} = -32e^{-2x+2y}.$

(10) $2u_{xx} + u_{xy} - 6u_{yy} = 98e^{7x}.$

(11) $6u_{xx} + u_{xy} - 2u_{yy} + 42u_x - 21u_y = 0.$

(12) $u_{xx} + 2u_{xy} - 3u_{yy} + 4u_x - 4u_y = 32(3x - y).$

(13) $3u_{xx} + u_{xy} - 2u_{yy} - 30u_x + 20u_y = 25(2x + 3y).$

(14) $2u_{xx} - 3u_{xy} - 2u_{yy} + 10u_x - 20u_y = -25(4x + 2y + 5).$

(15) $4u_{xx} + 14u_{xy} + 6u_{yy} - 10u_x - 5u_y = 25(4y - 7x - 2).$

(16) $2u_{xx} - 2u_{xy} - 4u_{yy} - 9u_x + 18u_y = 9(24x - 6y - 1).$

In (17)–(22) verify that the given PDE is parabolic everywhere in the (x, y)-plane, reduce it to its canonical form, and find its general solution.

(17) $u_{xx} + 8u_{xy} + 16u_{yy} + 64u = 16.$

(18) $16u_{xx} - 24u_{xy} + 9u_{yy} + 36u_x - 27u_y = 9.$

(19) $25u_{xx} + 30u_{xy} + 9u_{yy} - 45u_x - 27u_y + 18u = 18(3xy - 5y^2).$

(20) $9u_{xx} - 6u_{xy} + u_{yy} + 12u_x - 4u_y + 3u = 9x + 21y + 8.$

(21) $4u_{xx} + 4u_{xy} + u_{yy} - 2u_x - u_y - 2u = 3y - x.$

(22) $u_{xx} + 6u_{xy} + 9u_{yy} + 9u_x + 27u_y + 18u = 27(4x - 2y - 1).$

In (23)–(26) verify that the given PDE is elliptic everywhere in the (x,y)-plane and reduce it to its real canonical form.

(23) $u_{xx} + 4u_{xy} + 5u_{yy} - 2u_x + u_y = 3.$

(24) $u_{xx} + 2u_{xy} + 10u_{yy} + 3u_x - 2u_y + 2u = x + y.$

(25) $u_{xx} + 6u_{xy} + 10u_{yy} - u_x + u_y - 3u = 2x - y.$

(26) $u_{xx} + 4u_{xy} + 13u_{yy} + 2u_x - u_y + u = 3x + 2y.$

In (27)–(30) identify the type of the PDE, reduce the equation to its canonical form, and compute the solution of the given IVP.

(27) $4u_{xx} + 12u_{xy} + 9u_{yy} - 6u_x - 9u_y = 27(3x - 2y),\ -\infty < x < \infty,\ y > 0,$
      $u(x,0) = -3x,\ \ u_y(x,0) = 1,\ \ -\infty < x < \infty.$

(28) $9u_{xx} + 6u_{xy} + u_{yy} + 3u_x + u_y = 2x - 6y + 1,\ \ x > 0,\ -\infty < y < \infty,$
      $u(0,y) = y,\ \ u_x(0,y) = 2y,\ \ -\infty < y < \infty.$

(29) $u_{xx} + 6u_{xy} + 8u_{yy} = 8(10x - 3y),\ \ -\infty < x < \infty,\ y > 0,$
      $u(x,0) = -10(8x^3 + x),\ \ u_y(x,0) = 84x^2 + 3,\ \ -\infty < x < \infty.$

(30) $u_{xx} - u_{xy} - 6u_{yy} - 10u_x + 30u_y = 50,\ \ -\infty < x < \infty,\ y > 0,$
      $u(x,0) = 4x + e^{6x},\ \ u_y(x,0) = 3 + 2e^{6x},\ \ -\infty < x < \infty.$

In (31)–(36) identify the type of the PDE and reduce the equation to its canonical form.

(31) $u_{xx} + (x + 1)u_{xy} + xu_{yy} + (2y - x^2)u = y + x - x^2,\ \ x \neq 1.$

(32) $u_{xx} + 2xu_{xy} + (x^2 + 1)u_{yy} = x + y.$

(33) $u_{xx} + 2yu_{xy} + y^2 u_{yy} + 2u_y = x + y.$

(34) $u_{xx} + 2u_{xy} + (x^2 + 1)u_{yy} = 4(y - x),\ \ x \neq 0.$

(35) $u_{xx} + xu_{xy} + (2x - 4)u_{yy} + (4y - x^2)u = (y - 2x)(2y + 4x - x^2),$
      $x \neq 4.$

(36) $u_{xx} + 2\sqrt{x}\,u_{xy} + xu_{yy} - yu_y = 2y + 1,\ x > 0.$

Chapter 12
The Method of Characteristics

The equations in the problems we have investigated so far are all linear and the terms containing the unknown function and its derivatives have constant coefficients. The only exception is the type of problem where we need to make use of polar coordinates, but in such problems the polar radius is present in some of the coefficients in a very specific way, which does not disturb the solution scheme. Below we discuss a procedure for solving first-order linear PDEs with more general variable coefficients, and first-order nonlinear PDEs of a particular form. We also re-examine the one-dimensional wave equation from the perspective of this new technique.

12.1. First-Order Linear Equations

Consider the IVP

$$u_t(x,t) + cu_x(x,t) = 0, \quad -\infty < x < \infty, \ t > 0, \tag{PDE}$$
$$u(x,0) = f(x), \quad -\infty < x < \infty, \tag{IC}$$

where $c = $ const. If we measure the rate of change of u from a moving position given by $x = x(t)$, then, by the chain rule,

$$\frac{d}{dt} u(x(t),t) = u_t(x(t),t) + u_x(x(t),t)x'(t).$$

The first term on the right-hand side above is the change in u at a fixed point x, while the second one is the change in u resulting from the movement of the observation position.

Assuming that $x'(t) = c$, from the PDE we see that

$$\frac{d}{dt} u(x(t),t) = u_t(x(t),t) + cu_x(x(t),t) = 0;$$

that is, $u = $ const as perceived from the moving observation point. The position of this point is obtained by integrating its velocity $x'(t) = c$:

$$x = ct + x_0, \quad x_0 = x(0). \tag{12.1}$$

This formula defines a family of lines in the (x,t)-plane, which are called *characteristics* (see Fig. 12.1). As mentioned above, the characteristics have the property that $u(x,t)$ takes a constant value along each one of them (but, in general, different constant values on different characteristics).

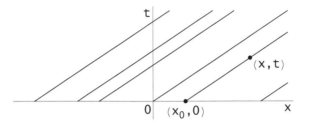

Fig. 12.1. Characteristic lines.

Hence, to find the value of the solution u at (x,t), we consider the characteristic through (x,t), of equation $x = ct + x_0$, which intersects the x-axis at $(x_0,0)$. Since u is constant on this line, its value at (x,t) is the same as at $(x_0,0)$. But the latter is known from the IC, so

$$u(x,t) = u(x_0,0) = f(x_0). \tag{12.2}$$

The parameter x_0 is now replaced from the equation (12.1) of the characteristic line: $x_0 = x - ct$. So, by (12.2), the solution of the given IVP is

$$u(x,t) = f(x - ct).$$

This formula shows that at a fixed time t, the shape of the solution is the same as at $t = 0$, but is shifted by ct along the x-axis. In other words, the shape of the initial data function travels in the positive (negative) x-direction with velocity c if $c > 0$ ($c < 0$), which means that the solution is a wave.

12.1. Example. In the IVP

$$u_t(x,t) + \tfrac{1}{2} u_x(x,t) = 0, \quad -\infty < x < \infty, \ t > 0,$$

$$u(x,0) = \begin{cases} \sin x, & 0 \le x \le \pi, \\ 0 & \text{otherwise,} \end{cases}$$

the ODE of the characteristics is $x'(t) = 1/2$, so the characteristic passing through $x = x_0$ at $t = 0$ has equation $x = t/2 + x_0$. The dotted lines in Fig. 12.2 are the characteristics passing through $x = 0$ and $x = \pi$ at $t = 0$, that is, the lines of equations $x = \frac{1}{2}t$ and $x = \frac{1}{2}t + \pi$, respectively.

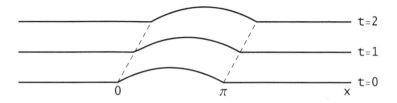

Fig. 12.2. The characteristic lines through $(0,0)$ and $(\pi,0)$.

Since

$$\frac{du}{dt} = u_t + u_x x' = u_t + \frac{1}{2}u_x = 0,$$

the solution u is constant along the characteristics:

$$u(x,t) = u(x_0,0) = \begin{cases} \sin x_0, & 0 \le x_0 \le \pi, \\ 0 & \text{otherwise;} \end{cases}$$

therefore, since $x_0 = x - \frac{1}{2}t$ on the characteristic through (x,t), it follows that

$$u(x,t) = \begin{cases} \sin\left(x - \frac{1}{2}t\right), & 0 \le x - \frac{1}{2}t \le \pi, \\ 0 & \text{otherwise.} \end{cases}$$

This can also be written as

$$u(x,t) = \begin{cases} \sin\left(x - \frac{1}{2}t\right), & \frac{1}{2}t \le x \le \frac{1}{2}t + \pi, \\ 0 & \text{otherwise.} \quad \blacksquare \end{cases}$$

12.2. Example. The velocity of the observation point in the IVP

$$u_t(x,t) + 3t u_x(x,t) = u, \quad -\infty < x < \infty, \ t > 0,$$

$$u(x,0) = \cos x, \quad -\infty < x < \infty,$$

is $x'(t) = 3t$, so the characteristic through (x,t) is $x = \frac{3}{2}t^2 + x_0$, $x_0 = x(0)$. Along this chàracteristic we have

$$\frac{du}{dt} = u_t + u_x x' = u_t + 3t u_x = u,$$

with solution $u(x,t) = Ce^t$, $C = $ const. Since the characteristic through (x,t) also passes through $(x_0,0)$ and $ue^{-t} = C$ is constant on this curve, we use the IC to write

$$C = u(x,t)e^{-t} = u(x_0,0)e^0 = u(x_0,0) = \cos x_0.$$

But $x_0 = x - \frac{3}{2}t^2$ on the characteristic; consequently,

$$u(x,t) = Ce^t = e^t \cos x_0 = e^t \cos(x - \frac{3}{2}t^2). \quad \blacksquare$$

12.3. Example. In the IVP

$$u_t(x,t) + xu_x(x,t) = 1, \quad -\infty < x < \infty, \ t > 0,$$
$$u(x,0) = x^2, \quad -\infty < x < \infty,$$

the velocity of the observation point satisfies the ODE $x'(t) = x$, with general solution $x = ce^t$, $c = $ const. Thus, the characteristic through (x,t) that also passes through $(x_0,0)$ has equation $x = x_0e^t$. On this characteristic,

$$\frac{du}{dt} = u_t + u_x x' = u_t + xu_x = 1,$$

that is, $u(x,t) = t + C$, $C = $ const. Using the IC, we see that

$$C = u(x,t) - t = u(x_0,0) - 0 = u(x_0,0) = x_0^2.$$

The solution of the IVP is now obtained by replacing $x_0 = xe^{-t}$ from the equation of the characteristic:

$$u(x,t) = t + C = t + x_0^2 = t + x^2 e^{-2t}. \quad \blacksquare$$

12.4. Example. The IBVP

$$u_t(x,t) + u_x(x,t) = x, \quad x > 0, \ t > 0,$$
$$u(0,t) = t, \quad t > 0,$$
$$u(x,0) = \sin x, \quad x > 0,$$

needs slightly different handling since here x is restricted to nonnegative values and we also have a BC at $x = 0$. First, using the standard argument, we see that the velocity of the moving observation point satisfies $x'(t) = 1$; therefore, the equation of the family of characteristics is $x = t + c$, $c = $ const. Since this problem is defined in the first quadrant of the (x,t)-plane, we notice (see Fig. 12.3) that if the point (x,t) is above the line $x = t$, then the characteristic passing through this point never reaches the x-axis, so the IC cannot be used for it. However, this characteristic reaches the t-axis, and we can use the BC instead. We split the discussion into three parts.

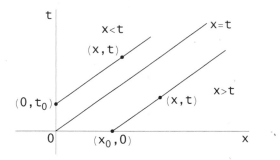

Fig. 12.3. Characteristic lines in the first quadrant.

(i) Let $x > t$. Since the characteristic through (x,t) also passes through $(x_0, 0)$, its equation is written in the form $x = t + c = t + x_0$, and the PDE shows that on this line we have

$$\frac{du}{dt} = u_t + u_x x' = u_t + u_x = x = t + x_0;$$

hence, $u(x,t) = \frac{1}{2}t^2 + x_0 t + C$, $C = $ const. Consequently,

$$C = u(x,t) - \tfrac{1}{2}t^2 - x_0 t = u(x_0, 0) - 0 - 0 = \sin x_0,$$

which, on substituting $x_0 = x - t$ from the equation of the characteristic, yields

$$u(x,t) = \tfrac{1}{2}t^2 + x_0 t + \sin x_0$$
$$= \tfrac{1}{2}t^2 + t(x - t) + \sin(x - t) = tx - \tfrac{1}{2}t^2 + \sin(x - t).$$

(ii) Let $x < t$. Then, since the characteristic through (x,t) also passes through $(0, t_0)$, its equation is written as $x = t + c = t - t_0$ and on it,

$$\frac{du}{dt} = x = t - t_0,$$

with solution

$$u(x,t) = \tfrac{1}{2}t^2 - t_0 t + C, \quad C = \text{const};$$

therefore, using the BC, we deduce that

$$C = u(x,t) - \tfrac{1}{2}t^2 + t_0 t = u(0,t_0) - \tfrac{1}{2}t_0^2 + t_0^2 = t_0 + \tfrac{1}{2}t_0^2.$$

To find the solution at (x,t), we now need to replace the parameter $t_0 = t - x$ from the equation of the characteristic; thus,

$$u(x,t) = \tfrac{1}{2}t^2 - t_0 t + t_0 + \tfrac{1}{2}t_0^2$$
$$= \tfrac{1}{2}t^2 - t(t - x) + t - x + \tfrac{1}{2}(t - x)^2 = \tfrac{1}{2}x^2 - x + t.$$

(iii) We see that as the point (x,t) approaches the line $x = t$ from either side, we obtain the same limiting value $u(x,t) = x^2/2$. Hence, the solution is continuous across this line and we can write

$$u(x,t) = \begin{cases} \tfrac{1}{2}x^2 - x + t, & x \le t, \\ xt - \tfrac{1}{2}t^2 + \sin(x - t), & x > t. \end{cases} \blacksquare$$

12.5. Remark. The continuity of u across the line $x = t$ is due to the continuity of the data at $(0,0)$; that is,

$$\lim_{x \to 0} \sin x = \lim_{t \to 0} t = 0.$$

When this condition is not satisfied in an IBVP of this type, then the solution u is discontinuous across the corresponding dividing line in the (x,t)-plane. Discontinuities—and, in general, any perturbations—in the solution always propagate along the characteristic lines. $\blacksquare$

12.6. Example. The problem

$$u_x(x,y) + u_y(x,y) + 2u(x,y) = 0, \quad -\infty < x, y < \infty,$$
$$u(x,y) = x + 1 \quad \text{on the line } 2x + y + 1 = 0$$

is neither an IVP nor a BVP. However, the method of characteristics works in this case as well. Thus, assuming that $x = x(y)$, we can write

$$\frac{d}{dy} u(x(y),y) = u_y(x(y),y) + u_x(x(y),y)x'(y);$$

so, if $x'(y) = 1$—that is, $x = y + c$, $c = \text{const}$—then the PDE becomes

$$\frac{du}{dy} + 2u = 0,$$

with general solution

$$u(x,y) = Ce^{-2y}, \quad C = \text{const}.$$

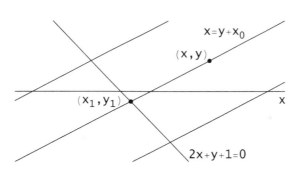

Fig. 12.4. Characteristics and the data line.

The equation of the characteristic through (x,y) and (x_1, y_1) (see Fig. 12.4) is $x = y + x_1 - y_1$ and, in view of the prescribed data, on this line we have

$$u(x,y)e^{2y} = C = u(x_1,y_1)e^{2y_1} = (x_1 + 1)e^{2y_1}.$$

Since the point (x_1, y_1) lies on both the characteristic and the data lines, its coordinates satisfy the system

$$x_1 - y_1 = x - y,$$
$$2x_1 + y_1 = -1,$$

with solution

$$x_1 = \tfrac{1}{3}(x - y - 1),$$
$$y_1 = \tfrac{1}{3}(-2x + 2y - 1).$$

Consequently, the solution of the given problem is

$$u(x,y) = Ce^{-2y} = (x_1 + 1)e^{2(y_1 - y)}$$
$$= \tfrac{1}{3}(x - y + 2)e^{-2(2x+y+1)/3}. \quad \blacksquare$$

12.2. First-Order Quasilinear Equations

A PDE of the form

$$u_t(x,t) + c(x,t,u)u_x(x,t) = q(x,t,u)$$

is called *quasilinear*. Although technically nonlinear, it is linear in the first-order derivatives of u. Such equations arise in the modeling of a variety of phenomena (for example, traffic flow) and can be solved by the method of characteristics.

12.7. Example. Consider the IVP

$$u_t(x,t) + u^3(x,t)u_x(x,t) = 0, \quad -\infty < x < \infty, \ t > 0,$$
$$u(x,0) = x^{1/3}, \quad -\infty < x < \infty.$$

If $x = x(t)$, then the usual procedure leads to the conclusion that the characteristic line through (x,t) and $(x_0,0)$ satisfies

$$x'(t) = u^3(x(t),t), \quad x(0) = x_0.$$

On this line,

$$\frac{du}{dt} = u_t + u_x x' = u_t + u^3 u_x = 0,$$

which, in view of the IC, yields

$$u(x,t) = C = u(x_0,0) = x_0^{1/3}.$$

Hence, the ODE problem for the characteristic line becomes

$$x'(t) = x_0, \quad x(0) = x_0,$$

with solution $x = x_0 t + x_0 = x_0(t+1)$. Since on this line we have $x_0 = x/(t+1)$, we can now write the solution of the given IVP as

$$u(x,t) = x_0^{1/3} = \left(\frac{x}{t+1}\right)^{1/3}. \quad \blacksquare$$

12.8. Example. A similar procedure is used to solve the IVP

$$u_t(x,t) + u(x,t)u_x(x,t) = 2t, \quad -\infty < x < \infty, \ t > 0,$$
$$u(x,0) = x, \quad -\infty < x < \infty.$$

If $x = x(t)$ satisfies $x'(t) = u(x(t),t)$, $x(0) = x_0$, then on the characteristic curve through (x,t) and $(x_0,0)$,

$$\frac{du}{dt} = u_t + u_x x' = u_t + uu_x = 2t.$$

Therefore, $u(x,t) = t^2 + C$, or

$$u(x,t) - t^2 = C = u(x_0,0) - 0 = x_0,$$

so $u(x,t) = t^2 + x_0$. This means that the characteristic curve satisfies

$$x'(t) = t^2 + x_0, \quad x(0) = x_0,$$

with solution

$$x = \tfrac{1}{3}t^3 + x_0 t + x_0 = \tfrac{1}{3}t^3 + x_0(t+1).$$

Since on this curve we have $x_0 = (x - \tfrac{1}{3}t^3)/(t+1) = (3x - t^3)/(3(t+1))$, the solution of the IBVP is

$$u(x,t) = t^2 + \frac{3x - t^3}{3(t+1)}. \quad \blacksquare$$

12.3. The One-Dimensional Wave Equation

We now reconsider the wave equation in terms of details provided by the method of characteristics.

The d'Alembert solution. Formally, we can write the one-dimensional wave equation as

$$u_{tt}(x,t) - c^2 u_{xx}(x,t) = \left(\frac{\partial}{\partial t} + c\frac{\partial}{\partial x}\right)\left(\frac{\partial}{\partial t} - c\frac{\partial}{\partial x}\right)u(x,t)$$

$$= \left(\frac{\partial}{\partial t} + c\frac{\partial}{\partial x}\right)w(x,t)$$

$$= w_t(x,t) + cw_x(x,t) = 0. \tag{12.3}$$

We know from Section 12.1 that the general solution of the equation satisfied by w in (12.3) is

$$w(x,t) = u_t(x,t) - cu_x(x,t) = P(x - ct), \tag{12.4}$$

where P is an arbitrary one-variable function. On the other hand, if we write the wave equation in the alternative form

$$
\begin{aligned}
u_{tt}(x,t) - c^2 u_{xx}(x,t) &= \left(\frac{\partial}{\partial t} - c\frac{\partial}{\partial x}\right)\left(\frac{\partial}{\partial t} + c\frac{\partial}{\partial x}\right)u(x,t) \\
&= \left(\frac{\partial}{\partial t} - c\frac{\partial}{\partial x}\right)v(x,t) \\
&= v_t(x,t) - cv_x(x,t) = 0,
\end{aligned}
$$

then, as above,

$$
v(x,t) = u_t(x,t) + cu_x(x,t) = Q(x + ct), \tag{12.5}
$$

where Q is another arbitrary one-variable function. Adding (12.4) and (12.5) side by side, we find that

$$
u_t(x,t) = \tfrac{1}{2}[P(x - ct) + Q(x + ct)].
$$

Direct integration now yields

$$
u(x,t) = F(x - ct) + G(x + ct), \tag{12.6}
$$

where F and G are arbitrary one-variable functions.

According to the explanation given in Section 12.1, $F(x - ct)$ is a fixed-shape wave traveling to the right with velocity c, and is constant on the characteristics $x - ct = \text{const}$. Similarly, $G(x + ct)$ is a fixed-shape wave traveling to the left with velocity $-c$, and is constant on the characteristics $x + ct = \text{const}$ (see Fig. 12.5). Through every point (x,t) in the $t > 0$ half plane there pass two characteristics, one from each family.

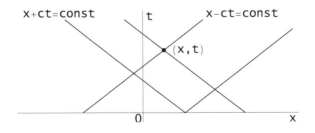

Fig. 12.5. The two families of characteristics.

Consider an infinite vibrating string, which is modeled by the IVP

$$u_{tt}(x,t) = c^2 u_{xx}(x,t), \quad -\infty < x < \infty, \ t > 0,$$
$$u(x,0) = f(x), \quad u_t(x,0) = g(x), \quad -\infty < x < \infty.$$

Differentiating the solution (12.6) of the PDE with respect to t and recalling that F and G are one-variable functions, we obtain

$$u_t(x,t) = -cF'(x - ct) + cG'(x + ct).$$

From the ICs it now follows that

$$f(x) = u(x,0) = F(x) + G(x), \quad g(x) = u_t(x,0) = -cF'(x) + cG'(x).$$

We solve the above equations for F and G. Thus, $F' = f' - G'$, so that $g/c = -f' + 2G'$, from which we easily find that

$$G' = \frac{1}{2}\left(f' + \frac{1}{c}g\right), \quad F' = \frac{1}{2}\left(f' - \frac{1}{c}g\right);$$

hence, by integration,

$$F(x) = \frac{1}{2}f(x) - \frac{1}{2c}\int_0^x g(y)\,dy,$$

$$G(x) = \frac{1}{2}f(x) + \frac{1}{2c}\int_0^x g(y)\,dy.$$

From these expressions and (12.6) we conclude that

$$u(x,t) = \frac{1}{2}[f(x + ct) + f(x - ct)] + \frac{1}{2c}\int_{x-ct}^{x+ct} g(y)\,dy. \tag{12.7}$$

This is *d'Alembert's solution*, which was derived earlier in Section 10.3 by means of the Green's function for the wave equation.

12.9. Example. Suppose that in the above IVP we have

$$u(x,0) = f(x) = \begin{cases} 1, & |x| < h, \\ 0, & |x| > h, \end{cases} \quad u_t(x,0) = g(x) = 0.$$

By (12.7), the solution is

$$u(x,t) = \tfrac{1}{2}[f(x - ct) + f(x + ct)],$$

where

$$\tfrac{1}{2}f(x - ct) = \begin{cases} \tfrac{1}{2}, & |x - ct| < h, \text{ or } -h + ct < x < h + ct, \\ 0 & \text{otherwise,} \end{cases}$$

$$\tfrac{1}{2}f(x + ct) = \begin{cases} \tfrac{1}{2}, & |x + ct| < h, \text{ or } -h - ct < x < h - ct, \\ 0 & \text{otherwise.} \end{cases}$$

Thus, the solution is the sum of two pulses of amplitude $1/2$, which move away from each other with velocity $2c$. Since their endpoints are initially $2h$ apart, they separate after $t = h/c$. ∎

12.10. Example. In the case of the IVP

$$u_{tt}(x,t) = c^2 u_{xx}(x,t), \quad -\infty < x < \infty, \ t > 0,$$
$$u(x,0) = \sin x, \quad u_t(x,0) = 0, \quad -\infty < x < \infty,$$

d'Alembert's solution (12.7) yields

$$u(x,t) = \tfrac{1}{2}\big[\sin(x + ct) + \sin(x - ct)\big] = \sin x \cos(ct).$$

Waves represented by solutions like this, where the variables separate, are called *standing waves*. ∎

12.11. Example. Also by (12.7), the solution of the IVP

$$u_{tt}(x,t) = c^2 u_{xx}(x,t), \quad -\infty < x < \infty, \ t > 0,$$
$$u(x,0) = 0, \quad u_t(x,0) = \sin x, \quad -\infty < x < \infty,$$

is

$$u(x,t) = \frac{1}{2c} \int_{x-ct}^{x+ct} \sin y \, dy$$

$$= \frac{1}{2c}\big[\cos(x - ct) - \cos(x + ct)\big] = \frac{1}{c}\sin x \sin(ct).$$

This solution also represents standing waves. ∎

12.12. Example. For the IVP

$$u_{tt}(x,t) = 4u_{xx}(x,t), \quad -\infty < x < \infty, \ t > 0,$$
$$u(x,0) = x, \quad u_t(x,0) = 2x^2, \quad -\infty < x < \infty,$$

d'Alembert's formula with $c = 2$ produces the solution

$$u(x,t) = \tfrac{1}{2}\left[(x - 2t) + (x + 2t)\right] + \tfrac{1}{4}\int_{x-2t}^{x+2t} 2y^2\, dy$$

$$= x + 2x^2 t + \tfrac{8}{3}t^3. \ \blacksquare$$

The semi-infinite vibrating string. Consider the IBVP

$$u_{tt}(x,t) = c^2 u_{xx}(x,t), \quad x > 0, \ t > 0,$$
$$u(0,t) = 0, \quad t > 0,$$
$$u(x,0) = f(x), \quad u_t(x,0) = g(x), \quad x > 0.$$

As in the preceding case, from the PDE we obtain

$$u(x,t) = F(x - ct) + G(x + ct),$$

where

$$F(x) = \frac{1}{2}f(x) - \frac{1}{2c}\int_0^x g(y)\, dy, \quad x > 0,$$

$$G(x) = \frac{1}{2}f(x) + \frac{1}{2c}\int_0^x g(y)\, dy, \quad x > 0.$$

Since $x > 0$ in this problem, the functions F and G are determined only for positive values of their arguments. This does not affect $G(x + ct)$, since $t > 0$. But the argument of $F(x - ct)$ is negative if $0 < x < ct$. To obtain $F(x - ct)$ for $x - ct < 0$, we use the BC. Thus,

$$u(0,t) = 0 = F(-ct) + G(ct), \quad t > 0,$$

so for $\xi < 0$ we have $F(\xi) = -G(-\xi)$, which means that the solution for $0 < x < ct$ is

$$u(x,t) = F(x - ct) + G(x + ct) = -G(ct - x) + G(x + ct)$$

$$= \frac{1}{2}[f(x + ct) - f(ct - x)] + \frac{1}{2c}\left[\int_0^{x+ct} g(y)\,dy - \int_0^{ct-x} g(y)\,dy\right]$$

$$= \frac{1}{2}[f(ct + x) - f(ct - x)] + \frac{1}{2c}\int_{ct-x}^{ct+x} g(y)\,dy. \tag{12.8}$$

The term $-G(ct - x)$ is a fixed-shape wave that travels to the right and is called the *reflected wave*.

For $x > ct$, the solution of the problem is given by d'Alembert's formula, as before.

12.13. Example. The solution of the IBVP

$$u_{tt}(x,t) = 4u_{xx}(x,t), \quad x > 0, \ t > 0,$$
$$u(0,t) = 0, \quad t > 0,$$
$$u(x,0) = 4x, \quad u_t(x,0) = 2x + 6, \quad x > 0,$$

is computed in two stages. First, for $0 < x < 2t$ we use (12.8) with $c = 2$ to find that

$$u(x,t) = \tfrac{1}{2}[4(2t + x) - 4(2t - x)] + \tfrac{1}{4}\int_{2t-x}^{2t+x}(2y + 6)\,dy$$

$$= 2xt + 7x;$$

then, by (12.7), for $x > 2t$ we have

$$u(x,t) = \tfrac{1}{2}[4(x + 2t) + 4(x - 2t)] + \tfrac{1}{4}\int_{x-2t}^{x+2t}(2y + 6)\,dy$$

$$= 2xt + 4x + 6t.$$

The solution is continuous across the characteristic line $x = 2t$ since

$$\lim_{x \to 0} u(x,0) = \lim_{t \to 0} u(0,t) = 0.$$

Thus, we can write

$$u(x,t) = \begin{cases} 2xt + 7x, & 0 \le x \le 2t, \\ 2xt + 4x + 6t, & x > 2t. \end{cases} \ \blacksquare$$

The finite string. Consider the IBVP

$$u_{tt}(x,t) = c^2 u_{xx}(x,t), \quad 0 < x < L, \ t > 0,$$

$$u(0,t) = 0, \quad u(L,t) = 0, \quad t > 0,$$

$$u(x,0) = f(x), \quad u_t(x,0) = g(x), \quad 0 < x < L.$$

Using separation of variables, in Section 5.2 we obtained the solution

$$u(x,t) = \sum_{n=1}^{\infty} \sin \frac{n\pi x}{L} \left(b_{1n} \cos \frac{n\pi ct}{L} + b_{2n} \sin \frac{n\pi ct}{L} \right),$$

where the coefficients b_{1n} and b_{2n} are determined from the expansions

$$f(x) = \sum_{n=1}^{\infty} b_{1n} \sin \frac{n\pi x}{L},$$

$$g(x) = \sum_{n=1}^{\infty} b_{2n} \frac{n\pi c}{L} \sin \frac{n\pi x}{L}.$$

Suppose that $f \neq 0$ and $g = 0$; then

$$b_{2n} = 0, \quad n = 1, 2, \ldots.$$

Using the formula

$$\sin \alpha \cos \beta = \tfrac{1}{2} [\sin(\alpha + \beta) + \sin(\alpha - \beta)],$$

we write the solution in the form

$$u(x,t) = \tfrac{1}{2} \sum_{n=1}^{\infty} b_{1n} \left[\sin \frac{n\pi(x+ct)}{L} + \sin \frac{n\pi(x-ct)}{L} \right]$$

$$= \tfrac{1}{2} [f(x+ct) + f(x-ct)].$$

Suppose now that $f = 0$ and $g \neq 0$; then $b_{1n} = 0$, $n = 1, 2, \ldots$. Using the formula

$$\sin \alpha \sin \beta = \tfrac{1}{2} [\cos(\alpha - \beta) - \cos(\alpha + \beta)],$$

we rewrite the solution as

$$u(x,t) = \tfrac{1}{2} \sum_{n=1}^{\infty} b_{2n} \left[\cos \frac{n\pi(x-ct)}{L} - \cos \frac{n\pi(x+ct)}{L} \right].$$

On the other hand,

$$
\int_{x-ct}^{x+ct} g(y)\,dy = \sum_{n=1}^{\infty} cb_{2n} \int_{x-ct}^{x+ct} \frac{n\pi}{L} \sin \frac{n\pi x}{L}\,dx
$$

$$
= c \sum_{n=1}^{\infty} b_{2n} \left[\cos \frac{n\pi(x-ct)}{L} - \cos \frac{n\pi(x+ct)}{L} \right];
$$

hence,

$$
u(x,t) = \frac{1}{2c} \int_{x-ct}^{x+ct} g(y)\,dy.
$$

Combining the two separate solutions and using the superposition principle, we now regain d'Alembert's formula (12.7).

12.4. Other Hyperbolic Equations

The method set out in the preceding section can be extended to more general hyperbolic equations. We illustrate how this is done in two particular cases.

One-dimensional waves. The solution of the type of problem discussed here is based on a decomposition of the PDE operator which is similar to that performed for the wave equation.

12.14. Example. The IVP

$$
u_{tt}(x,t) + 5u_{xt}(x,t) + 6u_{xx}(x,t) = 0, \quad -\infty < x < \infty,\ t > 0,
$$
$$
u(x,0) = x + 2, \quad u_t(x,0) = 2x, \quad -\infty < x < \infty,
$$

is hyperbolic because $B^2 - 4AC = 25 - 24 = 1 > 0$. It is easily verified that the PDE can be rewritten alternatively as

$$
\left(\frac{\partial}{\partial t} + 2\frac{\partial}{\partial x} \right)\left(\frac{\partial}{\partial t} + 3\frac{\partial}{\partial x} \right) u(x,t) = \left(\frac{\partial}{\partial t} + 3\frac{\partial}{\partial x} \right)\left(\frac{\partial}{\partial t} + 2\frac{\partial}{\partial x} \right) u(x,t) = 0.
$$

Setting

$$
u_t + 3u_x = w, \quad u_t + 2u_x = v,
$$

we have

$$
w_t + 2w_x = 0, \quad v_t + 3v_x = 0.
$$

By the argument developed in Section 12.1, the general solutions of these equations are, respectively,

$$w(x,t) = P(x - 2t),$$
$$v(x,t) = Q(x - 3t),$$

where P and Q are arbitrary one-variable functions; hence,

$$u_t + 3u_x = P(x - 2t),$$
$$u_t + 2u_x = Q(x - 3t).$$

The elimination of u_x between these equations now yields

$$u_t(x,t) = 3Q(x - 3t) - 2P(x - 2t),$$

from which, by integration, we find that

$$u(x,t) = F(x - 2t) + G(x - 3t),$$

where F and G are a new pair of arbitrary one-variable functions. We remark that, by the chain rule, the time derivative of u is

$$u_t(x,t) = -2F'(x - 2t) - 3G'(x - 3t). \tag{12.9}$$

At this point we apply the ICs and arrive at the equations

$$F(x) + G(x) = x + 2,$$
$$-2F'(x) - 3G'(x) = 2x.$$

Differentiating the first equation term by term, we obtain $F'(x) + G'(x) = 1$, which, combined with the second equation above, leads to $G'(x) = -2x - 2$. This allows us to determine G and then F:

$$G(x) = -x^2 - 2x + c, \quad c = \text{const},$$
$$F(x) = x + 2 - G(x) = x^2 + 3x + 2 - c.$$

Replacing in (12.9), we conclude that the solution of the given IVP is

$$u(x,t) = \left[(x - 2t)^2 + 3(x - 2t) + 2 - c\right] + \left[-(x - 3t)^2 - 2(x - 3t) + c\right]$$
$$= 2xt - 5t^2 + x + 2. \quad \blacksquare$$

12.15. Example. Consider the IBVP

$$u_{tt}(x,t) - 3u_{xt}(x,t) - 4u_{xx}(x,t) = 0, \quad x > 0, \ t > 0,$$

$$u(0,t) = 0, \quad t > 0,$$

$$u(x,0) = 1 - x^2, \quad u_t(x,0) = x + 2, \quad x > 0.$$

Since $B^2 - 4AC = 9 + 16 = 25 > 0$, the PDE is hyperbolic. Rewriting the equation as

$$\left(\frac{\partial}{\partial t} + \frac{\partial}{\partial x} \right) \left(\frac{\partial}{\partial t} - 4\frac{\partial}{\partial x} \right) u(x,t) = \left(\frac{\partial}{\partial t} - 4\frac{\partial}{\partial x} \right) \left(\frac{\partial}{\partial t} + \frac{\partial}{\partial x} \right) u(x,t) = 0$$

and proceeding as in Example 12.14, we find that

$$u(x,t) = F(x - t) + G(x + 4t), \tag{12.10}$$

where

$$\begin{aligned} F(x) &= \tfrac{1}{10}(-9x^2 - 4x + 10) - c, \quad c = \text{const}, \\ G(x) &= \tfrac{1}{10}(-x^2 + 4x) + c \end{aligned} \tag{12.11}$$

are defined for positive values of their arguments.

If $x > t$, then $x - t > 0$ and we replace (12.11) in (12.10):

$$\begin{aligned} u(x,t) &= \tfrac{1}{10}\big[-9(x - t)^2 - 4(x - t) + 10 - (x + 4t)^2 + 4(x + 4t) \big] \\ &= -x^2 + xt - \tfrac{5}{2}t^2 + 2t + 1. \end{aligned}$$

If $0 \le x < t$, then $x - t < 0$ and we make use of the BC:

$$0 = u(0,t) = F(-t) + G(4t) = F(-t) + G(-4(-t)),$$

which implies that $F(z) = -G(-4z)$ for $z < 0$; so, by (12.10) and (12.11),

$$\begin{aligned} u(x,t) &= -G(-4x + 4t) + G(x + 4t) \\ &= \tfrac{1}{10}\big[(-4x + 4t)^2 - 4(-4x + 4t) - (x + 4t)^2 + 4(x + 4t)\big] \\ &= \tfrac{3}{2}x^2 - 4xt + 2x. \end{aligned}$$

The two cases can be written together in the form

$$u(x,t) = \begin{cases} \tfrac{3}{2}x^2 - 4xt + 2x, & 0 \le x < t, \\ -x^2 + xt - \tfrac{5}{2}t^2 + 2t + 1, & x > t. \end{cases}$$

We notice that the solution is discontinuous across the line $x = t$. ∎

12.16. Remark. A d'Alembert-type formula can also be derived for PDEs of the kind studied in Examples 12.14 and 12.15. ∎

Spherical waves. The three-dimensional wave equation is

$$u_{tt} = c^2 \Delta u, \tag{12.12}$$

where the Laplacian Δ is expressed in terms of a system of coordinates appropriate for the geometry of the problem. If the amplitude of the waves depends only on the distance r of the wave front from a point source, then we choose spherical coordinates, in which case $u = u(r,t)$ and, by Remark 4.11(v),

$$\Delta u = r^{-2}(r^2 u_r)_r = r^{-2}(2r u_r + r^2 u_{rr}) = u_{rr} + 2r^{-1} u_r.$$

Therefore, the equation governing the propagation of spherical waves is

$$u_{tt}(r,t) = c^2 \big[u_{rr}(r,t) + 2r^{-1} u_r(r,t) \big].$$

Multiplying this equality by r and manipulating the right-hand side, we find that

$$r u_{tt} = c^2 \big[(r u_{rr} + u_r) + u_r \big] = c^2 \big[(r u_r)_r + u_r \big]$$
$$= c^2 (r u_r + u)_r = c^2 ((ru)_r)_r = c^2 (ru)_{rr}.$$

The substitution $ru(r,t) = v(r,t)$ now reduces the above equation to

$$v_{tt}(r,t) = c^2 v_{rr}(r,t),$$

so, by analogy with the argument developed in Section 12.1,

$$v(r,t) = F(r - ct) + G(r + ct).$$

Consequently, the general solution of (12.12) is

$$u(r,t) = \frac{1}{r} \big[F(r - ct) + G(r + ct) \big],$$

where F and G are arbitrary one-variable functions.

Exercises

In (1)–(10) use the method of characteristics to solve the IVP

$$u_t(x,t) + c(x,t)u_x(x,t) = q(x,t,u), \quad -\infty < x < \infty, \ t > 0,$$
$$u(x,0) = f(x), \quad -\infty < x < \infty,$$

with the functions c, q, and f as indicated. In each case sketch the family of characteristics in the (x,t)-plane.

(1) $c(x,t) = 2, \quad q(x,t,u) = t, \quad f(x) = 1 - x.$

(2) $c(x,t) = -2t, \quad q(x,t,u) = 2, \quad f(x) = x^3.$

(3) $c(x,t) = 2t + 1, \quad q(x,t,u) = 2u, \quad f(x) = \sin x.$

(4) $c(x,t) = t + 2, \quad q(x,t,u) = u + 2t, \quad f(x) = x + 2.$

(5) $c(x,t) = x, \quad q(x,t,u) = e^{-t}, \quad f(x) = x^2 + 1.$

(6) $c(x,t) = x - 1, \quad q(x,t,u) = 3t - 2, \quad f(x) = 2x - 1.$

(7) $c(x,t) = -3, \quad q(x,t,u) = 1 - x, \quad f(x) = e^{-2x}.$

(8) $c(x,t) = x + 2, \quad q(x,t,u) = 2x + t^2, \quad f(x) = 2x + 3.$

(9) $c(x,t) = 4, \quad q(x,t,u) = u + x + t, \quad f(x) = \cos(2x).$

(10) $c(x,t) = -2x, \quad q(x,t,u) = u - 2x, \quad f(x) = 2 - x.$

In (11)–(16) use the method of characteristics to solve the IBVP

$$u_t(x,t) + cu_x(x,t) = q(x,t,u), \quad x > 0, \ t > 0,$$
$$u(0,t) = g(t), \quad t > 0,$$
$$u(x,0) = f(x), \quad x > 0,$$

with the constant c and functions q, f, and g as indicated. In each case sketch the family of characteristics in the (x,t)-plane.

(11) $c = 1/2, \quad q(x,t,u) = u, \quad g(t) = 1, \quad f(x) = e^x.$

(12) $c = 1, \quad q(x,t,u) = u - 1, \quad g(t) = t - 1, \quad f(x) = x^2.$

(13) $c = 2, \quad q(x,t,u) = x^2, \quad g(t) = t^2 + 1, \quad f(x) = x.$

(14) $c = 3/2, \quad q(x,t,u) = x + t, \quad g(t) = e^t, \quad f(x) = 1 - x.$

(15) $c = 1, \quad q(x,t,u) = u + x, \quad g(t) = 2t + 1, \quad f(x) = \cos x.$

(16) $c = 2, \quad q(x,t,u) = 2u - t, \quad g(t) = t^2, \quad f(x) = 2x - 1.$

In (17)–(22) use the method of characteristics to solve the problem

$$u_y(x,y) + cu_x(x,y) = q(x,y,u), \quad -\infty < x, y < \infty,$$
$$u(x,y) = f(x,y) \quad \text{on the line } ax + by + d = 0,$$

with the constant c, functions q and f, and line $ax+by+d = 0$ as indicated. In each case sketch the family of characteristics and the data line in the (x,y)-plane.

(17) $c = 2$, $q(x,y,u) = u$, $f(x,y) = 2x - y$, $x + y - 1 = 0$.

(18) $c = -1$, $q(x,y,u) = 2y$, $f(x,y) = xy$, $x - 2y - 1 = 0$.

(19) $c = 1/2$, $q(x,y,u) = 2u - y$, $f(x,y) = x + y$, $2x + 3y - 2 = 0$.

(20) $c = 1$, $q(x,y,u) = u + x$, $f(x,y) = 2x - y - 1$, $2x + y - 3 = 0$.

(21) $c = -2$, $q(x,y,u) = u - x - y$, $f(x,y) = x - 2xy$, $x - y - 2 = 0$.

(22) $c = -1/2$, $q(x,y,u) = 2x + y$, $f(x,y) = e^{x-y}$, $3x - y - 3 = 0$.

In (23)–(28) use the method of characteristics to solve the quasilinear IVP

$$u_t(x,t) + c(x,t,u)u_x(x,t) = q(x,t,u), \quad -\infty < x < \infty, \ t > 0,$$
$$u(x,0) = f(x), \quad -\infty < x < \infty,$$

with the functions c, q, and f as indicated.

(23) $c(x.t,u) = -2u - 1$, $q(x,t,u) = 3$, $f(x) = 1 - x$.

(24) $c(x.t,u) = u - 1$, $q(x,t,u) = t + 1$, $f(x) = 2x$.

(25) $c(x.t,u) = u + t$, $q(x,t,u) = 1$, $f(x) = x + 1$.

(26) $c(x.t,u) = 1 + 2t - u$, $q(x,t,u) = 4t - 1$, $f(x) = 2 - 3x$.

(27) $c(x.t,u) = 1$, $q(x,t,u) = 2tu^2$, $f(x) = -e^{-x}$.

(28) $c(x.t,u) = (t + 2)u$, $q(x,t,u) = u + 1$, $f(x) = x$.

In (29)–(36) use the method of operator decomposition to solve the IVP

$$au_{tt}(x,t) + bu_{xt}(x,t) + cu_{xx}(x,t) = 0, \quad -\infty < x < \infty, \ t > 0,$$
$$u(x,0) = f(x), \quad u_t(x,0) = g(x), \quad -\infty < x < \infty,$$

with the constants a, b, and c and the functions f and g as indicated. (Do not use the d'Alembert formula directly.)

(29) $a = 1$, $b = 0$, $c = -1$, $f(x) = 2x + 3$, $g(x) = x^2 + 1$.

(30) $a = 1$, $b = 0$, $c = -9$, $f(x) = x^2 + x$, $g(x) = 2 - x$.

(31) $a = 1$, $b = 0$, $c = -1/4$, $f(x) = e^{-2x}$, $g(x) = x - x^2$.

(32) $a = 1$, $b = 0$, $c = -4$, $f(x) = x^2 - 3x + 1$, $g(x) = \sin x$.

(33) $a = 1$, $b = 1$, $c = -2$, $f(x) = x^2$, $g(x) = 2x + 1$.

(34) $a = 1$, $b = 1$, $c = -6$, $f(x) = 1 - 2x$, $g(x) = 3x^2$.

(35) $a = 1$, $b = 4$, $c = 3$, $f(x) = x^2 - x$, $g(x) = 4x - 2$.

(36) $a = 2$, $b = -7$, $c = -4$, $f(x) = 3x + 1$, $g(x) = 1 - 2x$.

In (37)–(44) use the method of operator decomposition to solve the IBVP

$$au_{tt}(x,t) + bu_{xt}(x,t) + cu_{xx}(x,t) = 0, \quad x > 0, \ t > 0,$$
$$u(0,t) = 0, \quad t > 0,$$
$$u(x,0) = f(x), \quad u_t(x,0) = g(x), \quad x > 0,$$

with the constants a, b, and c and the functions f and g as indicated. (Do not use the d'Alembert formula directly.)

(37) $a = 1$, $b = 0$, $c = -1$, $f(x) = 2x^2 - x$, $g(x) = 4x + 1$.

(38) $a = 1$, $b = 0$, $c = -1/4$, $f(x) = x + 2$, $g(x) = 3x^2$.

(39) $a = 1$, $b = 0$, $c = -9$, $f(x) = 1 - x^2$, $g(x) = \cos x$.

(40) $a = 1$, $b = 0$, $c = -4$, $f(x) = e^{-x} + 2$, $g(x) = 3x^2 - 2x$.

(41) $a = 1$, $b = 2$, $c = -3$, $f(x) = x + 1$, $g(x) = 2x$.

(42) $a = 1$, $b = 2$, $c = -8$, $f(x) = 2x - 1$, $g(x) = 4x + 1$.

(43) $a = 2$, $b = 1$, $c = -1$, $f(x) = x^2$, $g(x) = 2x + 3$.

(44) $a = 2$, $b = -3$, $c = -1$, $f(x) = x^2 + 1$, $g(x) = x - 3$.

Chapter 13
Perturbation and Asymptotic Methods

If nothing else works do this.

Owing to the complexity of the PDEs involved in some mathematical models, it is not always possible to find an exact solution to an initial/boundary value problem. The next best thing in such situations is to compute an approximate solution instead. This is the idea behind the method of asymptotic expansion, which is applicable to problems that depend on a small positive parameter and relies on the expansion of the solution in a series of powers of the parameter. If the series converges, then the technique is called a *perturbation method;* when the series diverges but is asymptotic (in a sense that will be explained below), we have an *asymptotic method.*

In what follows we discuss only the formal construction of the series solution and obtain the first few terms, without considering the question of convergence.

13.1. Asymptotic Series

In order to state what an asymptotic series is, we need a mechanism to compare the "magnitude" of functions.

13.1. Definition. Let f and g be two functions of a real variable x. We say that f *is of order* g *near* $x = a$, and write

$$f(x) = O\big(g(x)\big) \quad \text{as } x \to a,$$

if

$$\left| \frac{f(x)}{g(x)} \right| \text{ is bounded as } x \to a.$$

We also write

$$f(x) = o\big(g(x)\big) \quad \text{as } x \to a$$

if

$$\frac{f(x)}{g(x)} \to 0 \quad \text{as } x \to a. \ \blacksquare$$

13.2. Examples. (i) The function $\sin x$ is of order x near $x = 0$ since $\lim_{x \to 0} (\sin x)/x = 1$; therefore, $(\sin x)/x$ is bounded for x close to 0.

(ii) We have $x^2 \ln |x| = o(x)$ near $x = 0$ since $\lim_{x \to 0} (x^2 \ln |x|)/x = 0$.

(iii) Similarly, $e^{-1/|x|} = o(x^n)$ near $x = 0$ for any positive integer n because $\lim_{x \to 0} (e^{-1/|x|})/x^n = 0$. ∎

13.3. Definition. A function $f(x, \varepsilon)$, where $0 < \varepsilon \ll 1$ is a small parameter, is said to have the *asymptotic (power) series*

$$f(x, \varepsilon) \approx \sum_{n=0}^{\infty} f_n(x) \varepsilon^n \quad \text{as } \varepsilon \to 0+$$

if for any positive integer N

$$f(x, \varepsilon) = \sum_{n=0}^{N-1} f_n(x) \varepsilon^n + O(\varepsilon^N) \quad \text{as } \varepsilon \to 0+, \tag{13.1}$$

uniformly for x in some interval. ∎

13.4. Remarks. (i) Equality (13.1) means that the remainder after N terms is of order ε^N as $\varepsilon \to 0+$; it can also be written as

$$f(x, \varepsilon) = \sum_{n=0}^{N-1} f_n(x) \varepsilon^n + o(\varepsilon^{N-1}) \quad \text{as } \varepsilon \to 0+ .$$

(ii) The right-hand side of (13.1) may diverge as $N \to \infty$, but it yields a good approximation of $f(x, \varepsilon)$ when N is fixed and $\varepsilon > 0$ is very small. We do not need convergence for the result to be acceptable. Also, the approximation may not get better if we take additional terms because, as mentioned above, the series may be divergent.

(iii) In an asymptotic series it is assumed that the terms containing higher powers of ε are much smaller than those with lower powers.

(iv) In this chapter we assume that asymptotic series may be differentiated and integrated term by term.

(v) If $k \gg 1$ is a large parameter, then we can reduce the problem to one with a small parameter by setting $k = 1/\varepsilon$. ∎

13.5. Definition. Consider an initial/boundary value problem that depends (smoothly) on a small parameter $\varepsilon > 0$. The problem obtained by setting $\varepsilon = 0$ in the equation and data functions is called the *reduced (unperturbed) problem*. If the reduced problem is of the same type and order as the given one and both have unique solutions, then the given problem is called a *regular perturbation problem*; otherwise, it is called a *singular perturbation problem*. ■

13.6. Example. The IBVP

$$u_t(x,t) = k u_{xx}(x,t) + \varepsilon u_x(x,t), \quad 0 < x < L, \ t > 0,$$
$$u(0,t) = 0, \quad u(L,t) = 0, \quad t > 0,$$
$$u(x,0) = f(x), \quad 0 < x < L,$$

is a regular perturbation problem since both this IBVP and its reduced version (involving the heat equation) are second-order parabolic problems with unique solutions. ■

13.7. Example. The signalling (hyperbolic) problem

$$\varepsilon u_{tt}(x,t) - c^2 u_{xx}(x,t) + u_t(x,t) = 0, \quad x > 0, \ -\infty < t < \infty,$$
$$u(0,t) = f(t), \quad -\infty < t < \infty,$$

reduces to a parabolic one when we set $\varepsilon = 0$. Therefore, although both the given and reduced problems have unique solutions, the given IVP is a singular perturbation problem. ■

13.8. Example. Let D be a finite region bounded by a smooth, closed, simple curve ∂D in the (x,y)-plane, and let n be the unit outward normal to ∂D. The fourth-order BVP

$$(\Delta\Delta u)(x,y) = 0, \quad (x,y) \text{ in } D,$$
$$u(x,y) = f(x,y), \quad \varepsilon u_n(x,y) + u(x,y) = g(x,y), \quad (x,y) \text{ on } \partial D,$$

where $u_n = \partial u/\partial n$, reduces, when we set $\varepsilon = 0$, to the BVP

$$(\Delta\Delta u)(x,y) = 0, \quad (x,y) \text{ in } D,$$
$$u(x,y) = f(x,y), \quad u(x,y) = g(x,y), \quad (x,y) \text{ on } \partial D.$$

If $f \neq g$, then the reduced problem has no solution. On the other hand, if $f = g$, then the solution is not unique, since we have lost one of the BCs. Hence, the given BVP is a singular perturbation problem. ■

13.9. Definition. If the solution u of a perturbation problem has the asymptotic series $u \approx \sum_{n=0}^{\infty} u_n \varepsilon^n$, then $u - u_0$ is called a *perturbation* of the solution of the reduced problem. ■

13.2. Regular Perturbation Problems

The best way to see how the method works in this case is to examine a few specific models.

13.10. Example. Let

$$
D = \{(r,\theta) : 0 \leq r < 1, \ -\pi \leq \theta < \pi\},
$$
$$
\partial D = \{(r,\theta) : r = 1, \ -\pi \leq \theta < \pi\}
$$

be the unit disk (with the center at the origin) and its circular boundary, and consider the Dirichlet problem for the two-dimensional Helmholtz equation

$$
(\Delta u)(r,\theta) + \varepsilon u(r,\theta) = 0, \quad (r,\theta) \text{ in } D, \ r \neq 0,
$$
$$
u(r,\theta) = 1, \quad (r,\theta) \text{ on } \partial D,
$$

where $0 < \varepsilon \ll 1$, which ensures that the above BVP has a unique solution. The reduced problem (for $\varepsilon = 0$) is the Dirichlet problem for the Laplace equation, also uniquely solvable. Both the reduced and the perturbed problems are elliptic and of second order, so the given BVP is a regular perturbation problem.

Assuming a perturbation series of the form $u(r,\theta) \approx \sum_{n=0}^{\infty} u_n(r,\theta) \varepsilon^n$, from the PDE we find that

$$
\Delta u + \varepsilon u \approx \sum_{n=0}^{\infty} (\Delta u_n) \varepsilon^n + \sum_{n=0}^{\infty} u_n \varepsilon^{n+1}
$$

$$
= \Delta u_0 + \sum_{n=1}^{\infty} (\Delta u_n + u_{n-1}) \varepsilon^n = 0 \quad \text{in } D, \ r \neq 0.
$$

At the same time, the BC yields

$$u \approx u_0 + \sum_{n=1}^{\infty} u_n \varepsilon^n = 1 \quad \text{on } \partial D.$$

Equating the coefficients of every power of ε on both sides, we then obtain the BVPs

$$\Delta u_0 = 0 \quad \text{in } D, \ r \neq 0,$$
$$u_0 = 1 \quad \text{on } \partial D,$$

and, for $n \geq 1$,

$$\Delta u_n = -u_{n-1} \quad \text{in } D, \ r \neq 0,$$
$$u_n = 0 \quad \text{on } \partial D.$$

Since the region where the problem is posed and the boundary data are independent of the polar angle θ, we may assume that $u_n = u_n(r)$; consequently, by Remark 4.11(ii), the problem for u_0 can be written as

$$(\Delta u_0)(r) = u_0''(r) + r^{-1} u_0'(r) = 0, \quad 0 < r < 1,$$
$$u_0(1) = 1.$$

Multiplying the differential equation by r and noting that the left-hand side of the new ODE is, in fact, $(r u_0')'$, we find that

$$u_0(r) = C_1 \ln r + C_2, \quad C_1, C_2 = \text{const}.$$

We know (see Section 5.3) that for this type of problem we also have additional conditions, generated by physical considerations. Since we have assumed that the solution is independent of θ, the only other condition of this kind to be satisfied here is that the solution and its derivative be continuous (hence, bounded) in D. This implies that $C_1 = 0$; the value of C_2 is then found from the BC at $r = 1$, which yields $u_0(r) = 1$.

Next, applying the same argument to the BVP satisfied by u_1, namely,

$$(\Delta u_1)(r) = u_1''(r) + r^{-1} u_1'(r) = -u_0(r) = -1, \quad 0 < r < 1,$$
$$u_1(1) = 0,$$

we find that $u_1(r) = \frac{1}{4}(1 - r^2)$; hence,

$$u(r) = 1 + \frac{1}{4}\varepsilon(1 - r^2) + O(\varepsilon^2). \tag{13.2}$$

For this problem we can actually see how good an approximation (13.2) is. The given (perturbed) equation can be written in the form

$$u''(r) + r^{-1}u'(r) + \varepsilon u(r) = 0,$$

which is Bessel's equation of order zero (see (3.12) with $m = 0$, $\lambda = \varepsilon$, and x replaced by r). Its bounded solution satisfying $u(1) = 1$ is

$$u(r) = J_0(\sqrt{\varepsilon}r)/J_0(\sqrt{\varepsilon}),$$

where J_0 is the Bessel function of the first kind and order zero. This formula makes sense because $0 < \varepsilon \ll 1$; that is, $\sqrt{\varepsilon}$ is smaller than the first zero ($\xi \cong 2.4$) of $J_0(\xi)$. Since for small values of ξ

$$J_0(\xi) = 1 - \tfrac{1}{4}\xi^2 + O(\xi^4),$$

we use the formula for the sum of an infinite geometric progression with ratio q, $0 < |q| < 1$, namely,

$$1 + q + q^2 + \cdots + q^n + \cdots = \frac{1}{1-q},$$

to find that

$$\frac{J_0(\sqrt{\varepsilon}r)}{J_0(\sqrt{\varepsilon})} = \frac{1 - \tfrac{1}{4}\varepsilon r^2 + O(\varepsilon^2)}{1 - \tfrac{1}{4}\varepsilon + O(\varepsilon^2)} = \frac{1 - \tfrac{1}{4}\varepsilon r^2 + O(\varepsilon^2)}{1 - \left(\tfrac{1}{4}\varepsilon + O(\varepsilon^2)\right)}$$

$$= \left[1 - \tfrac{1}{4}\varepsilon r^2 + O(\varepsilon^2)\right]\left[1 + \left(\tfrac{1}{4}\varepsilon + O(\varepsilon^2)\right) + O(\varepsilon^2)\right]$$

$$= 1 + \tfrac{1}{4}\varepsilon(1 - r^2) + O(\varepsilon^2).$$

Thus, the perturbation solution (13.2) coincides with the exact solution to $O(\varepsilon)$-terms throughout D. ∎

13.11. Example. The elliptic nonlinear BVP

$$\Delta u(r,\theta) + \varepsilon u^2(r,\theta) = 36r, \quad (r,\theta) \text{ in } D,$$

$$u(r,\theta) = 4, \quad (r,\theta) \text{ on } \partial S,$$

where $0 < \varepsilon \ll 1$ and D and ∂D are the same as in Example 13.10, is a regular perturbation problem. Since we again notice that the solution u is expected to depend only on r, we assume for the solution a series of the form $u(r) \approx \sum\limits_{n=0}^{\infty} u_n(r)\varepsilon^n$. Then, in the usual way, taking into account the asymptotic expansion

$$u^2 = [u_0 + \varepsilon u_1 + O(\varepsilon^2)]^2 = u_0^2 + 2\varepsilon u_0 u_1 + O(\varepsilon^2)$$

and recalling the boundedness condition mentioned in Example 13.10, we see that u_0 and u_1 are, respectively, the solutions of the linear BVPs

$$u_0''(r) + r^{-1}u_0'(r) = 36r, \quad 0 < r < 1,$$
$$u_0(r), \ u_0'(r) \text{ bounded as } r \to 0+, \quad u_0(1) = 4,$$

$$u_1''(r) + r^{-1}u_1'(r) = -u_0^2(r), \quad 0 < r < 1,$$
$$u_1(r), \ u_1'(r) \text{ bounded as } r \to 0+, \quad u_1(1) = 0,$$

with solutions $u_0(r) = 4r^3$ and $u_1(r) = \frac{1}{4}(1 - r^8)$. Consequently,

$$u(r) = 4r^3 + \tfrac{1}{4}\varepsilon(1 - r^8) + O(\varepsilon^2).$$

This example shows how, in certain cases, the perturbation method reduces a nonlinear problem to a sequence of linear ones. ∎

13.12. Example. In the IVP

$$w_{tt}(x,t) - w_{xx}(x,t) + (3 + \varepsilon)w(x,t) = 0,$$
$$-\infty < x < \infty, \ t > 0,$$
$$w(x,0) = \varepsilon \cos x, \quad w_t(x,0) = 0, \quad -\infty < x < \infty,$$

the small parameter $0 < \varepsilon \ll 1$ occurs not only in the PDE, but also in one of the ICs. Both the given IVP and the reduced one are second-order hyperbolic problems with unique solutions, so this is a regular perturbation problem.

Since the equation is homogeneous and the data functions are uniformly small, we expect the solution to have the same property. Hence, we seek a solution of the form $w(x,y) = \varepsilon u(x,y)$. Substituting above, we arrive at the new IVP

$$u_{tt}(x,t) - u_{xx}(x,t) + (3 + \varepsilon)u(x,t) = 0,$$
$$-\infty < x < \infty, \ t > 0,$$
$$u(x,0) = \cos x, \quad u_t(x,0) = 0, \quad -\infty < x < \infty,$$

in which the small parameter now occurs only in the PDE. Replacing $u(x,t) \approx \sum_{n=0}^{\infty} u_n(x,t)\varepsilon^n$ in the PDE and ICs and equating the coefficients of each power of ε on both sides, we find that the functions u_n, $n = 0,1,\ldots,$ are, respectively, the solutions of the IVPs

$$(u_0)_{tt}(x,t) - (u_0)_{xx}(x,t) + 3u_0(x,t) = 0, \quad -\infty < x < \infty, \ t > 0,$$
$$u_0(x,0) = \cos x, \quad (u_0)_t(x,0) = 0, \quad -\infty < x < \infty,$$

$$(u_1)_{tt}(x,t) - (u_1)_{xx}(x,t) + 3u_1(x,t) = -u_0(x,t),$$
$$-\infty < x < \infty, \ t > 0,$$
$$u_1(x,0) = 0, \quad (u_1)_t(x,0) = 0, \quad -\infty < x < \infty,$$

and so on.

We solve the IVP for u_0 by means of separation of variables. Thus, we seek a solution of the form

$$u_0(x,t) = X(x)T(t),$$

where, as remarked in Chapter 5, neither X nor T is the zero function. Replaced in the ICs, this yields

$$X(x)T(0) = \cos x, \quad X(x)T'(0) = 0.$$

Clearly, $T(0) \neq 0$. Since, as already noted, $X \neq 0$, it follows that

$$X(x) = \frac{1}{T(0)}\cos x, \quad T'(0) = 0.$$

Then the form of the solution is

$$u_0(x,t) = \left(\frac{1}{T(0)}\cos x\right)T(t),$$

and the PDE for u_0 becomes

$$\left(\frac{1}{T(0)}\cos x\right)T''(t) + \left(\frac{1}{T(0)}\cos x\right)T(t) + 3\left(\frac{1}{T(0)}\cos x\right)T(t) = 0.$$

Hence, T is the solution of the IVP

$$T''(t) + 4T(t) = 0, \quad t > 0,$$
$$T'(0) = 0,$$

which is $T(t) = T(0)\cos(2t)$; so,

$$u_0(x,t) = \cos(2t)\cos x.$$

The PDE for u_1 is now written as

$$(u_1)_{tt}(x,t) - (u_1)_{xx}(x,t) + 3u_1(x,t) = -\cos(2t)\cos x. \tag{13.3}$$

Guided by the function on the right-hand side, we seek its solution in the form

$$u_1(x,t) = T(t)\cos x, \quad T \neq 0.$$

Substituting in (13.3), equating the coefficients of $\cos x$ on both sides, and making use of the ICs as above, we find that T is the solution of the IVP

$$T''(t) + 4T(t) = -\cos(2t), \quad t > 0,$$
$$T(0) = 0, \quad T'(0) = 0,$$

which is $T(t) = -\frac{1}{4}t\sin(2t)$. Thus,

$$u_1(x,t) = -\frac{1}{4}t\sin(2t)\cos x.$$

Combining u_0 and u_1, we conclude that

$$u(x,t) = \cos(2t)\cos x - \frac{1}{4}\varepsilon t\sin(2t)\cos x + O(\varepsilon^2). \tag{13.4}$$

This series is a good asymptotic approximation for the solution of the given IVP as $\varepsilon \to 0$ if t is restricted to any fixed finite interval $[0,t_0]$. But in the absence of such a restriction, we see that when $t = O(\varepsilon^{-1})$, the term $\frac{1}{4}\varepsilon t\sin(2t)\cos x$ is of the same order of magnitude as the leading $O(1)$-term. Such a term is called *secular*, and its presence means that the perturbation series is not valid for very large values of t. To extend the validity of the series for all $t > 0$, we use the power series expansions of $\sin\alpha$ and $\cos\alpha$ and write

$$\cos\left(2t + \tfrac{1}{4}\varepsilon t\right) = \cos(2t)\cos\left(\tfrac{1}{4}\varepsilon t\right) - \sin(2t)\sin\left(\tfrac{1}{4}\varepsilon t\right)$$
$$= \cos(2t)\left[1 + O(\varepsilon^2)\right] - \sin(2t)\left[\tfrac{1}{4}\varepsilon t + O(\varepsilon^3)\right]$$
$$= \cos(2t) - \tfrac{1}{4}\varepsilon t\sin(2t) + O(\varepsilon^2).$$

Then the asymptotic solution series (13.4) can be rearranged in the form

$$u(x,t) = \cos x \cos\left(2t + \tfrac{1}{4}\varepsilon t\right) + O(\varepsilon^2),$$

which does not contain the secular term identified earlier. Higher-order terms in ε may, however, contain further secular terms. ∎

13.13. Example. The first-order IVP

$$u_t(x,t) + u_x(x,t) + \varepsilon u(x,t) = 0, \quad -\infty < x < \infty, \ t > 0,$$
$$u(x,0) = \cos x, \quad -\infty < x < \infty,$$

where $0 < \varepsilon \ll 1$, is a regular perturbation problem because the reduced problem is also of first order and both problems have unique solutions. Writing $u(x,t) \approx \sum_{n=0}^{\infty} u_n(x,t)\varepsilon^n$, we are led to the sequence of IVPs

$$(u_0)_t(x,t) + (u_0)_x(x,t) = 0, \quad -\infty < x < \infty, \ t > 0,$$
$$u_0(x,0) = \cos x, \quad -\infty < x < \infty,$$

$$(u_1)_t(x,t) + (u_1)_x(x,t) = -u_0(x,t), \quad -\infty < x < \infty, \ t > 0,$$
$$u_1(x,0) = 0, \quad -\infty < x < \infty,$$

and so on. Using the method of characteristics (see Section 12.1), we find that $u_0(x,t) = \cos(x - t)$ and $u_1(x,t) = -t\cos(x - t)$; hence,

$$u(x,t) = \cos(x - t) - \varepsilon t \cos(x - t) + O(\varepsilon^2).$$

As explained in Example 13.12, $\varepsilon t \cos(x - t)$ is a secular term. Since the procedure that removed such a term in the preceding example does not work here, we try another technique, called the *method of multiple scales*, which consists in introducing an additional variable $\tau = \varepsilon t$. Denoting by $v(x,t,\tau)$ the new unknown function and remarking that v depends on t both directly and through τ, we arrive at the problem

$$v_t(x,t,\tau) + \varepsilon v_\tau(x,t,\tau) + v_x(x,t,\tau) + \varepsilon v(x,t,\tau) = 0,$$
$$- \infty < x < \infty, \ t, \tau > 0,$$
$$v(x,0,0) = \cos x, \quad -\infty < x < \infty.$$

We now set $v(x,t,\tau) \approx \sum_{n=0}^{\infty} v_n(x,t,\tau)\varepsilon^n$ and deduce that v_0 and v_1 are, respectively, the solutions of the problems

$$(v_0)_t(x,t,\tau) + (v_0)_x(x,t,\tau) = 0, \quad -\infty < x < \infty, \; t, \tau > 0,$$
$$v_0(x,0,0) = \cos x,$$

$$(v_1)_t(x,t,\tau) + (v_1)_x(x,t,\tau) = -(v_0)_\tau(x,t,\tau) - v_0(x,t,\tau),$$
$$-\infty < x < \infty, \; t, \tau > 0,$$
$$v_1(x,0,0) = 0, \quad -\infty < x < \infty.$$

We seek the solution of the former as $v_0(x,t,\tau) = f(x,t)\varphi(\tau)$, where φ satisfies the condition $\varphi(0) = 1$ but is otherwise arbitrary. Then the problem for v_0 reduces to the IVP

$$f_t(x,t) + f_x(x,t) = 0, \quad -\infty < x < \infty, \; t > 0,$$
$$f(x,0) = \cos x, \quad -\infty < x < \infty,$$

with solution $f(x,t) = \cos(x - t)$, so $v_0(x,t,\tau) = \varphi(\tau)\cos(x - t)$. Replacing this on the right-hand side in the PDE for v_1, we arrive at the equation

$$(v_1)_t(x,t,\tau) + (v_1)_x(x,t,\tau) = -[\varphi'(\tau) + \varphi(\tau)]\cos(x - t),$$

whose solution satisfying the condition $v_1(x,0,0) = 0$ is

$$v_1(x,t,\tau) = -[\varphi'(\tau) + \varphi(\tau)]t\cos(x - t).$$

Consequently, the solution of the modified problem is

$$v(x,t,\tau) = \varphi(\tau)\cos(x - t) - \varepsilon[\varphi'(\tau) + \varphi(\tau)]t\cos(x - t) + O(\varepsilon^2).$$

To eliminate the secular term on the right-hand side above, we now choose φ so that $\varphi' + \varphi = 0$. In view of the earlier condition $\varphi(0) = 1$, we find that $\varphi(\tau) = e^{-\tau}$. Hence, $v(x,t,\tau) = e^{-\tau}\cos(x - t) + O(\varepsilon^2)$, which, since $\tau = \varepsilon t$, means that

$$u(x,t) = e^{-\varepsilon t}\cos(x - t) + O(\varepsilon^2).$$

Using the method of characteristics on the original IVP, we see that its exact solution is, in fact, $u(x,t) = e^{-\varepsilon t}\cos(x-t)$. The additional $O(\varepsilon^2)$-term represents "noise" generated by the approximating nature of the asymptotic expansion technique. ∎

13.3. Singular Perturbation Problems

In this type of problem we need to construct different solutions that are valid
in different regions, and then to match them in an appropriate manner for
overall uniform validity.

13.14. Example. Consider the IVP

$$\varepsilon\big[u_t(x,t) + 2u_x(x,t)\big] + u(x,t) = \sin t, \quad -\infty < x < \infty,\ t > 0,$$
$$u(x,0) = x, \quad -\infty < x < \infty,$$

where $0 < \varepsilon \ll 1$. This is a singular perturbation problem since the reduced
(unperturbed) problem contains no derivatives of u.

Following the standard procedure, we try to seek a solution of the form
$u(x,t) \approx \sum_{n=0}^{\infty} u_n(x,t)\varepsilon^n$. Replacing in the PDE, we obtain

$$u_0(x,t) = \sin t,$$
$$u_n(x,t) = -(u_{n-1})_t(x,t) - 2(u_{n-1})_x(x,t), \quad n = 1,2,\dots.$$

From the above recurrence relation it is easily seen that

$$u_{2n} = (-1)^n \sin t, \quad u_{2n+1} = (-1)^{n+1} \cos t, \quad n \geq 0;$$

therefore,

$$u(x,t) = \left[\sum_{n=0}^{\infty}(-1)^n \varepsilon^{2n}\right] \sin t - \varepsilon\left[\sum_{n=0}^{\infty}(-1)^n \varepsilon^{2n}\right] \cos t$$
$$= \frac{1}{1+\varepsilon^2}\,(\sin t - \varepsilon \cos t),$$

where we have used the formula for the sum of an infinite geometric progres-
sion with ratio $-\varepsilon^2$. However, it is immediately obvious that this function
does not satisfy the initial condition. Also, when $t \approx \varepsilon$, using the power
series for $\cos\varepsilon$ and $\sin\varepsilon$, we have

$$\sin t - \varepsilon \cos t \approx \sin\varepsilon - \varepsilon\cos\varepsilon = \big(\varepsilon + O(\varepsilon^3)\big) - \varepsilon\big(1 + O(\varepsilon^2)\big) = O(\varepsilon^2);$$

in other words, for $t = O(\varepsilon)$ the two terms in $u(x,t)$ are of the same order,
which is not allowed in an asymptotic series. Consequently, the series is not
well ordered in the region where $t = O(\varepsilon)$, so it is not a valid representation
of the solution in that region. This means that we must seek a different series

in a *boundary layer* of width $O(\varepsilon)$ near the x-axis ($t = 0$). To construct the new series, we make the change of variables

$$\tau = \frac{t}{\varepsilon}, \quad u(x,t) = u(x,\varepsilon\tau) = u^i(x,\tau),$$

which amounts to a stretching of the time argument near $t = 0$. Under this transformation, the IVP becomes

$$u^i_\tau(x,\tau) + 2\varepsilon u^i_x(x,\tau) + u^i(x,\tau) = \sin\varepsilon\tau = \varepsilon\tau + O(\varepsilon^3),$$

$$-\infty < x < \infty, \ \tau > 0,$$

$$u^i(x,0) = x, \quad -\infty < x < \infty.$$

For the *inner solution* u^i of this boundary layer IVP we assume an asymptotic expansion of the form $u^i(x,\tau) \approx \sum_{n=0}^{\infty} u^i_n(x,\tau)\varepsilon^n$. Then the new PDE and IC yield, in the usual way, the sequence of IVPs

$$(u^i_0)_\tau(x,\tau) + u^i_0(x,\tau) = 0, \quad -\infty < x < \infty, \ \tau > 0,$$
$$u^i_0(x,0) = x, \quad -\infty < x < \infty,$$

$$(u^i_1)_\tau(x,\tau) + u^i_1(x,\tau) = \tau - 2(u^i_0)_x(x,\tau), \quad -\infty < x < \infty, \ \tau > 0,$$
$$u^i_1(x,0) = 0, \quad -\infty < x < \infty,$$

and so on. Restricting our attention to the first two terms, from the first problem we easily obtain

$$u^i_0(x,\tau) = xe^{-\tau},$$

and from the second problem (seeking a particular integral of the form $a\tau e^{-\tau}$, $a = \text{const}$),

$$u^i_1(x,\tau) = \tau - 1 + (1 - 2\tau)e^{-\tau};$$

hence,

$$u^i(x,\tau) = xe^{-\tau} + \varepsilon\left[\tau - 1 + (1 - 2\tau)e^{-\tau}\right] + O(\varepsilon^2).$$

We rename the first solution the *outer solution* and denote it by u^o:

$$u^o(x,t) = \frac{1}{1 + \varepsilon^2}(\sin t - \varepsilon\cos t) = \sin t - \varepsilon\cos t + O(\varepsilon^2).$$

Now we need to match u^o (valid for $t = O(1)$) and u^i (valid for $t = O(\varepsilon)$) up to the order of ε considered (here, it is $O(\varepsilon)$), in some common region of validity. To this end, first we write the outer solution in terms of the inner variable $\tau = t/\varepsilon$ and expand it for τ fixed and ε small, listing the terms up to $O(\varepsilon)$, after which we revert to t:

$$(u^o)^i \approx \sin t - \varepsilon \cos t + \cdots = \sin(\varepsilon\tau) - \varepsilon\cos(\varepsilon\tau) + \cdots$$
$$= \varepsilon(\tau - 1) + \cdots = t - \varepsilon + \cdots.$$

Next, we write the inner solution in terms of the outer variable $t = \varepsilon\tau$ and expand it for t fixed and ε small to the same order:

$$(u^i)^o \approx xe^{-t/\varepsilon} + \varepsilon\left[\frac{t}{\varepsilon} - 1 + \left(1 - 2\frac{t}{\varepsilon}\right)e^{-t/\varepsilon}\right] + \cdots = t - \varepsilon + \cdots$$

(the rest of the terms are $o(\varepsilon^n)$ for any positive integer n). Finally, we impose the condition $(u^o)^i = (u^i)^o$ up to $O(\varepsilon)$ terms. In our case this is already satisfied, so the two solutions match.

Since the common region of validity is not clear, it is useful to consider a *composite solution* of the form

$$u^c = u^o + u^i - (u^o)^i = u^o + u^i - (u^i)^o.$$

This is valid uniformly for $t > 0$ since

$$(u^c)^o = (u^o)^o + (u^i)^o - ((u^i)^o)^o = u^o + (u^i)^o - (u^i)^o = u^o,$$
$$(u^c)^i = (u^o)^i + (u^i)^i - ((u^o)^i)^i = (u^o)^i + u^i - (u^o)^i = u^i.$$

Here we have

$$u(x,t) \approx u^c(x,t) = \sin t - \varepsilon\cos t + (x - 2t + \varepsilon)e^{-t/\varepsilon} + \cdots. \quad \blacksquare$$

13.15. Example. Consider the elliptic problem in a semi-infinite strip

$$\varepsilon(\Delta u)(x,y) + u_x(x,y) + u_y(x,y) = 0, \quad x > 0, \; 0 < y < 1,$$
$$u(x,0) = e^{-x}, \quad u(x,1) = p(x), \quad x > 0,$$
$$u(0,y) = y, \quad u(x,y) \text{ bounded as } x \to \infty, \quad 0 < y < 1,$$

where $0 < \varepsilon \ll 1$ and

$$p(x) = \begin{cases} 1 - x, & 0 < x < 1, \\ 0, & x \geq 1. \end{cases}$$

This is a singular perturbation problem because the perturbed BVP is of second order whereas the reduced BVP is of first order.

Let $u(x,y) \approx \sum_{n=0}^{\infty} u_n(x,y)\varepsilon^n$. Then it is readily seen that u_0 is the solution of the BVP

$$(u_0)_x(x,y) + (u_0)_y(x,y) = 0, \quad x > 0, \; 0 < y < 1,$$
$$u_0(x,0) = e^{-x}, \quad u_0(x,1) = p(x), \quad x > 0,$$
$$u_0(0,y) = y, \quad u_0(x,y) \text{ bounded as } x \to \infty, \quad 0 < y < 1,$$

u_1 is the solution of the BVP

$$(u_1)_x(x,y) + (u_1)_y(x,y) = -(\Delta u_0)(x,y), \quad x > 0, \; 0 < y < 1,$$
$$u_1(x,0) = 0, \quad u_1(x,1) = 0, \quad x > 0,$$
$$u_1(0,y) = 0, \quad u_1(x,y) \text{ bounded as } x \to \infty, \quad 0 < y < 1,$$

and so on.

We find u_0 by the method of characteristics (see Chapter 12), using $y = 1$ as the data line. If $x = x(y)$ is a characteristic curve, then on it

$$\frac{d}{dy} u_0(x(y),y) = (u_0)_x(x(y),y)x'(y) + (u_0)_y(x(y),y).$$

Hence, $x'(y) = 1$ implies that $du_0/dy = 0$, so

$$x = y + c, \quad u_0(x,y) = c', \quad c, c' = \text{const.}$$

The equation of the characteristic through the points (x,y) and $(x_0,1)$ is $x = y + x_0 - 1$, and on this line

$$u_0(x,y) = c' = u_0(x_0,1) = p(x_0) = p(x - y + 1);$$

therefore, the solution of the given BVP is

$$u(x,y) = u_0(x,y) + O(\varepsilon) = p(x - y + 1) + O(\varepsilon)$$
$$= \begin{cases} y - x + O(\varepsilon), & 0 < x < y, \\ O(\varepsilon), & x \geq y. \end{cases} \tag{13.5}$$

Since it is clear that this solution does not satisfy the other BCs, we need to introduce two boundary layers.

In the boundary layer near $y = 0$ we make the substitution $\eta = y/\delta(\varepsilon)$, write $u(x,y) = u(x, \delta(\varepsilon)\eta) = v(x,\eta)$, and arrive at the new BVP

$$\varepsilon v_{xx} + \frac{\varepsilon}{\delta^2(\varepsilon)} v_{\eta\eta} + v_x + \frac{1}{\delta(\varepsilon)} v_\eta = 0, \quad v(x,0) = e^{-x}.$$

The unspecified boundary layer width $\delta(\varepsilon)$ must be chosen so that the $v_{\eta\eta}$ term is one of the dominant terms in the above equation. Comparing the coefficients of all the terms in the new PDE, we see that there are three possibilities:

$$\text{(i)} \ \ \frac{\varepsilon}{\delta^2(\varepsilon)} \approx \varepsilon; \quad \text{(ii)} \ \ \frac{\varepsilon}{\delta^2(\varepsilon)} \approx 1; \quad \text{(iii)} \ \ \frac{\varepsilon}{\delta^2(\varepsilon)} \approx \frac{1}{\delta(\varepsilon)}.$$

It is not difficult to check that cases (i) and (ii) do not satisfy our magnitude requirement, whereas (iii) does. Hence, we can take $\delta(\varepsilon) = \varepsilon$, for simplicity, and the BVP becomes

$$\varepsilon^2 v_{xx} + v_{\eta\eta} + \varepsilon v_x + v_\eta = 0, \quad v(x,0) = e^{-x}.$$

Writing $v(x,\eta) = \sum\limits_{n=0}^{\infty} v_n(x,\eta)\varepsilon^n$, from the PDE and BC satisfied by v we find that

$$(v_0)_{\eta\eta} + (v_0)_\eta = 0, \quad v_0(x,0) = e^{-x},$$

with solution

$$v_0(x,\eta) = \alpha(x) + \left[e^{-x} - \alpha(x)\right]e^{-\eta},$$

where α is an arbitrary function; therefore,

$$
\begin{aligned}
u(x,y) = v(x,\eta) &= v_0(x,\eta) + O(\varepsilon) \\
&= \alpha(x) + \left[e^{-x} - \alpha(x)\right]e^{-\eta} + O(\varepsilon). \quad (13.6)
\end{aligned}
$$

We rename (13.5) the outer solution u^o and (13.6) the first inner solution u^{i_1}, and match them by the method used in Example 13.14. Thus, expanding p in powers of ε, we have

$$
\begin{aligned}
(u^o)^{i_1} &= p(x - \varepsilon\eta + 1) + O(\varepsilon) = p(x+1) + O(\varepsilon) = O(\varepsilon), \\
(u^{i_1})^o &= \alpha(x) + \left[e^{-x} - \alpha(x)\right]e^{-y/\varepsilon} + O(\varepsilon) = \alpha(x) + O(\varepsilon),
\end{aligned}
\quad (13.7)
$$

so $(u^o)^{i_1} = (u^{i_1})^o$ to $O(1)$ terms if

$$\alpha(x) = 0, \quad x > 0. \quad (13.8)$$

The second boundary layer needs to be constructed near $x = 0$. Proceeding as above (this time with the transformed u_{xx} term as one of the dominant terms in the new PDE), we again conclude that the correct layer width is $\delta(\varepsilon) \approx \varepsilon$, so we set

$$\xi = \frac{x}{\varepsilon}.$$

Writing $u(x,y) = u(\varepsilon\xi, y) = w(\xi, y)$, we arrive at the new BVP

$$w_{\xi\xi} + \varepsilon^2 w_{yy} + w_\xi + \varepsilon w_y = 0, \quad w(0,y) = y.$$

If we assume that $w(\xi, y) \approx \sum_{n=0}^{\infty} w_n(\xi, y)\varepsilon^n$, then

$$(w_0)_{\xi\xi} + (w_0)_\xi = 0, \quad w_0(0,y) = y,$$

with solution

$$w_0(\xi, y) = \beta(y) + [y - \beta(y)]e^{-\xi},$$

where β is another arbitrary function; hence,

$$u(x,y) = w(\xi, y) = w_0(\xi, y) + O(\varepsilon)$$
$$= \beta(y) + [y - \beta(y)]e^{-\xi} + O(\varepsilon). \tag{13.9}$$

We call (13.9) the second inner solution u^{i_2} and match it with u^o up to $O(1)$ terms. As above, we have

$$(u^o)^{i_2} = p(\varepsilon\xi - y + 1) + O(\varepsilon) = p(-y + 1) + O(\varepsilon) = y + O(\varepsilon),$$
$$(u^{i_2})^o = \beta(y) + [y - \beta(y)]e^{-x/\varepsilon} + O(\varepsilon) = \beta(y) + O(\varepsilon), \tag{13.10}$$

so

$$\beta(y) = y. \tag{13.11}$$

It can be verified that the composite solution in this case is

$$u^c = u^o + u^{i_1} + u^{i_2} - (u^{i_1})^o - (u^{i_2})^o$$
$$= u^o + u^{i_1} + u^{i_2} - (u^o)^{i_1} - (u^o)^{i_2},$$

since $(u^{i_1})^{i_2} = (u^{i_1})^o$ and $(u^{i_2})^{i_1} = (u^{i_2})^o$. Thus, by (13.5)–(13.11), the asymptotic solution of the given BVP to $O(1)$ terms is

$$u(x,y) \approx u^c(x,y)$$
$$= p(x - y + 1) + \left[e^{-x} - p(x+1)\right]e^{-y/\varepsilon}$$
$$+ \left[y - p(1 - y)\right]e^{-x/\varepsilon} + O(\varepsilon)$$
$$= \begin{cases} y - x + e^{-x-y/\varepsilon} + O(\varepsilon), & 0 < x < y, \\ e^{-x-y/\varepsilon} + O(\varepsilon), & x \geq y. \end{cases} \blacksquare$$

13.16. Remarks. (i) We did not consider effects caused by the incompatibility of the boundary values at the "corner points" $(0,0)$ and $(0,1)$.

(ii) If we had tried to construct the first boundary layer solution near $y = 1$ instead of $y = 0$ (and, thus, use $y = 0$ as the data line for computing the first term in u^o), we would have failed. In that region we would need to substitute $\eta = (1 - y)/\varepsilon$, and setting $u(x,y) = u(x, 1 - \varepsilon\eta) = v(x,\eta)$, we would obtain the BVP

$$\varepsilon^2 v_{xx} + v_{\eta\eta} + \varepsilon v_x - v_\eta = 0, \quad v(x,0) = p(x),$$

from which

$$(v_0)_{\eta\eta} - (v_0)_\eta = 0, \quad v_0(x,0) = p(x).$$

This would yield

$$v_0(x,\eta) = \alpha(x) + \left[p(x) - \alpha(x)\right]e^\eta,$$

which is not good for matching since $e^\eta = e^{y/\varepsilon} \to \infty$ as $\varepsilon \to 0+$ with y fixed. $\blacksquare$

Exercises

In (1)–(4) use the method of asymptotic expansion to compute a formal approximate solution (to $O(\varepsilon)$ terms) for the regular perturbation problem

$$u''(r) + r^{-1}u'(r) + \varepsilon f(u) = q(r), \quad 0 < r < 1,$$
$$u(r), \ u'(r) \text{ bounded as } r \to 0+, \quad u(1) = g,$$

with the functions f and q and the number g as indicated, where $0 < \varepsilon \ll 1$.

(1) $f(u) = 4u$, $q(r) = 1$, $g = 1 + \varepsilon$.

(2) $f(u) = -2u$, $q(r) = -4 + 2\varepsilon(r^2 + 9r)$, $g = 1 + 3\varepsilon$.

(3) $f(u) = -u^3$, $q(r) = \varepsilon$, $g = -3$.

(4) $f(u) = u - u^2$, $q(r) = 6(3r - 2) + \varepsilon(12 - 3r^2 + 2r^3 - 9r^4 + 12r^5 - 4r^6)$,

$g = -1 + 3\varepsilon$.

In (5)–(14) use the method of asymptotic expansion to compute a formal approximate solution (to $O(\varepsilon)$ terms) for the regular perturbation problem

$$u_t(x,t) + au_x(x,t) + bu(x,t) = q(x,t), \quad -\infty < x < \infty, \ t > 0,$$

$$u(x,0) = f(x), \quad -\infty < x < \infty,$$

with the functions q and f and the coefficients a and b as indicated, where $0 < \varepsilon \ll 1$.

(5) $q(x,t) = 0$, $f(x) = \sin x$, $a = \varepsilon$, $b = 1 + \varepsilon$.

(6) $q(x,t) = 1$, $f(x) = -2x$, $a = -\varepsilon$, $b = 1 - \varepsilon$.

(7) $q(x,t) = \varepsilon$, $f(x) = e^{-x}$, $a = 2\varepsilon$, $b = 2 - \varepsilon$.

(8) $q(x,t) = -2 + \varepsilon e^{-t}$, $f(x) = \varepsilon$, $a = -2\varepsilon$, $b = 1 + 2\varepsilon$.

(9) $q(x,t) = 0$, $f(x) = x$, $a = 3$, $b = -2\varepsilon$.

(10) $q(x,t) = 0$, $f(x) = 2x - 1$, $a = -1$, $b = 1 + \varepsilon$.

(11) $q(x,t) = 2$, $f(x) = -x$, $a = 2 + \varepsilon$, $b = \varepsilon - 1$.

(12) $q(x,t) = x + 2\varepsilon$, $f(x) = 3 + x$, $a = 1 - \varepsilon$, $b = 1 + 2\varepsilon$.

(13) $q(x,t) = 2t$, $f(x) = e^x$, $a = 2 + \varepsilon$, $b = 0$.

(14) $q(x,t) = 6xt + \varepsilon x$, $f(x) = 1 + x + \varepsilon x$, $a = 1 - 3\varepsilon$, $b = 0$.

In (15)–(24) use the method of asymptotic expansion to compute a formal approximate solution (to $O(\varepsilon)$ terms) for the regular perturbation problem

$$u_{xx}(x,y) + au_x(x,y) + bu_y(x,y) + cu(x,y) = q(x,y),$$

$$0 < x < 1, \ -\infty < y < \infty,$$

$$u(0,y) = f(y), \quad u(1,y) = g(y), \quad -\infty < y < \infty,$$

with the functions q, f, and g and the coefficients a, b, and c as indicated, where $0 < \varepsilon \ll 1$.

(15) $q(x,y) = -xy^2$, $f(y) = \varepsilon y$, $g(y) = y^2 + \varepsilon(e - 2)y$,

$a = 0$, $b = -\varepsilon$, $c = -1$.

(16) $q(x,y) = 2 - \varepsilon(x^2 + 2y + 4)$, $f(y) = 2y$, $g(y) = 1 + 2y + \varepsilon y^2$,
      $a = 0$, $b = -2\varepsilon$, $c = -\varepsilon$.

(17) $q(x,y) = -4 - 2(1 + \varepsilon)x^2$, $f(y) = \varepsilon y$, $g(y) = -2 + \varepsilon y \cos 1$,
      $a = 0$, $b = \varepsilon$, $c = 1 + \varepsilon$.

(18) $q(x,y) = 4x^2 y - 2y - \varepsilon[x^2 y - x^2 + 3(y + 1)e^x]$, $f(y) = \varepsilon(y + 1)$,
      $g(y) = -y + \varepsilon e(y + 1)$, $a = 0$, $b = -\varepsilon$, $c = \varepsilon - 4$.

(19) $q(x,y) = -y$, $f(y) = 0$, $g(y) = y$, $a = -\varepsilon$, $b = -3\varepsilon$, $c = 0$.

(20) $q(x,y) = \varepsilon(y - 2)e^x$, $f(y) = 2y$, $g(y) = (2 + \varepsilon)ey$,
      $a = -1$, $b = -\varepsilon$, $c = 0$.

(21) $q(x,y) = 2\varepsilon - 6$, $f(y) = (2 + 3\varepsilon)y$, $g(y) = 2y + 6 + 3\varepsilon ey$,
      $a = \varepsilon - 1$, $b = -2\varepsilon$, $c = 0$.

(22) $q(x,y) = 2\varepsilon(y + e^{2x})$, $f(y) = (\varepsilon - 1)y$, $g(y) = -e^2 y + \varepsilon(e^2 + y)$,
      $a = -3$, $b = -\varepsilon$, $c = 2$.

(23) $q(x,y) = 2y - 4x - 2 + \varepsilon(3 - 5e^{-x})$, $f(y) = -y - 1$,
      $g(y) = 2 - e^{-1} - y + 2\varepsilon e^{-1}$, $a = \varepsilon - 1$, $b = -\varepsilon$, $c = -2$.

(24) $q(x,y) = 2\varepsilon[2y + 3\cos(2x) - y\sin(2x)]$, $f(y) = (1 + \varepsilon)y$,
      $g(y) = y\cos 2 + \varepsilon(y + \sin 2)$, $a = \varepsilon$, $b = 2\varepsilon$, $c = 4$.

In (25)–(34) use the method of asymptotic expansion (in conjunction with the method of separation of variables) to compute a formal approximate solution (to $O(\varepsilon)$ terms) for the regular perturbation problem

$$u_{tt}(x,t) - a^2 u_{xx}(x,t) + bu_t(x,t) + cu_x(x,t) + du(x,t) = q(x,t),$$
$$-\infty < x < \infty, \ t > 0,$$
$$u(x,0) = f(x), \quad u_t(x,0) = g(x), \quad -\infty < x < \infty,$$

with the functions q, f, and g and the coefficients a, b, c, and d as indicated, where $0 < \varepsilon \ll 1$.

(25) $q(x,t) = 0$, $f(x) = -x$, $g(x) = x$,
      $a = 2$, $b = 0$, $c = 0$, $d = 1 + \varepsilon$.

(26) $q(x,t) = x - 2$, $f(x) = x - 2$, $g(x) = \frac{7}{2}(x - 2)$,
      $a = 3$, $b = 2$, $c = 0$, $d = 2\varepsilon$.

(27) $q(x,t) = -9\sin(2x)$, $f(x) = \sin(2x)$, $g(x) = -3\sin(2x)$,
      $a = 1$, $b = 0$, $c = 0$, $d = 5 + \varepsilon$.

(28) $q(x,t) = 0$, $f(x) = \cos(3x)$, $g(x) = 4\cos(3x)$,
      $a = 2$, $b = 0$, $c = 0$, $d = -32 - \varepsilon$.

(29) $q(x,t) = 0$, $f(x) = e^{-2x}$, $g(x) = 2e^{-2x}$,
     $a = 2$, $b = 0$, $c = 0$, $d = 12 + 3\varepsilon$.

(30) $q(x,t) = 0$, $f(x) = e^{2x}$, $g(x) = -e^{2x}$,
     $a = 1$, $b = -1$, $c = 0$, $d = 2 + 2\varepsilon$.

(31) $q(x,t) = 0$, $f(x) = e^{x}$, $g(x) = 2e^{x}$,
     $a = 1$, $b = -3$, $c = 0$, $d = 3 + \varepsilon$.

(32) $q(x,t) = e^{-x}$, $f(x) = e^{-x}$, $g(x) = -\frac{4}{3}e^{-x}$,
     $a = 2$, $b = -2$, $c = 0$, $d = 1 - \varepsilon$.

(33) $q(x,t) = 0$, $f(x) = 2x - 1$, $g(x) = 4x - 2$,
     $a = 2$, $b = 0$, $c = 0$, $d = 3\varepsilon - 4$.

(34) $q(x,t) = -2x$, $f(x) = -2x$, $g(x) = \frac{9}{2}x$,
     $a = 1$, $b = -2$, $c = \varepsilon$, $d = \varepsilon - 8$.

In (35)–(40) use the method of matched asymptotic expansions to compute a formal composite solution (to $O(\varepsilon)$ terms) for the singular perturbation problem

$$\varepsilon u_{xx}(x,y) + a u_x(x,y) + b u_y(x,y) + c u(x,y) = 0,$$

$$0 < x < L, \quad -\infty < y < \infty,$$

$$u(0,y) = f(y), \quad u(L,y) = g(y), \quad -\infty < y < \infty,$$

with the functions f and g, the coefficients a, b, and c, and the number L as indicated, where $0 < \varepsilon \ll 1$. (The location of the boundary layer is specified in each case.)

(35) $f(y) = y + 1$, $g(y) = 2y$, $a = 1 + 2\varepsilon$, $b = 2\varepsilon$, $c = 1$, $L = 1$;
     boundary layer at $x = 0$.

(36) $f(y) = y$, $g(y) = 2y - 1$, $a = 1 - \varepsilon$, $b = 0$, $c = 1$, $L = 2$;
     boundary layer at $x = 0$.

(37) $f(y) = 2y + 3$, $g(y) = -y$, $a = \varepsilon - 1$, $b = \varepsilon$, $c = 0$, $L = 2$;
     boundary layer at $x = 2$.

(38) $f(y) = -3y$, $g(y) = y + 2$, $a = -1 - \varepsilon$, $b = 2\varepsilon$, $c = 1$, $L = 1$;
     boundary layer at $x = 1$.

(39) $f(y) = y^2$, $g(y) = 1 - y$, $a = 1 - 2\varepsilon$, $b = -\varepsilon$, $c = -1$, $L = 1$;
     boundary layer at $x = 0$.

(40) $f(y) = 2$, $g(y) = 2 - 3y$, $a = 1$, $b = -2\varepsilon$, $c = 1$, $L = 2$;
     boundary layer at $x = 0$.

In (41)–(46) use the method of matched asymptotic expansions (in conjunction with the method of characteristics) to compute a formal composite solution (to $O(\varepsilon)$ terms) for the singular perturbation problem

$$\varepsilon\left[u_{xx}(x,y) + au_{yy}(x,y)\right] + bu_x(x,y) + cu_y(x,y) = 0,$$

$$x > 0, \ 0 < y < L,$$

$$u(x,0) = f(x), \quad u(x,L) = g(x), \quad x > 0,$$

$$u(0,y) = h(y), \quad 0 < y < L.$$

with the functions f, g, and h, the coefficients a, b, and c, and the number L as indicated, where $0 < \varepsilon \ll 1$. (The locations of the boundary layers are specified in each case.)

(41) $f(x) = x^2 + 1$, $g(x) = 2x - 3$, $h(y) = y^2$,
　　$a = 1$, $b = 1$, $c = 2$, $L = 1$; boundary layers at $x = 0$, $y = 0$.

(42) $f(x) = x^2$, $g(x) = 1 - x$, $h(y) = y^2 - 1$,
　　$a = 2$, $b = 1$, $c = 1$, $L = 2$; boundary layers at $x = 0$, $y = 0$.

(43) $f(x) = 2x^2 - 1$, $g(x) = x + 2$, $h(y) = y^2 + 1$,
　　$a = 2$, $b = 1$, $c = -1$, $L = 1$; boundary layers at $x = 0$, $y = 1$.

(44) $f(x) = 1 - x^2$, $g(x) = 2x + 4$, $h(y) = 2y^2$,
　　$a = 1$, $b = 2$, $c = -2$, $L = 2$; boundary layers at $x = 0$, $y = 2$.

(45) $f(x) = x + 1$, $g(x) = x^2 - 2$, $h(y) = 1 - 2y$,
　　$a = 4$, $b = 2$, $c = 1$, $L = 1$; boundary layers at $x = 0$, $y = 0$.

(46) $f(x) = x^2 + x$, $g(x) = -3x$, $h(y) = y^2 - 2y$,
　　$a = 3$, $b = 1$, $c = -1$, $L = 2$; boundary layers at $x = 0$, $y = 2$.

Chapter 14
Complex Variable Methods

Certain linear two-dimensional elliptic problems turn out to be difficult to solve in a Cartesian coordinate setup. In many such cases it is advisable to go over to equivalent formulations in terms of complex variables, which may be able to help us find the solutions much more readily, and elegantly. Although complex numbers have already been mentioned in Chapters 1, 3, 8, 9, and 11, we start by giving a brief presentation of their basic rules of manipulation and a few essential details about complex functions.

14.1. Elliptic Equations

A complex number is an expression of the form

$$z = x + iy, \quad x, y \text{ real}, \quad i^2 = -1,$$

where x and y are called, respectively, the *real part* and the *imaginary part* of z. The number $\bar{z} = x - iy$ is the *complex conjugate* of z and

$$r = |z| = (z\bar{z})^{1/2} = (x^2 + y^2)^{1/2}$$

is the *modulus* of z.

A complex number can also be written in polar form as

$$z = re^{i\theta} = r(\cos\theta + i\sin\theta),$$

where θ, $-\pi < \theta \leq \pi$, called the *argument* of z, is determined from the equalities

$$\cos\theta = \frac{x}{r}, \quad \sin\theta = \frac{y}{r}.$$

Obviously,

$$\bar{z} = r(\cos\theta - i\sin\theta) = re^{-i\theta}.$$

Addition and multiplication of complex numbers are performed according to the usual algebraic rules for real numbers.

A complex function of a complex variable has the general form

$$f(z) = (\operatorname{Re} f)(x,y) + i(\operatorname{Im} f)(x,y),$$

where $\operatorname{Re} f$ and $\operatorname{Im} f$ are its real and imaginary parts. Such a function f is called *holomorphic* if its derivative $f'(z)$ exists at all points in its domain of definition. A holomorphic function is *analytic*—that is, it can be expanded in a convergent power series.

14.1. Example. If $f(z) = z^2$, then

$$(\operatorname{Re} f)(x,y) = x^2 - y^2, \quad (\operatorname{Im} f)(x,y) = 2xy. \quad \blacksquare$$

14.2. Theorem. *A function f is holomorphic if and only if it satisfies the Cauchy–Riemann relations*

$$(\operatorname{Re} f)_x = (\operatorname{Im} f)_y, \quad (\operatorname{Im} f)_x = -(\operatorname{Re} f)_y. \tag{14.1}$$

In this case, both $\operatorname{Re} f$ and $\operatorname{Im} f$ are solutions of the Laplace equation.

14.3. Remarks. (i) Suppose that a smooth function f of real variables x and y is expressed in terms of the complex variables z and $\bar{z}$; that is,

$$f(x,y) = g(z,\bar{z}).$$

From the chain rule of differentiation it follows that

$$\begin{aligned}
f_x &= g_z + g_{\bar{z}}, \\
f_y &= i(g_z - g_{\bar{z}}), \\
f_{xx} &= g_{zz} + 2g_{z\bar{z}} + g_{\bar{z}\bar{z}}, \\
f_{yy} &= -g_{zz} + 2g_{z\bar{z}} - g_{\bar{z}\bar{z}}, \\
f_{xy} &= f_{yx} = i(g_{zz} - g_{\bar{z}\bar{z}}).
\end{aligned} \tag{14.2}$$

(ii) The Laplace equation, which is not easily integrated in terms of real variables, has a very simple solution in terms of complex variables. Thus, if $u(x,y) = v(z,\bar{z})$, then, by (14.2),

$$\Delta u = u_{xx} + u_{yy} = 4v_{z\bar{z}}, \tag{14.3}$$

so $\Delta u(x,y) = 0$ is equivalent to

$$v_{z\bar{z}}(z,\bar{z}) = 0.$$

It is now trivial to see that the latter has the general solution

$$v(z,\bar{z}) = \varphi(z) + \bar{\psi}(\bar{z}),$$

where φ and ψ are arbitrary analytic functions of z.

If we want the *real* general solution, then we must have $v(z,\bar{z}) = \bar{v}(\bar{z}, z)$; that is,

$$\varphi(z) + \bar{\psi}(\bar{z}) = \bar{\varphi}(\bar{z}) + \psi(z),$$

or

$$\varphi(z) - \bar{\varphi}(\bar{z}) = \psi(z) - \bar{\psi}(\bar{z}),$$

which means that $\operatorname{Im}\varphi = \operatorname{Im}\psi$. Using this equality and (14.1), we find that $\operatorname{Re}\varphi = \operatorname{Re}\psi$ as well, so $\psi = \varphi$; therefore, the real general solution of the two-dimensional Laplace equation is

$$v(z,\bar{z}) = \varphi(z) + \bar{\varphi}(\bar{z}), \tag{14.4}$$

where φ is an arbitrary analytic function.

(iii) A similar treatment can be applied to the biharmonic equation

$$\Delta\Delta u(x,y) = 0.$$

Since, as we have seen, $\Delta v = 4v_{z\bar{z}}$, where $v(z,\bar{z}) = u(x,y)$, we easily deduce that $\Delta\Delta v(x,y) = 16v_{zz\bar{z}\bar{z}}(z,\bar{z})$; hence, the biharmonic equation is equivalent to

$$v_{zz\bar{z}\bar{z}}(z,\bar{z}) = 0.$$

Then, by (ii) above,

$$\Delta v(z,\bar{z}) = 4v_{z\bar{z}}(z,\bar{z}) = \Phi(z) + \bar{\Phi}(\bar{z}).$$

Integrating and applying the argument in (ii), we arrive at the real general solution

$$v(z,\bar{z}) = \bar{z}\Phi(z) + z\bar{\Phi}(\bar{z}) + \varphi(z) + \bar{\varphi}(\bar{z}),$$

where Φ and φ are arbitrary analytic functions of z. ∎

14.4. Example. Consider the BVP

$$\Delta u = 8 \quad \text{in } D,$$
$$u = 1 - 2\cos\theta - \cos(2\theta) + 2\sin(2\theta) \quad \text{on } \partial D,$$

where D and ∂D are the unit circular disk and its boundary, respectively, and r, θ are polar coordinates with the pole at the center of the disk. On ∂D we have

$$z = e^{i\theta} = \sigma, \quad \bar{z} = e^{-i\theta} = \sigma^{-1},$$
$$\cos(n\theta) = \tfrac{1}{2}(e^{in\theta} + e^{-in\theta}) = \tfrac{1}{2}(\sigma^n + \sigma^{-n}), \quad n = 1, 2, \dots, \quad (14.5)$$
$$\sin(n\theta) = -\tfrac{1}{2}i(e^{in\theta} - e^{-in\theta}) = -\tfrac{1}{2}i(\sigma^n - \sigma^{-n});$$

hence, using (14.3), we bring the given BVP to the equivalent form

$$v_{z\bar{z}} = 2 \quad \text{in } D,$$
$$v = -(\tfrac{1}{2} - i)\sigma^{-2} - \sigma^{-1} + 1 - \sigma - (\tfrac{1}{2} + i)\sigma^2 \quad \text{on } \partial D.$$

It is easily seen that $2z\bar{z}$ is a particular solution of the PDE, so, by (14.4), the general solution of the equation is

$$v(z, \bar{z}) = \varphi(z) + \bar{\varphi}(\bar{z}) + 2z\bar{z}. \quad (14.6)$$

Since the arbitrary function φ is analytic, it admits a series expansion of the form

$$\varphi(z) = \sum_{n=0}^{\infty} a_n z^n. \quad (14.7)$$

Replacing (14.7) in (14.6) and then v with $z = \sigma$ and $\bar{z} = \sigma^{-1}$ in the BC and performing the usual comparison of coefficients, we find that

$$a_0 + \bar{a}_0 + 2 = 1, \quad a_1 = -1, \quad a_2 = -(\tfrac{1}{2} + i), \quad a_n = 0 \ (n \neq 0, 1, 2).$$

Therefore, $a_0 + \bar{a}_0 = -1$, and (14.6) and (14.7) yield the solution

$$v(z, \bar{z}) = a_0 + \bar{a}_0 + a_1 z + \bar{a}_1 \bar{z} + a_2 z^2 + \bar{a}_2 \bar{z}^2 + 2z\bar{z}$$
$$= -1 - z - \bar{z} - (\tfrac{1}{2} + i)z^2 - (\tfrac{1}{2} - i)\bar{z}^2 + 2z\bar{z}.$$

In Cartesian coordinates with the origin at the center of the disk, this becomes

$$u(x, y) = -1 - 2x + x^2 + 4xy + 3y^2. \quad \blacksquare$$

14.5. Example. To solve the elliptic BVP

$$u_{xx} - 2u_{xy} + 2u_{yy} = 4 \quad \text{in } D,$$

$$u = \tfrac{3}{2} - \tfrac{1}{2}\cos(2\theta) + \tfrac{3}{2}\sin(2\theta) \quad \text{on } \partial D,$$

where the notation is the same as in Example 14.4, we use a slightly altered procedure. First, by (14.2), we easily see that $v(z, \bar{z}) = u(x, y)$ is the solution of the problem

$$(1 + 2i)v_{zz} - 6v_{z\bar{z}} + (1 - 2i)v_{\bar{z}\bar{z}} = -4 \quad \text{in } D,$$

$$v = -\tfrac{1}{4}(1 - 3i)\sigma^{-2} + \tfrac{3}{2} - \tfrac{1}{4}(1 + 3i)\sigma^2 \quad \text{on } \partial D. \tag{14.8}$$

We now perform a simple transformation of the form

$$\zeta = z + \bar{\alpha}\bar{z}, \quad \bar{\zeta} = \bar{z} + \alpha z, \quad v(z, \bar{z}) = w(\zeta, \bar{\zeta}),$$

where α is a complex number to be chosen so that the left-hand side of the PDE for w consists only of the mixed second-order derivative. Thus, by the chain rule,

$$v_{zz} = w_{\zeta\zeta} + 2\alpha w_{\zeta\bar{\zeta}} + \alpha^2 w_{\bar{\zeta}\bar{\zeta}},$$

$$v_{z\bar{z}} = \bar{\alpha}w_{\zeta\zeta} + (1 + \alpha\bar{\alpha})w_{\zeta\bar{\zeta}} + \alpha w_{\bar{\zeta}\bar{\zeta}},$$

$$v_{\bar{z}\bar{z}} = \bar{\alpha}^2 w_{\zeta\zeta} + 2\bar{\alpha}w_{\zeta\bar{\zeta}} + w_{\bar{\zeta}\bar{\zeta}}.$$

When we replace this in the PDE in (14.8), we see that the coefficients of both $w_{\zeta\zeta}$ and $w_{\bar{\zeta}\bar{\zeta}}$ vanish if α is a root of the quadratic equation

$$(1 + 2i)\alpha^2 - 6\alpha + 1 - 2i = 0;$$

that is, if $\alpha = 1 - 2i$ or $\alpha = \tfrac{1}{5}(1 - 2i)$. Choosing, say, the first root, we have the transformation

$$\zeta = z + (1 + 2i)\bar{z}, \quad \bar{\zeta} = \bar{z} + (1 - 2i)z, \tag{14.9}$$

which leads to the equation

$$w_{\zeta\bar{\zeta}} = \tfrac{1}{4},$$

with general solution

$$w(\zeta, \bar{\zeta}) = \varphi(\zeta) + \bar{\varphi}(\bar{\zeta}) + \tfrac{1}{4}\zeta\bar{\zeta}, \tag{14.10}$$

where $\varphi(\zeta)$ is an arbitrary analytic function. As in (14.7), and taking (14.9) into account, we now write

$$\varphi(\zeta) = \sum_{n=0}^{\infty} a_n \zeta^n = \sum_{n=0}^{\infty} a_n \big(z + (1+2i)\bar{z}\big)^n,$$

so, by (14.9) and (14.10),

$$v(z,\bar{z}) = \sum_{n=0}^{\infty} \Big[a_n \big(z + (1+2i)\bar{z}\big)^n + \bar{a}_n \big(\bar{z} + (1-2i)z\big)^n \Big]$$
$$+ \tfrac{1}{4}\big(z + (1+2i)\bar{z}\big)\big(\bar{z} + (1-2i)z\big). \qquad (14.11)$$

On the boundary ∂D, this and the BC in (14.8) give rise to the equality

$$\sum_{n=0}^{\infty} \Big[a_n \big(\sigma + (1+2i)\sigma^{-1}\big)^n + \bar{a}_n \big(\sigma^{-1} + (1-2i)\sigma\big)^n \Big]$$
$$+ \tfrac{1}{4}\Big[(1+2i)\sigma^{-2} + 6 + (1-2i)\sigma^2 \Big]$$
$$= -\tfrac{1}{4}(1-3i)\sigma^{-2} + \tfrac{3}{2} - \tfrac{1}{4}(1+3i)\sigma^2.$$

Expanding the binomials on the left-hand side and equating the coefficients of each power of σ on both sides, we immediately note that

$$a_n = 0, \quad n = 3, 4, \ldots,$$

and that, in this case,

$$a_0 + \bar{a}_0 + 2(1+2i)a_2 + 2(1-2i)\bar{a}_2 + \tfrac{3}{2} = \tfrac{3}{2},$$
$$a_1 + (1-2i)\bar{a}_1 = 0,$$
$$a_2 - (3+4i)\bar{a}_2 + \tfrac{1}{4}(1-2i) = -\tfrac{1}{4}(1+3i).$$

The second and third equalities, taken together with their conjugates, yield the systems

$$a_1 + (1-2i)\bar{a}_1 = 0,$$
$$(1+2i)a_1 + \bar{a}_1 = 0,$$

$$a_2 - (3+4i)\bar{a}_2 = -\tfrac{1}{4}(2+i),$$
$$(-3+4i)a_2 + \bar{a}_2 = -\tfrac{1}{4}(2-i),$$

from which $a_1 = 0$ and $a_2 = \frac{1}{16}(2 + i)$. Replacing a_2 in the first equality, we find that $a_0 + \bar{a}_0 = 0$. Hence, by (14.11),

$$v(z, \bar{z}) = a_0 + \bar{a}_0 + a_1\left(z + (1 + 2i)\bar{z}\right) + \bar{a}_1\left(\bar{z} + (1 - 2i)z\right)$$
$$+ a_2\left(z + (1 + 2i)\bar{z}\right)^2 + \bar{a}_2\left(\bar{z} + (1 - 2i)z\right)^2$$
$$= -\tfrac{1}{4}(1 + 3i)z^2 + \tfrac{3}{2}z\bar{z} - \tfrac{1}{4}(1 - 3i)\bar{z}^2,$$

or, in Cartesian coordinates,

$$u(x, y) = x^2 + 3xy + 2y^2. \quad \blacksquare$$

14.2. Systems of Equations

As an illustration of the efficiency of the complex variable method in solving linear two-dimensional systems of partial differential equations, we consider the mathematical model of plane deformation of an elastic body. This state is characterized by a two-component displacement vector $u = (u_1, u_2)$ defined in the two-dimensional domain D occupied by the body. In the absence of body forces, u satisfies the system of PDEs

$$\begin{aligned}(\lambda + \mu)\left[(u_1)_{xx} + (u_2)_{xy}\right] + \mu\Delta u_1 = 0,\\(\lambda + \mu)\left[(u_1)_{xy} + (u_2)_{yy}\right] + \mu\Delta u_2 = 0,\end{aligned} \quad (14.12)$$

where λ and μ are physical constants and Δ is the Laplacian. We assume that D is finite, simply connected (roughly, this means that D has no "holes"), and bounded by a simple, smooth, closed contour ∂D, and consider the Dirichlet problem for (14.12); that is, the BVP with the displacement components prescribed on the boundary:

$$u_1\big|_{\partial D} = f_1, \quad u_2\big|_{\partial D} = f_2,$$

where f_1 and f_2 are known functions. Our aim is to find u at every point (x, y) in D.

14.6. Example. Using the same notation as in Example 14.4, consider the BVP

$$\begin{aligned}2\left[(u_1)_{xx} + (u_2)_{xy}\right] + \Delta u_1 = 0,\\2\left[(u_1)_{xy} + (u_2)_{yy}\right] + \Delta u_2 = 0\end{aligned} \quad \text{in } D, \quad (14.13)$$

$$u_1\big|_{\partial D} = \tfrac{1}{2}\sin(2\theta), \quad u_2\big|_{\partial D} = \cos\theta. \quad (14.14)$$

To treat this problem in terms of complex variables, we define the complex displacement

$$U = u_1 + iu_2$$

and see that

$$
\begin{aligned}
(u_1)_x + (u_2)_y &= u_{1,z} + u_{1,\bar{z}} + iu_{2,z} - iu_{2,\bar{z}} \\
&= (u_1 + iu_2)_z + (u_1 - iu_2)_{\bar{z}} = U_z + \bar{U}_{\bar{z}}.
\end{aligned}
\tag{14.15}
$$

Next, we rewrite system (14.13) in the form

$$2[(u_1)_x + (u_2)_y]_x + \Delta u_1 = 0,$$
$$2[(u_1)_x + (u_2)_y]_y + \Delta u_2 = 0$$

and remark that, by (14.2), we have the operational equality $\partial_x + i\partial_y = 2\partial_{\bar{z}}$. Multiplying the second equation above by i, adding it to the first one, and using (14.15) and (14.3), we deduce that

$$
\begin{aligned}
0 &= 2(\partial_x + i\partial_y)\big[(u_1)_x + (u_2)_y\big] + \Delta(u_1 + iu_2) \\
&= 4\partial_{\bar{z}}(U_z + \bar{U}_{\bar{z}}) + \Delta U = 4\big[(U_z + \bar{U}_{\bar{z}})_{\bar{z}} + U_{z\bar{z}}\big],
\end{aligned}
$$

or

$$(2U_z + \bar{U}_{\bar{z}})_{\bar{z}} = 0,$$

with general solution

$$2U_z + \bar{U}_{\bar{z}} = \tfrac{1}{2}\alpha'(z),$$

where α' is an analytic function of z. The algebraic system formed by this equation and its conjugate now yields

$$U_z = \tfrac{1}{3}\alpha'(z) - \tfrac{1}{6}\bar{\alpha}'(\bar{z}); \tag{14.16}$$

therefore, by integration,

$$U(z,\bar{z}) = \tfrac{1}{3}\alpha(z) - \tfrac{1}{6}z\bar{\alpha}'(\bar{z}) + \bar{\beta}(\bar{z}), \tag{14.17}$$

where β is another analytic function of z.

To investigate the arbitrariness of α and β, let p and q be functions of z such that $\alpha + p$ and $\beta + q$ generate the same displacement U as α and β. Then, by (14.16),

$$\tfrac{1}{3}\big[\alpha'(z) + p'(z)\big] - \tfrac{1}{6}\big[\bar{\alpha}'(\bar{z}) + \bar{p}'(\bar{z})\big] = \tfrac{1}{3}\alpha'(z) - \tfrac{1}{6}\bar{\alpha}'(\bar{z}),$$

from which $2p'(z) - \bar{p}'(\bar{z}) = 0$. This equation together with its conjugate yield $p'(z) = 0$, so $p(z) = c$, where c is a complex number. Using (14.17), we now see that

$$\tfrac{1}{3}\alpha(z) + \tfrac{1}{3}c - \tfrac{1}{6}z\bar{\alpha}'(\bar{z}) + \bar{\beta}(\bar{z}) + \bar{q}(\bar{z}) = \tfrac{1}{3}\alpha(z) - \tfrac{1}{6}z\bar{\alpha}'(\bar{z}) + \bar{\beta}(\bar{z}),$$

so $q(z) = -\tfrac{1}{3}\bar{c}$. The arbitrariness produced by the complex constant c in the functions α and β can be eliminated by imposing an additional condition. For example, if the origin is in D, we may ask that

$$\alpha(0) = 0. \tag{14.18}$$

As in the preceding section, let σ be a generic point on the circle ∂D. Then, by (14.5) and the fact that $\bar{\sigma} = \sigma^{-1}$, the BC (14.14) can be written as

$$U\big|_{\partial D} = (u_1 + iu_2)\big|_{\partial D} = \tfrac{1}{4}i(\sigma^{-2} + 2\sigma^{-1} + 2\sigma - \sigma^2);$$

hence, in view of (14.17), we must have

$$\tfrac{1}{3}\alpha(\sigma) - \tfrac{1}{6}\sigma\bar{\alpha}'(\sigma^{-1}) + \bar{\beta}(\sigma^{-1}) = \tfrac{1}{4}i(\sigma^{-2} + 2\sigma^{-1} + 2\sigma - \sigma^2). \tag{14.19}$$

Since D is finite and simply connected, we can consider series expansions of the analytic functions α and β, of the form

$$\alpha(z) = \sum_{n=0}^{\infty} a_n z^n, \quad \beta(z) = \sum_{n=0}^{\infty} b_n z^n.$$

Substituting these series in (14.19), we arrive at

$$\sum_{n=0}^{\infty} \left(\tfrac{1}{3}a_n\sigma^n - \tfrac{1}{6}n\bar{a}_n\sigma^{-n+2} + \bar{b}_n\sigma^{-n}\right) = \tfrac{1}{4}i(\sigma^{-2} + 2\sigma^{-1} + 2\sigma - \sigma^2).$$

The next step consists in equating the coefficients of each power of σ on both sides, and it is clear that we have $a_n = b_n = 0$ for all $n = 3, 4, \ldots$. Thus, since (14.18) implies that $a_0 = 0$, the only nonzero coefficients are given by the equalities

$$\bar{b}_2 = \tfrac{1}{4}i, \quad \bar{b}_1 = \tfrac{1}{2}i, \quad \bar{b}_0 - \tfrac{1}{3}\bar{a}_2 = 0, \quad \tfrac{1}{3}a_1 - \tfrac{1}{6}\bar{a}_1 = \tfrac{1}{2}i, \quad \tfrac{1}{3}a_2 = -\tfrac{1}{4}i.$$

Coefficient a_1 is computed by combining the equation that it satisfies with its complex conjugate form. In the end, we obtain

$$a_1 = i, \quad a_2 = -\tfrac{3}{4}i, \quad b_0 = -\tfrac{1}{4}i, \quad b_1 = -\tfrac{1}{2}i, \quad b_2 = -\tfrac{1}{4}i,$$

which generate the functions

$$\alpha(z) = iz - \tfrac{3}{4}iz^2, \quad \beta(z) = -\tfrac{1}{4}i - \tfrac{1}{2}iz - \tfrac{1}{4}iz^2$$

and, by (14.17), the complex displacement

$$U(z,\bar{z}) = \tfrac{1}{4}i(1 + 2z + 2\bar{z} - z^2 - z\bar{z} + \bar{z}^2).$$

Hence, in terms of Cartesian coordinates, the solution of the given BVP is

$$u_1(x,y) = \operatorname{Re}\left(U(z,\bar{z})\right) = xy,$$
$$u_2(x,y) = \operatorname{Im}\left(U(z,\bar{z})\right) = \tfrac{1}{4}\left(1 + 4x - x^2 - y^2\right). \quad \blacksquare$$

14.7. Remark. If the functions f_1 and f_2 prescribed on ∂D are not finite sums of integral powers of σ, they need to be expanded in full Fourier series. Using an argument similar to that in Section 8.1, we can write such a series in the form (see Chapter 8)

$$f(\theta) = \sum_{n=-\infty}^{\infty} c_n e^{-in\theta} = \sum_{n=-\infty}^{\infty} c_n \sigma^{-n},$$

where

$$c_n = \frac{1}{2\pi} \int_{-\pi}^{\pi} f(\theta) e^{in\theta}\, d\theta. \quad \blacksquare$$

Exercises

In (1)–(6) find the solution $u(x,y)$ of the BVP

$$\Delta u = q \quad \text{in } D,$$
$$u = f \quad \text{on } \partial D,$$

where D and ∂D are the disk with the center at the origin and radius 1 and its circular boundary, respectively, θ is the polar angle mentioned earlier in this chapter, and the number q and function f are as indicated.

(1) $q = -6$, $f(\theta) = \frac{1}{2}[1 + 5\cos(2\theta)]$.

(2) $q = 8$, $f(\theta) = 2 + 2\cos\theta - 3\cos(2\theta)$.

(3) $q = -2$, $f(\theta) = \frac{1}{2}[-5 + 2\sin\theta - 3\cos(2\theta) + 3\sin(2\theta)]$.

(4) $q = 2$, $f(\theta) = \frac{1}{2}[1 + 6\sin\theta - 5\cos(2\theta) + \sin(2\theta)]$.

(5) $q = 4$, $f(\theta) = 1 + 2\cos\theta - 3\sin\theta + 2\cos(2\theta) - 2\sin(2\theta)$.

(6) $q = -4$, $f(\theta) = 2 + 2\cos\theta - \sin\theta + 2\cos(2\theta) - 4\sin(2\theta)$.

In (7)–(14) find the solution $u(x,y)$ of the BVP

$$u_{xx} + au_{xy} + bu_{yy} = q \quad \text{in } D,$$
$$u = f \quad \text{on } \partial D,$$

where D and ∂D are the disk with the center at the origin and radius 1 and its circular boundary, respectively, θ is the polar angle, and the numbers a, b, and q and function f are as indicated.

(7) $a = -4$, $b = 5$, $q = 18$, $f(\theta) = \frac{1}{2}[3 + \cos(2\theta) - \sin(2\theta)]$.

(8) $a = -2$, $b = 5$, $q = -8$, $f(\theta) = 2\cos\theta + 2\cos(2\theta) + \sin(2\theta)$.

(9) $a = 2$, $b = 2$, $q = 2$, $f(\theta) = \frac{1}{2}[-1 - 2\sin\theta + 3\cos(2\theta) + 4\sin(2\theta)]$.

(10) $a = 4$, $b = 5$, $q = -28$,
$f(\theta) = 2 + \cos\theta - 2\sin\theta + 3\cos(2\theta) + \sin(2\theta)$.

(11) $a = -1$, $b = \frac{5}{4}$, $q = \frac{25}{2}$,
$f(\theta) = \frac{1}{2}[-4 + 4\cos\theta - 2\cos(2\theta) - 3\sin(2\theta)]$.

(12) $a = -4$, $b = \frac{17}{4}$, $q = -\frac{67}{2}$,
$f(\theta) = \frac{1}{2}[-5 + 2\cos\theta + 5\cos(2\theta) + 3\sin(2\theta)]$.

(13) $a = -2$, $b = 10$, $q = 66$,
$f(\theta) = 3 + \cos\theta - 2\sin\theta - \cos(2\theta) - \sin(2\theta)$.

(14) $a = -4$, $b = 13$, $q = -4$,
$f(\theta) = 1 + 2\cos\theta - 3\sin\theta + 2\cos(2\theta) - 2\sin(2\theta)$.

In (15)–(24) find the solution $(u_1(x,y), u_2(x,y))$ of the BVP

$$(\lambda + \mu)[(u_1)_{xx} + (u_2)_{xy}] + \mu\Delta u_1 = 0,$$
$$\quad \text{in } D,$$
$$(\lambda + \mu)[(u_1)_{xy} + (u_2)_{yy}] + \mu\Delta u_2 = 0$$
$$u_1 = f_1, \quad u_2 = f_2 \quad \text{on } \partial D,$$

where D and ∂D are the disk with the center at the origin and radius 1 and its circular boundary, respectively, Δ is the Laplacian, θ is the polar angle, and the numbers λ and μ and functions f_1 and f_2 are as indicated.

(15) $\lambda = 1$, $\mu = \frac{1}{2}$, $f_1(\theta) = 2\cos\theta - \cos(2\theta)$, $f_2(\theta) = \cos\theta + \sin(2\theta)$.

(16) $\lambda = 1$, $\mu = 1$, $f_1(\theta) = \sin\theta + \cos(2\theta)$, $f_2(\theta) = 1 + 2\sin(2\theta)$.

(17) $\lambda = 2$, $\mu = 1$, $f_1(\theta) = 2 - \cos\theta + 2\sin(2\theta)$, $f_2(\theta) = 1 - \cos(2\theta)$.

(18) $\lambda = -1$, $\mu = 2$,
$$f_1(\theta) = \cos\theta + 2\sin(2\theta), \quad f_2(\theta) = 1 + 2\cos(2\theta) + \sin(2\theta).$$

(19) $\lambda = -2$, $\mu = 3$,
$$f_1(\theta) = 1 + 2\cos(2\theta) - \sin(2\theta), \quad f_2(\theta) = \cos\theta - 2\sin(2\theta).$$

(20) $\lambda = 1$, $\mu = 2$,
$$f_1(\theta) = 1 - \sin\theta + \cos(2\theta), \quad f_2(\theta) = \cos(2\theta) - 2\sin(2\theta).$$

(21) $\lambda = \frac{1}{2}$, $\mu = 1$,
$$f_1(\theta) = 2 + \sin\theta + \sin(2\theta), \quad f_2(\theta) = 3\cos(2\theta) - \sin(2\theta).$$

(22) $\lambda = -\frac{1}{2}$, $\mu = 1$,
$$f_1(\theta) = -\cos\theta + 2\cos(2\theta), \quad f_2(\theta) = 2 - \cos(2\theta) - \sin(2\theta).$$

(23) $\lambda = 3$, $\mu = 1$,
$$f_1(\theta) = \cos(2\theta) - \sin(2\theta), \quad f_2(\theta) = 1 + \sin\theta - 2\cos(2\theta).$$

(24) $\lambda = -\frac{1}{2}$, $\mu = 2$,
$$f_1(\theta) = \cos\theta - 2\sin(2\theta), \quad f_2(\theta) = \cos(2\theta) + 3\sin(2\theta).$$

J

Answers to Odd-Numbered Exercises

CHAPTER 1

(1) $y = C(x^2 + 1)$.   (3) $y = (x-1)^{-2}(x^3/3 - x^2/2 + C)$.

(5) $y = Ce^{-5x/2}$.   (7) $y = C_1 e^x + C_2 e^{3x}$.

(9) $y = (C_1 + C_2 x)e^{-x/2}$.   (11) $y = e^{-x}[C_1 \cos(2x) + C_2 \sin(2x)]$.

(13) $y = Ce^{-2x} + x - 1/2 + e^{4x}/6$.   (15) $y = (C + x/2)e^{x/2}$.

(17) $y = C_1 \cosh x + C_2 \sinh x - x^2 + x - 4$.   (19) $y = C_1 e^{5x} + (C_2 - 3x)e^{-5x}$.

(21) $y = C_1 x^{3/2} + C_2 x^{-1}$.   (23) Linear.   (25) Nonlinear.

CHAPTER 2

(1) $f(x) \sim 1/2 + \sum\limits_{n=1}^{\infty} [(-1)^n - 1](1/(n\pi)) \sin(n\pi x)$.

(3) $f(x) \sim 1/2 + \sum\limits_{n=1}^{\infty} [1 - (-1)^n](5/(n\pi)) \sin(n\pi x)$.

(5) $f(x) \sim 1 + \sum\limits_{n=1}^{\infty} (-1)^{n+1}(2/(n\pi)) \sin(n\pi x)$.

(7) $f(x) \sim \sum\limits_{n=1}^{\infty} \{[1 - (-1)^n](2/(n^2\pi^2)) \cos(n\pi x/2)$
$$+ [3 - (-1)^n](1/(n\pi)) \sin(n\pi x/2)\}.$$

(9) $f(x) \sim -1/4 + \sum\limits_{n=1}^{\infty} \{[(-1)^n - 1](1/(n^2\pi^2)) \cos(n\pi x)$
$$- [2(-1)^n + 1](1/(n\pi)) \sin(n\pi x)\}.$$

(11) $f(x) \sim 10/3 + \sum\limits_{n=1}^{\infty} (-1)^n(4/(n^2\pi^2))[\cos(n\pi x) + n\pi \sin(n\pi x)]$.

(13) $f(x) \sim (e^4 - 1)/(4e^2) + \sum\limits_{n=1}^{\infty} (-1)^n[(e^4 - 1)/(e^2(4 + n^2\pi^2))]$
$$\times [2\cos(n\pi x/2) - n\pi \sin(n\pi x/2)].$$

(15) $f(x) \sim 9/8 + \sum\limits_{n=1}^{\infty} \{(2/(n^2\pi^2))[(-1)^n - \cos(n\pi/2)] \cos(n\pi x/2)$
$$+ [(-1)^{n+1}(3/(n\pi)) + (2/(n^2\pi^2)) \sin(n\pi/2)] \sin(n\pi x/2)\}.$$

(17) $f(x) \sim \sum\limits_{n=1}^{\infty} (2/(n\pi))[\cos(n\pi/2) - (-1)^n] \sin(n\pi x/2)$;

$f(x) \sim 1/2 - \sum\limits_{n=1}^{\infty} (2/(n\pi)) \sin(n\pi/2) \cos(n\pi x/2)$;

(19) $f(x) \sim \sum\limits_{n=1}^{\infty} (2/(n\pi))[1 + (-1)^n - 2\cos(n\pi/2)] \sin(n\pi x/2)$;

$f(x) \sim \sum\limits_{n=1}^{\infty} (4/(n\pi)) \sin(n\pi/2) \cos(n\pi x/2)$.

(21) $f(x) \sim \sum\limits_{n=1}^{\infty} [2-(-1)^n](2/(n\pi))\sin(n\pi x)$;

$f(x) \sim 3/2 + \sum\limits_{n=1}^{\infty} [1-(-1)^n](2/(n^2\pi^2))\cos(n\pi x)$.

(23) $f(x) \sim \sum\limits_{n=1}^{\infty} (2/(n\pi))[-3\cos(n\pi/2)+(2/(n\pi))\sin(n\pi/2)+(-1)^n 2]\sin(n\pi x/2)$;

$f(x) \sim -3/4 + \sum\limits_{n=1}^{\infty} (2/(n\pi))[3\sin(n\pi/2) + (2/(n\pi))\cos(n\pi/2) - 2/(n\pi)]$
$$\times \cos(n\pi x/2).$$

(25) $f(x) \sim \sum\limits_{n=1}^{\infty} (2/(n\pi))[2+(-1)^n-3\cos(n\pi/2)+(4/(n\pi))\sin(n\pi/2)]\sin(n\pi x/2)$;

$f(x) \sim 1 + \sum\limits_{n=1}^{\infty} \{(6/(n\pi))\sin(n\pi/2) + (8/(n^2\pi^2))\cos(n\pi/2)$
$$+[(-1)^{n+1}-1](4/(n^2\pi^2))\}\cos(n\pi x/2).$$

(27) $f(x) \sim \sum\limits_{n=1}^{\infty} (2/(n^3\pi^3))[8(-1)^n + n^2\pi^2 - (8+3n^2\pi^2)\cos(n\pi/2)$
$$+ 2n\pi\sin(n\pi/2)]\sin(n\pi x/2);$$

$f(x) \sim 5/12 + \sum\limits_{n=1}^{\infty} (2/(n^3\pi^3))\{[2(-1)^n - 1]2n\pi + 2n\pi\cos(n\pi/2)$
$$+(8+3n^2\pi^2)\sin(n\pi/2)\}\cos(n\pi x/2).$$

(29) $f(x) \sim 3\sin x + \sum\limits_{n=2}^{\infty} (-1)^{n+1}(2/n)\sin(nx)$;

$f(x) \sim 2/\pi+\pi/2-(4/\pi)\cos x+\sum\limits_{n=2}^{\infty} [2(1-(-1)^n-2n^2)/((n^2-1)n^2\pi)]\cos(nx)$.

CHAPTER 3

(1) Regular. (3) Singular. (5) Singular.

(7) $\lambda_n = (2n-1)^2/4$, $f_n(x) = \sin\left((2n-1)x/2\right)$, $n = 1, 2, \ldots$.

(9) $\lambda_n = \zeta_n^2$, where ζ_n are the roots of the equation $\tan\zeta = -\zeta$,

$f_n(x) = \zeta_n\cos(\zeta_n x) + \sin(\zeta_n x)$, $n = 1, 2, \ldots$.

(11) $\lambda_n = \zeta_n^2$, where ζ_n are the roots of the equation $\cot\zeta = \zeta$,

$f_n(x) = \zeta_n\cos(\zeta_n x) + \sin(\zeta_n x)$, $n = 1, 2, \ldots$.

(13) $\lambda_n = (4 + n^2\pi^2)/3$, $f_n(x) = e^{-2x}\sin(n\pi x)$, $n = 1, 2, \ldots$.

(15) $\lambda_n = 2\zeta_n^2 + 9/8$, where ζ_n are the roots of the equation $\tan\zeta = 4\zeta/3$,

$f_n(x) = e^{-3x/4}\sin(\zeta_n x)$, $n = 1, 2, \ldots$.

(17) $u(x) \sim \sum\limits_{n=1}^{\infty} [4/((2n-1)\pi)]\sin((2n-1)x/2)$.

(19) $u(x) \sim \sum\limits_{n=1}^{\infty} [8(2n-1-3(-1)^n)/((2n-1)^2\pi)]\sin((2n-1)x/2)$.

(21) $u(x) \sim \sum\limits_{n=1}^{\infty} \{[8-20\sin((3-2n)\pi/4)]/((2n-1)\pi)\}\sin((2n-1)x/2)$.

(23) $u(x) \sim \sum\limits_{n=1}^{\infty} [2/((2n-1)^2\pi)]\{(2n-1)(\pi-2)\sin((3-2n)\pi/4)$
$$- 4[1+(-1)^n - 2n + \sin((2n-1)\pi/4)]\}\sin((2n-1)x/2).$$

(25) $u(x) \sim \sum_{n=1}^{\infty} [2(-1)^{n+1}/((2n-1)\pi)]\cos((2n-1)\pi x/2)$.

(27) $u(x) \sim \sum_{n=1}^{\infty} [(8 - 4(-1)^n(2n-1)\pi)/((2n-1)^2\pi^2)]\cos((2n-1)\pi x/2)$.

(29) $u(x) \sim \sum_{n=1}^{\infty} \{4[(-1)^{n+1} - 3\sin((2n-1)\pi/4)]/((2n-1)\pi)\}\cos((2n-1)\pi x/2)$.

(31) $u(x) \sim \sum_{n=1}^{\infty} [4/((2n-1)^2\pi^2)]\{4(2n-1)\pi\sin((2n-1)\pi/4)$
$\quad - 4\sin((3-2n)\pi/4) + (2n-1)\pi[(-1)^n - \sin((2n-1)\pi/4)]\}\cos((2n-1)\pi x/2)$.

(33) $u(x) \sim 1.1892\sin(2.0288x) + 0.3134\sin(4.9132x) + 0.2776\sin(7.9787x)$
$\quad + 0.1629\sin(11.0855x) + 0.1499\sin(14.2074x) + \cdots$.

(35) $u(x) \sim 0.9639\sin(1.1444x) + 0.8718\sin(2.5435x) + 0.4227\sin(4.0481x)$
$\quad + 0.3836\sin(5.5863x) + 0.2583\sin(7.1382x) + \cdots$.

(37) $u(x) \sim 2.4222\sin(2.0288x) - 2.9144\sin(4.9132x) - 1.3509\sin(7.9787x)$
$\quad + 0.3704\sin(11.0855x) + 0.3322\sin(14.2074x) + \cdots$.

(39) $u(x) \sim 1.0549\sin(2.0288x) + 0.0227\sin(4.9132x) - 0.0157\sin(7.9787x)$
$\quad - 0.3785\sin(11.0855x) + 0.0242\sin(14.2074x) + \cdots$.

(41) $u(x) \sim e^{-2x}[3.8004\sin(\pi x) - 1.8466\sin(2\pi x) + 1.7035\sin(3\pi x)$
$\quad - 0.9917\sin(4\pi x) + 1.0511\sin(5\pi x) + \cdots]$.

(43) $u(x) \sim e^{-2x}[8.3031\sin(\pi x) - 5.7781\sin(2\pi x) + 4.5576\sin(3\pi x)$
$\quad - 3.2366\sin(4\pi x) + 2.8691\sin(5\pi x) + \cdots]$.

(45) $u(x) \sim e^{-2x}[-6.4533\sin(\pi x/2) + 9.8385\sin(\pi x) - 4.7668\sin(3\pi x/2)$
$\quad + 1.9083\sin(2\pi x) - 2.4786\sin(5\pi x/2) + \cdots]$.

(47) $u(x) \sim e^{-2x}[-3.3732\sin(\pi x) + 2.1406\sin(2\pi x) - 1.8243\sin(3\pi x)$
$\quad + 0.7873\sin(4\pi x) - 1.0104\sin(5\pi x) + \cdots]$.

(49) (i) $u(x) \sim 1.6020J_0(2.4048x) - 1.0463J_0(5.5201x) + 0.8514J_0(8.6537x)$
$\quad - 0.7296J_0(11.7915x) + 0.6485J_0(14.9309x) + \cdots$.

(ii) $u(x) \sim 2.2131J_1(3.8317x) - 0.5171J_1(7.0156x) + 1.1046J_1(10.1735x)$
$\quad - 0.4550J_1(13.3237x) + 0.8113J_1(16.4706x) + \cdots$.

(51) (i) $u(x) \sim -0.8503J_0(2.4048x) + 2.2951J_0(5.5201x) - 1.5435J_0(8.6537x)$
$\quad + 1.5114J_0(11.7915x) - 1.2461J_0(14.9309x) + \cdots$.

(ii) $u(x) \sim -1.6747J_1(3.8317x) + 2.3326J_1(7.0156x) - 1.2572J_1(10.1735x)$
$\quad + 1.6073J_1(13.3237x) - 1.0429J_1(16.4706x) + \cdots$.

(53) (i) $u(x) \sim 0.8947J_0(2.4048x) - 4.0540J_0(5.5201x) + 2.5517J_0(8.6537x)$
$\quad - 0.0663J_0(11.7915x) + 0.7009J_0(14.9309x) + \cdots$.

(ii) $u(x) \sim 2.2028 J_1(3.8317x) - 4.4956 J_1(7.0156x) + 0.8536 J_1(10.1735x)$
$$- 0.1883 J_1(13.3237x) + 1.4431 J_1(16.4706x) + \cdots.$$

(55) (i) $u(x) \sim 1.2920 J_0(2.4048x) - 0.2378 J_0(5.5201x) - 0.0983 J_0(8.6537x)$
$$+ 0.0356 J_0(11.7915x) + 0.0602 J_0(14.9309x) + \cdots.$$

(ii) $u(x) \sim 1.6299 J_1(3.8317x) + 0.4360 J_1(7.0156x) + 0.2091 J_1(10.1735x)$
$$+ 0.2123 J_1(13.3237x) + 0.2485 J_1(16.4706x) + \cdots.$$

(57) $u(x) = (13/3)P_0(x) - 3P_1(x) + (2/3)P_2(x).$

(59) $u(x) \sim -P_0(x) + 3P_1(x) - (7/4)P_3(x) + (11/8)P_5(x) + \cdots.$

(61) $u(x) \sim -(5/4)P_0(x) + (5/4)P_1(x) + (5/16)P_2(x) - (7/16)P_3(x) - (3/32)P_4(x)$
$$+ (11/32)P_5(x) + \cdots.$$

(63) $u(x) \sim P_0(x) - (3/2)P_1(x) - (5/4)P_2(x) + (7/8)P_3(x) + (3/8)P_4(x)$
$$- (11/16)P_5(x) + \cdots.$$

(65) $u(\theta,\varphi) = -(2\sqrt{\pi}/3)Y_{0,0}(\theta,\varphi) - \sqrt{2\pi/15}\, Y_{2,-2}(\theta,\varphi) - (2\sqrt{5\pi}/3)Y_{2,0}(\theta,\varphi)$
$$- \sqrt{2\pi/15}\, Y_{2,2}(\theta,\varphi).$$

(67) $u(\theta,\varphi) \sim \sqrt{\pi}\, Y_{0,0}(\theta,\varphi) + (\sqrt{3\pi}/2)Y_{1,0}(\theta,\varphi) + \cdots.$

(69) $u(\theta,\varphi) \sim 2\sqrt{\pi}\, Y_{0,0}(\theta,\varphi) + \sqrt{\pi/6}\,(1-i)Y_{1,-1}(\theta,\varphi) + \sqrt{3\pi}\, Y_{1,0}(\theta,\varphi)$
$$- \sqrt{\pi/6}\,(1+i)Y_{1,1}(\theta,\varphi) - (1/8)\sqrt{15\pi/2}\,(1-i)Y_{2,-1}(\theta,\varphi)$$
$$+ (1/8)\sqrt{15\pi/2}\,(1+i)Y_{2,1}(\theta,\varphi) + \cdots.$$

(71) $u(\theta,\varphi) \sim (\sqrt{\pi}/3)Y_{0,0}(\theta,\varphi) - (5/8)\sqrt{\pi/6}\,iY_{1,-1}(\theta,\varphi) - (\sqrt{3\pi}/8)Y_{1,0}(\theta,\varphi)$
$$- (5/8)\sqrt{\pi/6}\,iY_{1,1}(\theta,\varphi) + \sqrt{\pi/30}\, Y_{2,-2}(\theta,\varphi)$$
$$+ (83/8)\sqrt{\pi/30}\,iY_{2,-1}(\theta,\varphi) - (1/3)\sqrt{\pi/5}\, Y_{2,0}(\theta,\varphi)$$
$$+ (83/8)\sqrt{\pi/30}\,iY_{2,1}(\theta,\varphi) + \sqrt{\pi/30}\, Y_{2,2}(\theta,\varphi) + \cdots.$$

CHAPTER 4

(5) $\alpha = 3/4, \; \beta = -17/8.$

(7) $\alpha = 4, \; \beta = -10.$

(9) (i) $\alpha = 1, \; \beta = -1/2, \; v_{tt}(x,t) + (7/4)v(x,t) = v_{xx}(x,t);$

(ii) $\alpha = 1 - \sqrt{7}/2, \; \beta = -1/2, \; v_{tt}(x,t) = v_{xx}(x,t) - \sqrt{7}\,v_x(x,t);$

(11) (iii) $\alpha = \sqrt{2}/4, \; \beta = -1/2, \; v_{tt}(x,t) + v_t(x,t) = 2v_{xx}(x,t).$

(13) $\alpha = 1, \; \beta = 2, \; v_{xx}(x,y) + v_{yy}(x,y) + 2v(x,y) = 0.$

(15) $\alpha = 3/2, \; \beta = 1/2, \; v_{xx}(x,y) + v_{yy}(x,y) - (1/2)v(x,y) = 0.$

CHAPTER 5

(1) $u(x,t) = \sin(2\pi x)e^{-4\pi^2 t} - 3\sin(6\pi x)e^{-36\pi^2 t}$.

(3) $u(x,t) = \sum_{n=1}^{\infty} [(-1)^n - 1](4/(n\pi))\sin(n\pi x)e^{-n^2\pi^2 t}$.

(5) $u(x,t) = \sum_{n=1}^{\infty} [1 - (-1)^n 3](2/(n\pi))\sin(n\pi x)e^{-n^2\pi^2 t}$.

(7) $u(x,t) = \sum_{n=1}^{\infty} [(2/(n^2\pi^2))\sin(n\pi/2) - (1/(n\pi))\cos(n\pi/2)]\sin(n\pi x)e^{-n^2\pi^2 t}$.

(9) $u(x,t) = 3 - 2\cos(4\pi x)e^{-16\pi^2 t}$.

(11) $u(x,t) = 1/2 + \sum_{n=1}^{\infty} [1 - (-1)^n](6/(n^2\pi^2))\cos(n\pi x)e^{-n^2\pi^2 t}$;

(13) $u(x,t) = -1 - \sum_{n=1}^{\infty} (4/(n\pi))\sin(n\pi/2)\cos(n\pi x)e^{-n^2\pi^2 t}$.

(15) $u(x,t) = 3/4 + \sum_{n=1}^{\infty} \{-(2/(n\pi))\sin(n\pi/2) + (4/(n^2\pi^2))[(-1)^n - \cos(n\pi/2)]\}$
$$\times \cos(n\pi x)e^{-n^2\pi^2 t}.$$

(17) $u(x,t) = 3\sin(n\pi x/2)e^{-\pi^2 t/4} - \sin(5\pi x/2)e^{-25\pi^2 t/4}$.

(19) $u(x,t) = \sum_{n=1}^{\infty} [8/((2n-1)\pi) + (-1)^{n+1}8/((2n-1)^2\pi^2)]$
$$\times \sin((2n-1)\pi x/2)e^{-(2n-1)^2\pi^2 t/4}.$$

(21) $u(x,t) = 2\cos(5\pi x/2)e^{-25\pi^2 t/4}$.

(23) $u(x,t) = \sum_{n=1}^{\infty} [(-1)^n 4/((2n-1)\pi) - 16/((2n-1)^2\pi^2)]$
$$\times \cos((2n-1)\pi x/2)e^{-(2n-1)^2\pi^2 t/4}.$$

(25) $u(x,t) = 2\sin(2\pi x)e^{-4\pi^2 t} - \cos(5\pi x)e^{-25\pi^2 t}$.

(27) $u(x,t) = 3/2 + \sum_{n=1}^{\infty} [(-1)^n - 1](3/(n\pi))\sin(n\pi x)e^{-n^2\pi^2 t}$.

(29) $u(x,t) = -3\sin(2\pi x)\cos(2\pi t) + 4\sin(7\pi x)\cos(7\pi t)$
$$+ (1/(3\pi))\sin(3\pi x)\sin(3\pi t).$$

(31) $u(x,t) = 2\sin(3\pi x)\cos(3\pi t) + \sum_{n=1}^{\infty} [1 - (-1)^n](4/(n^2\pi^2))\sin(n\pi x)\sin(n\pi t)$.

(33) $u(x,t) = \sum_{n=1}^{\infty} \sin(n\pi x)[(4/(n\pi))(3 + 4\cos(n\pi/2))\sin^2(n\pi/4)\cos(n\pi t)$
$$+ (3/(2\pi))\sin(2\pi t)].$$

(35) $u(x,t) = \sum_{n=1}^{\infty} \sin(n\pi x)\{(1/(n^2\pi^2))[(-1)^{n+1}2n\pi + n\pi\cos(n\pi/2)$
$$+ 2\sin(n\pi/2)]\cos(n\pi t) + [(-1)^n - 1](2/(n^2\pi^2))\sin(n\pi t)\}.$$

(37) $u(x,t) = 2 - 3\cos(4\pi x)\cos(4\pi t) + (2/(3\pi))\cos(3\pi x)\sin(3\pi t)$.

(39) $u(x,t) = -3\cos(2\pi x)\cos(2\pi t)$
$$+ \sum_{n=1}^{\infty} [(-1)^n - 1](4/(n^3\pi^3))\cos(n\pi x)\sin(n\pi t).$$

(41) $u(x,t) = 5/2 + \sum_{n=1}^{\infty} -(2/(n\pi))\sin(n\pi/2)\cos(n\pi x)\cos(n\pi t)$
$$- (1/(3\pi))\cos(3\pi x)\sin(3\pi t).$$

(43) $u(x,t) = 13/8 + t/2$
$$+ \sum_{n=1}^{\infty} \cos(n\pi x)\{(1/(n^2\pi^2))[2\cos(n\pi/2) - n\pi\sin(n\pi/2) - 2]\cos(n\pi t)$$
$$+ [1 - (-1)^n](2/(n^3\pi^3))\sin(n\pi t)\}.$$

(45) $u(x,t) = (6/(5\pi))\sin(5\pi x/2)\sin(5\pi t/2)$
$$+ \sum_{n=1}^{\infty} [8/((2n-1)\pi)]\cos((2n-1)\pi/4)\sin((2n-1)\pi x/2)\cos((2n-1)\pi t/2).$$

(47) $u(x,y) = -3\operatorname{csch}(4\pi)\sin(2\pi x)\sinh(2\pi(y-2)) + \operatorname{csch}(6\pi)\sin(3\pi x)\sinh(3\pi y).$

(49) $u(x,y) = 3\operatorname{sech}(2\pi)\cosh(2\pi x)\cos(2\pi y) + 1 - x$
$$+ \sum_{n=1}^{\infty} [(-1)^n - 1](8/(n^3\pi^3))\operatorname{sech}(n\pi/2)\sinh(n\pi(x-1)/2)\cos(n\pi y/2).$$

(51) $u(x,y) = (6/\pi)\operatorname{sech}\pi\cos(\pi x/2)\sinh(\pi y/2)$
$$+ \sum_{n=1}^{\infty} (4/(2n-1)\pi)[(-1)^{n+1}2 - \sin((2n-1)\pi/4)]\operatorname{sech}((2n-1)\pi)$$
$$\times \cos((2n-1)\pi x/2)\cosh((2n-1)\pi(y-2)/2).$$

(53) $u(r,\theta) = 3 - 32r^3\cos(3\theta).$

(55) $u(r,\theta) = 1 + \sum_{n=1}^{\infty} [1 - (-1)^n](1/(2^{n-2}n\pi))r^n\sin(n\theta).$

(57) $u(x,t) = 3\sin(2\pi x)e^{-(1+16\pi^2)t/4+x/2}.$

(59) $u(x,t) = \sum_{n=1}^{\infty} (4/(1+4n^2\pi^2))$
$$\times \{(-1)^{n+1}2n\pi + e^{1/4}[2n\pi\cos(n\pi/2) + \sin(n\pi/2)]\}\sin(n\pi x)$$
$$\times e^{-1/2-(1+4n^2\pi^2)t/4+x/2}.$$

(61) $u(x,t) = (1 + 16\pi^2)^{-1/2}e^{(x-2t)/4}\sin(2\pi x)\sin((1+16\pi^2)^{1/2}t)$
$$- (1 + 36\pi^2)^{-1/2}e^{(x-2t)/4}\sin(3\pi x)$$
$$\times [2(1+36\pi^2)^{1/2}\cos((1+36\pi^2)^{1/2}t) + \sin((1+36\pi^2)^{1/2}t)].$$

(63) $u(x,t) = 4(1+4\pi^2)^{-1/2}e^{(x-2t)/4}\sin(\pi x)\sin((1+4\pi^2)^{1/2}t)$
$$+ \sum_{n=1}^{\infty} 4(1+4n^2\pi^2)^{-1/2}(1+16n^2\pi^2)^{-1}$$
$$\times [4n\pi\cos(n\pi/2) + \sin(n\pi/2) - 4e^{1/8}n\pi]e^{(2x-4t-1)/8}\sin(n\pi x)$$
$$\times [2(1+4n^2\pi^2)^{1/2}\cos((1+4n^2\pi^2)^{1/2}t) + \sin((1+4n^2\pi^2)^{1/2}t)].$$

(65) $u(x,y) = e^{y-1}[e\operatorname{csch}((4+\pi^2)^{1/2}/2)\sin(\pi x/2)\sinh((4+\pi^2)^{1/2}(y-1)/2)$
$$+2\operatorname{csch}((1+\pi^2)^{1/2})\sin(\pi x)\sinh((1+\pi^2)^{1/2}y)].$$

(67) $u(x,y) = -e^y[3\operatorname{csch}(2(1+4\pi^2)^{1/2})\sinh((1+4\pi^2)^{1/2}(x-2))\sin(2\pi y)$
$$+ 2\operatorname{csch}(2(1+\pi^2)^{1/2})\sinh((1+\pi^2)^{1/2}x)\sin(\pi y)].$$

(69) $u(x,y,t) = \sin(\pi x)\sin(2\pi y)\cos(\sqrt{5}\pi t)$
$$- (2/(\sqrt{5}\pi))\sin(2\pi x)\sin(\pi y)\sin(\sqrt{5}\pi t).$$

(71) $u(x,y,t) = -\sin(\pi x)\cos(2\pi y)\cos(\sqrt{5}\pi t) + (3/(2\pi))\sin(2\pi x)\sin(2\pi t).$

(73) $u(x,y,t) = (4/\pi^2)\sum_{m=1}^{\infty}\sum_{n=1}^{\infty} (1/(mn))\{[1 - (-1)^m][1 - (-1)^n]$
$$\times \cos((m^2+n^2)^{1/2}\pi t)$$
$$+(-1)^{m+n}(\pi(m^2+n^2))^{-1/2}\sin((m^2+n^2)^{1/2}\pi t)\}\sin(n\pi x)\sin(m\pi y).$$

(75) $u(r,\theta,t) = [5.2698J_2(5.1356r)\cos(5.1356t) - 0.3168J_2(8.4172r)\cos(8.4172t)$
$+ 2.6215J_2(11.6198r)\cos(11.6198t) - 0.4762J_2(14.7960r)\cos(14.7960t)$
$+ 1.9049J_2(17.9598r)\cos(17.9598t) + \cdots]\sin(2\theta).$

(77) $u(r,\theta,t) = [-0.3382J_1(3.8317r)\cos(3.8317t) + 0.1354J_1(7.0156r)\cos(7.0156t)$
$- 0.0774J_1(10.1735r)\cos(10.1735t) + 0.0516J_1(13.3237r)\cos(13.3237t)$
$- 0.0375J_1(16.4706r)\cos(16.4706t) + \cdots]\sin\theta.$

(79) $u(r,\theta,\varphi) = 1/3 + r^2(2/3 - 2\cos^2\varphi - \sin\theta\sin\varphi\cos\varphi + \cos(2\theta)\sin^2\varphi) + \cdots.$

CHAPTER 6

(1) $u_\infty(x) = 2x^3 - 3x^2 + 4x - 1.$ (3) $u_\infty(x) = x^4 + 2x^2 + 3x - 2.$

(5) $\gamma = -6,\ \ u_\infty(x) = x^3 + 2x^2 - 3x - 1.$ (7) $u_\infty(x) = x^2 + 2x - 1.$

(9) $u_\infty(x) = e^x - 2e^{2x} - x.$ (11) $u_\infty(x) = 3x^2 - 5x - 2.$

(13) $v(x,t) = u(x,t) - t - 1 + x(t^2 - t + 1),\ \ q(x,t) = 2xt + t - 3,\ \ f(x) = 3x - 1.$

(15) $v(x,t) = u(x,t) - 3xt - (1/2)x^2(t^2 - 2t),$
$q(x,t) = -x^2t + x^2 + t^2 - 2x - t - 2,\ \ f(x) = 2x.$

(17) $v(x,t) = u(x,t) - t/2 + 1 - x(t + 2),\ \ q(x,t) = -5/2 + t,\ \ f(x) = 1.$

(19) $v(x,t) = u(x,t) - 2t - 3 + x(7 + 2t - t^2),$
$q(x,t) = -xt,\ \ f(x) = 8x - 2,\ \ g(x) = 3x - 2.$

(21) $v(x,t) = u(x,t) - x(t - 1) - (1/2)x^2(t^2 - t + 3),$
$q(x,t) = -x^2 - xt + t^2 + 2x - t + 3,\ \ f(x) = 1 + 2x - 3x^2/2,\ \ g(x) = x^2/2.$

(23) $v(x,y) = u(x,y) - 2y^2 + 1 - x(1 + y - 3y^2),$
$q(x,y) = -4x - y + 4,\ \ f(x) = 2x + 1,\ \ g(x) = -x.$

(25) $v(x,y) = u(x,y) - x(2 - y) - x^2(4y - 1)/2,$
$q(x,y) = 2x + 3y - 1,\ \ f(x) = x^2/2 + x,\ \ g(x) = 1 - 3x - 3x^2/2.$

(27) $v(x,t) = u(x,t) - 2t - 3 - x(1 - 5t),$
$q(x,t) = xt + 4x + 10t - 3,\ \ f(x) = -1.$

(29) $v(x,t) = u(x,t) + x(2t - 1) - (1/2)x^2(3t + 2),$
$q(x,t) = -3x^2/2 - 5xt - 3x + 7t + 1,\ \ f(x) = 2 - x^2.$

CHAPTER 7

(1) $u(x,t) = [((8\pi^4 + 1)/(8\pi^4))e^{-4\pi^2 t} + (1/(2\pi^2))t - 1/(8\pi^4)]\sin(2\pi x)$
$- 5e^{-16\pi^2 t}\sin(4\pi x).$

(3) $u(x,t) = e^{-\pi^2 t}\sin(\pi x)$
$$+ [((18\pi^2 - 3)/(9\pi^2 - 1))e^{-9\pi^2 t} + (1/(9\pi^2 - 1))e^{-t}]\sin(3\pi x)$$
$$+ (1/(25\pi^2))(e^{-25\pi^2 t} - 1)\sin(5\pi x).$$

(5) $u(x,t) = [(1/\pi^2)(t - 1) - 1/\pi^4 + (1 + 1/\pi^2 + 1/\pi^4)e^{-\pi^2 t}]\sin(\pi x)$
$$+ 2e^{-4\pi^2 t}\sin(2\pi x).$$

(7) $u(x,t) = \sum\limits_{n=1}^{\infty} (2/(n^5\pi^5))\{[(-1)^{n+1} + 2n^4\pi^4\sin^2(n\pi/4)]e^{-n^2\pi^2 t}$
$$+ (-1)^n(1 - n^2\pi^2 t)\}\sin(n\pi x).$$

(9) $u(x,t) = 2t + 2e^{-\pi^2 t}\cos(\pi x) + [1/(4\pi^2) - ((4\pi^2 + 1)/(4\pi^2))e^{-4\pi^2 t}]\cos(2\pi x).$

(11) $u(x,t) = (1/2)t^2 + 1 + [1/\pi^4 - (1/\pi^2)t - (1/\pi^4)e^{-\pi^2 t}]\cos(\pi x)$
$$+ 3e^{-16\pi^2 t}\cos(4\pi x).$$

(13) $u(x,t) = 2 + (1/(\pi^2 - 1))(e^{-t} - e^{-\pi^2 t})\cos(\pi x) - e^{-9\pi^2 t}\cos(3\pi x).$

(15) $u(x,t) = 1 - t^2/2 - 3e^{-4\pi^2 t}\cos(2\pi x)$
$$+ \sum\limits_{n=1}^{\infty} (4/(n^6\pi^6))\{[1 + (-1)^n]e^{-n^2\pi^2 t} - 1 + (-1)^n$$
$$+ [1 - (-1)^n]n^2\pi^2 t\}\cos(n\pi x).$$

(17) $u(x,t) = [((9\pi^2 - 4)/(9\pi^2))e^{-9\pi^2 t/4} + 4/(9\pi^2)]\sin(3\pi x/2)$
$$+ (8/(25\pi^2))(e^{-25\pi^2 t/4} - 1)\sin(5\pi x/2).$$

(19) $u(x,t) = (t^2 - 2t)\sin(3\pi x/2) + 2e^{-t}\sin(5\pi x/2).$

(21) $u(x,t) = \cos(\pi t)\sin(\pi x)$
$$+ [1/(2\pi^2) - (1/(2\pi^2))\cos(2\pi t) - (3/(2\pi))\sin(2\pi t)]\sin(2\pi x).$$

(23) $u(x,t) = [\cos(\pi t) + ((2\pi^2 - 1)/\pi^3)\sin(\pi t) + (1/\pi^2)t]\sin(\pi x)$
$$+ (4/(3\pi))\sin(3\pi t)\sin(3\pi x).$$

(25) $u(x,t) = (1/(8\pi^3))[2\pi(1 + t) - 2\pi\cos(2\pi t) - \sin(2\pi t)]\sin(2\pi x)$
$$+ \sum\limits_{n=1}^{\infty} [(-1)^{n+1} - 1](2/(n^2\pi^2))\sin(n\pi t)\sin(n\pi x).$$

(27) $u(x,t) = (3/2)t^2 + 1 + 2\cos(2\pi t)\cos(2\pi x) + (1/(3\pi))\sin(3\pi t)\cos(3\pi x).$

(29) $u(x,t) = (1/2)t^2 + t + (1/\pi)\sin(\pi t)\cos(\pi x)$
$$+ [2\cos(2\pi t) - ((4\pi^2 + 1)/(8\pi^3))\sin(2\pi t) + (1/(4\pi^2))t]\cos(2\pi x).$$

(31) $u(x,t) = -1/2 - t^2/4$
$$+ \sum\limits_{n=1}^{\infty} [(-1)^n - 1](2/(n^4\pi^4))[(n^2\pi^2 + 1)\cos(n\pi t) - 1]\cos(n\pi x).$$

(33) $u(x,y) = -\operatorname{cosech}(2\pi)\sinh(\pi(y - 2))\sin(\pi x)$
$$+ [(1/(4\pi^2)) - 1)\operatorname{cosech}(4\pi)\sinh(2\pi y)$$
$$- (1/(4\pi^2))\operatorname{cosech}(4\pi)\sinh(2\pi(y - 2)) - 1/(4\pi^2)]\sin(2\pi x)$$
$$+ 2\operatorname{cosech}(6\pi)\sinh(3\pi(y - 2))\sin(3\pi x).$$

(35) $u(x,y) = (2/\pi^3)\operatorname{sech}\pi\sin(\pi x)[\sinh(\pi(y-1)) - \pi y\cosh\pi]$
$$+ \sum_{n=1}^{\infty} (2/(n^2\pi^2))\operatorname{csch}(2n\pi)\sin(n\pi x)\cosh(n\pi(y-2)).$$

(37) $u(x,y) = \operatorname{cosech}(\pi/2)\sinh(\pi x/2)\sin(\pi y/2)$
$$+ \{1 - (\operatorname{cosech}\pi)[\sinh(\pi x) + \sinh(\pi(x-1))]\}\sin(\pi y).$$

(39) $u(x,y) = (1/\pi^3)[\operatorname{sech}\pi\sinh(\pi(x-1)) + \pi(1-x)]\sin(\pi y)$
$$- \sum_{n=1}^{\infty} (4/(n\pi))\operatorname{sech}(n\pi/2)\cosh(n\pi x/2)\sin(n\pi y/2).$$

(41) $u(x,y) = 1 - y^2 + (2y-1)\cos(2\pi x).$

(43) $u(x,y) = 2\operatorname{sech}(2\pi)\cos(\pi x)\cosh(\pi(y-2)) + (1/(8\pi^4))\cos(2\pi x)$
$$\times\,[4\pi(1-\pi^2)\operatorname{sech}(4\pi)\sinh(2\pi y) + \operatorname{sech}(4\pi)\cosh(2\pi(y-2)) - 2\pi^2 y^2 - 1].$$

(45) $u(x,y) = 2\operatorname{csch}(\pi/2)\sinh(\pi x/2)\cos(\pi y/2)$
$$- (1/\pi^4)[\pi^2 - 2 - \pi^2 x^2 + 2\operatorname{csch}\pi\sinh(\pi x)$$
$$- (\pi^4 - \pi^2 + 2)\operatorname{csch}\pi\sinh(\pi(x-1))]\cos(\pi y).$$

(47) $u(x,y) = (4/(27\pi^3))\operatorname{sech}(3\pi/2)[3\pi\cosh(3\pi x/2) + 2\sinh(3\pi(x-1)/2)$
$$- 3\pi x\cosh(3\pi/2)]\cos(3\pi y/2)$$
$$+ y + \sum_{n=1}^{\infty}[1 - (-1)^n](4/(n^2\pi^2))\operatorname{sech}(n\pi/2)\cosh(n\pi x/2)\cos(n\pi y/2).$$

(49) $u(r,\theta) = 1 - 2r^2 + 2r\sin\theta + 2r^3\cos(3\theta).$

(51) $u(r,\theta) = 2r^3 - r^2 + r^3\cos\theta + (r^4 + r^2)\sin(3\theta).$

(53) $u(r,\theta) = (1/20)(-4r^3 + 4r^2 + 5r^2\ln r)\sin(2\theta)$
$$+ 1/2 + \sum_{n=1}^{\infty}[1 - (-1)^n](1/(n\pi))r^n\sin(n\theta).$$

(55) $u(x,t) = (1/(7 - 16\pi^2)^2)\{8[10 + 32\pi^2 + (21 - 48\pi^2)t]$
$$- (256\pi^4 + 32\pi^2 + 129)e^{(7-16\pi^2)t/8}\}\sin(\pi x).$$

(57) $u(x,t) = -e^{(7-64\pi^2)t/8 + x/4}\sin(2\pi x)$
$$+ \sum_{n=1}^{\infty}[(-1)^n - 1][16/(n\pi(16n^2\pi^2 - 7))][e^{(7-16n^2\pi^2)t/8} - 1]e^{x/4}\sin(n\pi x).$$

(59) $u(x,t) = te^x\sin(\pi x).$

(61) $u(x,t) = -e^{-t/2+x}[\cos((4\pi^2 + 7)^{1/2}t/2)$
$$+ (4\pi^2 + 7)^{-1/2}\sin((4\pi^2 + 7)^{1/2}t/2)]\sin(\pi x)$$
$$+ \sum_{n=1}^{\infty}[(-1)^n - 1]2[n\pi(n^2\pi^2 + 2)(4n^2\pi^2 + 7)]^{-1}\{-(7 + 4n^2\pi^2)$$
$$+ e^{-t/2}[(4n^2\pi^2 + 7)\cos((4n^2\pi^2 + 7)^{1/2}t/2)$$
$$+ (4n^2\pi^2 + 7)^{1/2}\sin((4n^2\pi^2 + 7)^{1/2}t/2)]\}e^x\sin(n\pi x).$$

(63) $u(x,y) = (2y + 1)e^x\sin(\pi x).$

(65) $u(x,y) = -\operatorname{csch}(4\pi)e^x \sin(2\pi x) \sinh(2\pi(y-2))$
$$+ \sum_{n=1}^{\infty} [8(1 - (-1)^n e)/(n^3\pi^3 + n\pi)] \operatorname{sech}(n\pi)$$
$$\times e^x \sin(n\pi x) \sinh(n\pi y/2) \sinh(n\pi(y-2)/2).$$

CHAPTER 8

(1) $u(x,t) = -3(1 + 4t)^{-1/2}e^{-x^2/(1+4t)}$.

(3) $u(x,t) = (1/(2\pi)) \int_{-\infty}^{\infty} ((\sin\omega)/\omega^3)(1 - e^{-2\omega^2 t})e^{-i\omega x}\, d\omega$.

(5) $u(x,t) = (1 - 2t)e^{-x^2}$.

(7) $u(x,t) = (4/\pi) \int_{0}^{\infty} \omega(\omega^2 + 1)^{-2}e^{-\omega^2 t} \sin(\omega x)\, d\omega$.

(9) $u(x,t) = (1/\pi) \int_{0}^{\infty} \omega^{-3}[(e^{-2\omega^2 t} - 1)\cos\omega + 2\omega^2 t] \sin(\omega x)\, d\omega$.

(11) $u(x,t) = (2t - x)e^{-x}$.

(13) $u(x,t) = -(2/\pi) \int_{0}^{\infty} (\omega^5 + \omega^3)^{-1}[(\omega^2 + 1)(\omega^2 t - 1) + (2\omega^4 + \omega^2 + 1)e^{-\omega^2 t}]$
$$\times \cos(\omega x)\, d\omega.$$

(15) $u(x,t) = (1/\pi) \int_{0}^{\infty} \omega^{-4}[-(\omega\sin\omega + 2\omega^2 + 1)e^{-2\omega^2 t}$
$$+ \omega\sin\omega + 2\omega^2(1 - t) + 1]\cos(\omega x)\, d\omega.$$

(17) $u(x,t) = (1 - 2xt)e^{-2x}$.

(19) $u(x,t) = (1/2)[(2x + 2t - 1)e^{-(x+t)^2} + (2x - 2t - 1)e^{-(x-t)^2}]$.

(21) $u(x,t) = (1/(2\pi)) \int_{-\infty}^{\infty} \omega^{-3} \sin\omega \sin^2(\omega t)e^{-i\omega x}\, d\omega$.

(23) $u(x,t) = 2e^{-t-x^2}$.

(25) $u(x,t) = (2/\pi) \int_{0}^{\infty} (\omega^3 + 4\omega)^{-1}[\omega^2 + 4 + 2(\omega^2 - 2)\cos(\omega t)]\sin(\omega x)\, d\omega$.

(27) $u(x,t) = (1/\pi) \int_{0}^{\infty} \omega^{-3} \sin^2(\omega/2)[1 - 2\omega\sin(2\omega t) - \cos(2\omega t)]\sin(\omega x)\, d\omega$.

(29) $u(x,t) = -e^{2t-x-1}$.

(31) $u(x,t) = (2/\pi) \int_{0}^{\infty} (\omega^4 + \omega^2)^{-1}[1 + \omega^2 + (2\omega^2 - 1)\cos(\omega t)]\cos(\omega x)\, d\omega$.

(33) $u(x,t) = (1/(4\pi)) \int_{0}^{\infty} \omega^{-4} \sin(2\omega)[2\omega t - (4\omega^2 + 1)\sin(2\omega t)]\cos(\omega x)\, d\omega$.

(35) $u(x,t) = (2 - xt)e^{-x}$.

(37) $u(x,y) = (1/\sqrt{3\pi}) \int_{-\infty}^{\infty} \operatorname{csch}\omega \sinh(\omega x)\, e^{-\omega^2/12 - i\omega y}\, d\omega$.

(39) $u(x,y) = (1/\pi) \int_{-\infty}^{\infty} \omega^{-3} \sin\omega[\operatorname{csch}\omega \sinh(\omega x) - x]e^{-i\omega y}\, d\omega$.

(41) $u(x,y) = (x^2 + 2)e^{-2y^2}$.

(43) $u(x,y) = (1/\pi) \int\limits_{0}^{\infty} (\omega^3 + 4\omega)^{-1} \{\operatorname{csch}\omega[-2(\omega^2 + 4)\sinh(\omega x)$

$$+ 2\omega^2 \sinh(\omega(x-1))] + 2(\omega^2 + 4)x\} \sin(\omega y)\, d\omega.$$

(45) $u(x,y) = (4/\pi) \int\limits_{0}^{\infty} \omega^{-3} \sin^2(\omega/2)$

$$\times \{-1 + \operatorname{csch}\omega[\sinh(\omega x) + (2\omega^2 - 1)\sinh(\omega(x-1))]\} \sin(\omega y)\, d\omega.$$

(47) $u(x,y) = (x^2 - y)e^{-y}.$

(49) $u(x,y) = (y+1)e^{x-2y}.$

(51) $u(x,y) = (2/\pi) \int\limits_{0}^{\infty} (\omega^4 + \omega^2)^{-1}[(1-\omega^2)\operatorname{csch}\omega\sinh(\omega(x-1))$

$$+ (\omega^2 + 1)(1-x)]\cos(\omega y)\, d\omega.$$

(53) $u(x,y) = (2/\pi) \int\limits_{0}^{\infty} \omega^{-3}(\omega - \sin\omega)$

$$\times \{\operatorname{csch}\omega[\sinh(\omega x) - \sinh(\omega(x-1))] - 1\}\cos(\omega y)\, d\omega.$$

(55) $u(x,y) = (2xy + y^2)e^{-2y}.$

(57) $u(x,y) = (x^2 + x - 1)e^{-y}.$

CHAPTER 9

(1) $3/(s^2 + 2s + 10) - 72/s^5.$

(3) $(s - 4)/(s^2 + 4s + 8) - 2((s^2 + 2s + 2)/s^3)e^{-s}.$

(5) $e^t[2\cos(5t) + (3/5)\sin(5t)].$

(7) $(3/(2\sqrt{2\pi}))t^{-3/2}e^{-1/(8t)} + \operatorname{erfc}(1/(2\sqrt{2t})).$

(9) $u(x,t) = t + \operatorname{erfc}(x/(2\sqrt{t})) + e^{-4t}\sin(2x).$

(11) $u(x,t) = e^{-3t} + 2t(x-1)^2 H(1-x).$

(13) $u(x,t) = (2t - 1)\cos x.$

(15) $u(x,t) = e^{-x}[\sin(2t) - 3\cos(2t)].$

(17) $u(x,t) = 2 - e^{-3t} + \operatorname{erfc}(x/(4\sqrt{t})).$

(19) $u(x,t) = (1/2)[t^2 - 2t - (x^2 - 2xt + t^2 + 4x - 4t)H(t - x)].$

(21) $u(x,t) = (1/9)(e^{-3t} - 6t - 1) - (1/2)(x - 2t)^2 H(t - x/2).$

(23) $u(x,t) = (2 - e^{-2t})e^{-x}.$

(25) $u(x,t) = 4e^{-t/2} - \sin(2x).$

(27) $u(x,t) = t^2 + t - 2H(t - x/2)\sin(t - x/2).$

(29) $u(x,t) = -2e^x \operatorname{erfc}(x/(2\sqrt{t})).$

(31) $u(x,t) = (3t + 1)\sin x.$

(33) $u(x,t) = 1 - 2(1+t)e^{-t} + (1 + t - x)(e^{-x} + 2e^{-t})H(t - x).$

(35) $u(x,t) = (t - 2)e^{-2x}.$

CHAPTER 10

(1) $G(x,\xi) = \begin{cases} x(1-\xi)/2, & x \le \xi, \\ \xi(1-x)/2, & x > \xi, \end{cases}$ $u(x) = (1/12)(12 - 37x + 3x^2 - 2x^3)$.

(3) $G(x,\xi) = \begin{cases} x, & x \le \xi, \\ \xi, & x > \xi, \end{cases}$ $u(x) = 2 + 5x - 2x^2$.

(5) $G(x,\xi) = \begin{cases} x/2, & x \le \xi, \\ \xi/2, & x > \xi, \end{cases}$ $u(x) = \begin{cases} x^2/4 + 7x/2 - 2, & 0 \le x \le 1, \\ -x^2/2 + 5x - 11/4, & 1 < x \le 2. \end{cases}$

(7) $G(x,\xi) = \begin{cases} (1-\xi)/3, & x \le \xi, \\ (1-x)/3, & x > \xi, \end{cases}$ $u(x) = (1/3)(20 - 9x + x^2)$.

(9) $G(x,\xi) = \begin{cases} 1-\xi, & x \le \xi, \\ 1-x, & x > \xi, \end{cases}$ $u(x) = \begin{cases} 59/8 - 2x - x^2, & 0 \le x \le 1/2, \\ 8 - 9x/2 + 3x^2/2, & 1/2 < x \le 1. \end{cases}$

(11) $G(x,t;\xi,\tau) = \sum_{n=1}^{\infty} 2e^{-2n^2\pi^2(t-\tau)}\sin(n\pi x)\sin(n\pi\xi)$,

$u(x,t) = (1/(4\pi^4))(e^{-2\pi^2 t} + 2t - 1)\sin(\pi x) - e^{-8\pi^2 t}\sin(2\pi x)$.

(13) $G(x,t;\xi,\tau) = \sum_{n=1}^{\infty} e^{-n^2\pi^2(t-\tau)/4}\sin(n\pi x/2)\sin(n\pi\xi/2)$,

$u(x,t) = \sum_{n=1}^{\infty} (2/(n^3\pi^3))\{[1 + (-1)^n]n^2\pi^2 + 4(e^{n^2\pi^2 t/4} - 1)$
$\times [1+(-1)^n - 2\cos(n\pi/2)]\}e^{-n^2\pi^2 t/4}\sin(n\pi x/2)$.

(15) $G(x,t;\xi,\tau) = 1 + \sum_{n=1}^{\infty} 2e^{-n^2\pi^2(t-\tau)}\cos(n\pi x)\cos(n\pi\xi)$,

$u(x,t) = 1/2 + t^2/2 + \sum_{n=1}^{\infty} [(-1)^n - 1](2/(n^6\pi^6))$
$\times [2 + n^4\pi^4 + 2(n^2\pi^2 t - 1)e^{n^2\pi^2 t}]e^{-n^2\pi^2 t}\cos(n\pi x)$.

(17) $G(x,t;\xi,\tau) = \sum_{n=1}^{\infty} e^{-(2n-1)^2\pi^2(t-\tau)/8}\sin((2n-1)\pi x/4)\sin((2n-1)\pi\xi/4)$,

$u(x,t) = \sum_{n=1}^{\infty} (4/((2n-1)^5\pi^5))$
$\times \{(2n-1)^4\pi^4 - 8[8 + ((2n-1)^2\pi^2 t - 8)e^{(2n-1)^2\pi^2 t/8}]\}$
$\times e^{-(2n-1)^2\pi^2 t/8}\sin((2n-1)\pi x/4)$.

(19) $G(x,t;\xi,\tau) = \sum_{n=1}^{\infty} 2e^{-(2n-1)^2\pi^2(t-\tau)/4}\cos((2n-1)\pi x/2)\cos((2n-1)\pi\xi/2)$,

$u(x,t) = \sum_{n=1}^{\infty} -[8/((2n-1)^3\pi^3)]\{(-1)^n(2n-1)^2\pi^2$
$- [2 - (2n-1)^2\pi^2 - 2e^{(2n-1)^2\pi^2 t/4}]\sin((2n-1)\pi/4)\}$
$\times e^{-(2n-1)^2\pi^2 t/4}\cos((2n-1)\pi x/2)$.

(21) $G(x,y;\xi,\eta) = \sum_{m=1}^{\infty}\sum_{n=1}^{\infty} -[8/((m^2 + 4n^2)\pi^2)]\sin(n\pi x)\sin(m\pi y/2)$
$\times \sin(n\pi\xi)\sin(m\pi\eta/2)$,

$u(x,y) = \sum_{n=1}^{\infty} [(-1)^n - 1]2[n(n^2 + 1)\pi^3]^{-1}\sin(n\pi x)\sin(\pi y)$.

(23) $G(x,y;\xi,\eta) = \sum_{m=1}^{\infty}\sum_{n=1}^{\infty} -[16/((4m^2 + 4n^2 - 4m + 1)\pi^2)]$
$\times \sin(n\pi x)\sin((2m-1)\pi y/2)\sin(n\pi\xi)\sin((2m-1)\pi\eta/2)$,

$u(x,y) = \sum_{n=1}^{\infty} (-1)^n 8[n(4n^2 + 9)\pi^3]^{-1}\sin(n\pi x)\sin(3\pi y/2)$.

(25) $G(x,y;\xi,\eta) = \sum\limits_{m=0}^{\infty} \sum\limits_{n=1}^{\infty} -[32/((16m^2 + 4n^2 - 4n + 1)\pi^2)]$

$\times \sin((2n-1)\pi x/4)\cos(m\pi y)\sin((2n-1)\pi\xi/4)\cos(m\pi\eta),$

$u(x,y) = \sum\limits_{n=1}^{\infty} -128[(2n-1)(4n^2-4n+145)\pi^3]^{-1}\sin((2n-1)\pi x/4)\cos(3\pi y).$

(27) $G(x,y;\xi,\eta) = \sum\limits_{m=1}^{\infty} \sum\limits_{n=1}^{\infty} -(4/((m^2 + n^2)\pi^2))$

$\times \sin(n\pi x)\sin(m\pi y)\sin(n\pi\xi)\sin(m\pi\eta),$

$u(x,y) = \sum\limits_{m=1}^{\infty} \sum\limits_{n=1}^{\infty} -[(-1)^m - 1][(-1)^n - 1]4[mn(m^2 + n^2)\pi^4]^{-1}$

$\times \sin(n\pi x)\sin(m\pi y).$

(29) $G(x,y;\xi,\eta) = \sum\limits_{m=1}^{\infty} \sum\limits_{n=1}^{\infty} -[8/((2m^2 + 2n^2 - 2m - 2n + 1)\pi^2)]$

$\times \cos((2n-1)\pi x/2)\sin((2m-1)\pi y/2)\cos((2n-1)\pi\xi/2)\sin((2m-1)\pi\eta/2),$

$u(x,y) = \sum\limits_{m=1}^{\infty} \sum\limits_{n=1}^{\infty} 32[2 + (-1)^n(2n-1)\pi]$

$\times [(2m-1)(2n-1)^2(2m^2 + 2n^2 - 2m - 2n + 1)\pi^5]^{-1}$

$\times \cos((2n-1)\pi x/2)\sin((2m-1)\pi y/2).$

(31) $u(-1,2) = -1/3.$ (33) $u(1,2) = -27/32.$ (35) $u(3,2) = 1/3.$

(37) $u(-3,4) = 8/3.$ (39) $u(4,2) = 133/24.$

CHAPTER 11

(1) $\{(x,y): -1 < x < 3\}$: elliptic; $\{(x,y): x < -1 \text{ or } x > 3\}$: hyperbolic;
    $\{(x,y): x = -1 \text{ or } x = 3\}$: parabolic.

(3) $\{(x,y): |x| < 2|y|\}$: elliptic; $\{(x,y): |x| > 2|y|\}$: hyperbolic;
    $\{(x,y): |x| = 2|y|\}$: parabolic.

(5) $\{(x,y): (x+1)^2 + (y-2)^2 < 1\}$: elliptic;
    $\{(x,y): (x+1)^2 + (y-2)^2 > 1\}$: hyperbolic;
    $\{(x,y): (x+1)^2 + (y-2)^2 = 1\}$: parabolic.

(7) $v_{rs} = 2r,$ $u(x,y) = (3x+y)^2(x+2y) + \varphi(3x+y) + \psi(x+2y).$

(9) $v_{rs} = 2e^{2s},$ $u(x,y) = (x+3y)e^{-2x+2y} + \varphi(x+3y) + \psi(y-x).$

(11) $v_{rs} + 3v_s = 0,$ $u(x,y) = \varphi(3y-2x)e^{-3(x+2y)} + \psi(x+2y).$

(13) $v_{rs} - 2v_s = -r,$ $u(x,y) = (1/4)(y-x) + (1/2)(2x+3y)(y-x)$
$+ \varphi(2x+3y) + \psi(y-x)e^{4x+6y}.$

(15) $2v_{rs} - v_s = 2 - r - 2s,$ $u(x,y) = (4x-3y)(3x-y) + \varphi(2y-x) + \psi(y-3x)e^{y-x/2}.$

(17) $v_{ss} + 4v = 1,$ $u(x,y) = 1/4 + \varphi(y-4x)\cos(2y) + \psi(y-4x)\sin(2y).$

(19) $v_{ss} - 3v_s + 2v = -2rs,$ $u(x,y) = (1/2)(2y+3)(3x-5y)$
$+ \varphi(5y-3x)e^y + \psi(5y-3x)e^{2y}.$

(21) $v_{ss} - v_s - 2v = r+s,$ $u(x,y) = (1/4)(1+2x-6y) + \varphi(2y-x)e^{-y} + \psi(2y-x)e^{2y}.$

(23) $w_{\alpha\alpha} + w_{\beta\beta} + 5w_\alpha - 2w_\beta = 3$.

(25) $w_{\alpha\alpha} + w_{\beta\beta} + 4w_\alpha - w_\beta - 3w = -\alpha - \beta$.

(27) Parabolic; $v_{ss} - v_s = -3r$,
$$u(x,y) = 1 - 12x + 8y - (1 - 9x + 6y)e^y - 9xy + 6y^2.$$

(29) Hyperbolic; $v_{rs} = 2r + 4s$, $u(x,y) = (3y - 10x)(8x^2 - 6xy + y^2 + 1) + 3y - 10x$.

(31) Hyperbolic; $2(2r - s + 1)v_{rs} - 2v_s + sv = s - r$.

(33) Parabolic; $s^2 v_{ss} + (2r/s - r)v_r + 2v_s = s - \ln(r/s)$.

(35) Hyperbolic; $2(2r - s - 16)v_{rs} - 2v_s + (2r + 2)v = rs$.

CHAPTER 12

(1) $x = 2t + x_0$, $u(x,t) = 1 - x + 2t + t^2/2$.

(3) $x = t^2 + t + x_0$, $u(x,t) = e^{2t}\sin(x - t^2 - t)$.

(5) $x = x_0 e^t$, $u(x,t) = x^2 e^{-2t} - e^{-t} + 2$.

(7) $x = -3t + x_0$, $u(x,t) = -xt - 3t^2/2 + t + e^{-2(x+3t)}$.

(9) $x = 4t + x_0$, $u(x,t) = [x - 4t + 5 + \cos(2x - 8t)]e^t - x - t - 5$.

(11) $x = \begin{cases} t/2 + x_0, & x \ge t/2, \\ (t - t_0)/2, & x < t/2, \end{cases}$ $u(x,t) = \begin{cases} e^{x + t/2}, & x \ge t/2, \\ e^{2x}, & x < t/2. \end{cases}$

(13) $x = \begin{cases} 2t + x_0, & x \ge 2t, \\ 2(t - t_0), & x < 2t, \end{cases}$

$u(x,t) = \begin{cases} x^2 t - 2xt^2 + 4t^3/3 + x - 2t, & x > 2t, \\ x^3/6 + x^2/4 - xt + t^2 + 1, & x < 2t. \end{cases}$

(15) $x = \begin{cases} t + x_0, & x \ge t, \\ t - t_0, & x < t, \end{cases}$

$u(x,t) = \begin{cases} [x - t + 1 + \cos(x - t)]e^t - x - 1, & x \ge t, \\ 2(t - x + 1)e^x - x - 1, & x < t. \end{cases}$

(17) $x = 2y + x_0$, $u(x,y) = (x - 2y + 1)e^{(x+y-1)/3}$.

(19) $x = y/2 + x_0$, $u(x,y) = (1/4)[(2x - y + 1)e^{(2x+3y-2)/2} + 2y + 1]$.

(21) $x = -2y + x_0$, $u(x,y) = -(1/9)(2x^2 + 8xy + 8y^2 + 7x + 14y - 31)e^{(y-x+2)/3}$
$$+ x + y - 1.$$

(23) $x = (2t + 1)x_0 - 3t^2 - t$, $u(x,t) = (3t^2 + 4t - x + 1)/(2t + 1)$.

(25) $x = t^2 + t + (t + 1)x_0$, $u(x,t) = (x + t + 1)/(t + 1)$.

(27) $x = t + x_0$, $u(x,t) = -1/(e^{x-t} + t^2)$.

(29) $u(x,t) = x^2 t + t^3/3 + 2x + t + 3$.

(31) $u(x,t) = (1/2)e^{t-2x} + (1/2)e^{-t-2x} - x^2 t - t^3/12 + xt$.

(33) $u(x,t) = x^2 + 2xt + t^2 + t$.

(35) $u(x,t) = x^2 + 4xt - 11t^2 - x - 2t$.

(37) $u(x,t) = \begin{cases} 8xt, & 0 \leq x < t, \\ 2x^2 + 4xt + 2t^2 - x + t, & x \geq t. \end{cases}$

(39) $u(x,t) = \begin{cases} -6xt + (1/3)\cos x \cos(3t), & 0 \leq x < 3t, \\ 1 - x^2 - 9t^2 + (1/3)\cos x \sin(3t), & x > 3t. \end{cases}$

(41) $u(x,t) = \begin{cases} 2x^2/9 + 2xt/3 + x, & 0 \leq x < 3t, \\ 2xt - 2t^2 + x + 1, & x > 3t. \end{cases}$

(43) $u(x,t) = \begin{cases} x^2 + 2xt + 3x, & 0 \leq x < t, \\ x^2 + 2xt + 3t, & x \geq t. \end{cases}$

CHAPTER 13

(1) $u(r) = (1/4)(3 + r^2) + (1/16)\varepsilon(29 - 12r^2 - r^4) + O(\varepsilon^2)$.

(3) $u(r) = -3 + (13/2)\varepsilon(r^2 - 1) + O(\varepsilon^2)$.

(5) $u(x,t) = e^{-t}\sin x - \varepsilon t e^{-t}(\cos x + \sin x) + O(\varepsilon^2)$.

(7) $u(x,t) = e^{-2t-x} + (1/2)\varepsilon(1 - e^{-2t} + 6te^{-2t-x}) + O(\varepsilon^2)$.

(9) $u(x,t) = x - 3t + 2\varepsilon t(x - 3t) + O(\varepsilon^2)$.

(11) $u(x,t) = (2 + 2t - x)e^t - 2 + \varepsilon[(xt - 2t^2 - t + 2)e^t - 2] + O(\varepsilon^2)$.

(13) $u(x,t) = e^{x-2t} + t^2 - \varepsilon t e^{x-2t} + O(\varepsilon^2)$.

(15) $u(x,y) = xy^2 + \varepsilon y(e^x - 2x) + O(\varepsilon^2)$.

(17) $u(x,y) = -2x^2 + \varepsilon y \cos x + O(\varepsilon^2)$.

(19) $u(x,y) = 3xy/2 - x^2y/2$
$$+ (1/24)\varepsilon(-15x + 18x^3 - 3x^4 - 14xy + 18x^2y - 4x^3y) + O(\varepsilon^2).$$

(21) $u(x,y) = 6x + 2y + 3\varepsilon y e^x + O(\varepsilon^2)$.

(23) $u(x,y) = 2x - y - e^{-x} + 2\varepsilon x e^{-x} + O(\varepsilon^2)$.

(25) $u(x,y) = x(\sin y - \cos y) + (1/2)\varepsilon x[y \cos y + (y - 1)\sin y] + O(\varepsilon^2)$.

(27) $u(x,y) = \sin(2x)[2\cos(3y) - \sin(3y) - 1]$
$$- (1/18)\varepsilon \sin(2x)[(3y + 2)\cos(3y) + (6y - 1)\sin(3y) - 2] + O(\varepsilon^2).$$

(29) $u(x,y) = e^{-2x+2y} - (3/16)\varepsilon e^{-2x}[e^{-2y} + (4y - 1)e^{2y}] + O(\varepsilon^2)$.

(31) $u(x,y) = e^{x+2y} - \varepsilon e^x[e^y + (y - 1)e^{2y}] + O(\varepsilon^2)$.

(33) $u(x,y) = (2x - 1)e^{2y} + (3/16)\varepsilon(2x - 1)[(1 - 4y)e^{2y} - e^{-2y}] + O(\varepsilon^2)$.

(35) $u^c(x,y) = 2ye^{1-x} + 2\varepsilon(xy - 2x - y + 2)e^{1-x}$
$$+ [1 - x + (1 - 2e)y - (1 - 2e)xy + 2\varepsilon e(y - 2)]e^{-x/\varepsilon} + O(\varepsilon^2).$$

(37) $u^c(x,y) = 2y + 3 + 2\varepsilon x + (3xy + 3x - 9y - 9 - 4\varepsilon)e^{-(2-x)/\varepsilon} + O(\varepsilon^2).$

(39) $u^c(x,y) = (1 - y)e^{x-1} + \varepsilon y(1 - x)e^{x-1}$
$$+ (exy^2 + xy + ey^2 - x + y - 1 - \varepsilon y)e^{-1-x/\varepsilon} + O(\varepsilon^2).$$

(41) $u^c(x,y) = 2x - y - 2 + (x^2 + xy - 2x - y + 3)e^{-2y/\varepsilon}$
$$+ (4xy + y^2 + 2x + y + 2)e^{-x/\varepsilon} + O(\varepsilon^2).$$

(43) $u^c(x,y) = 2x^2 + 4xy + 2y^2 - 1 + 12\varepsilon y$
$$+ (-2x^2 + 4xy - 7x + 3y - 2 - 12\varepsilon)e^{-(1-y)/(2\varepsilon)}$$
$$+ (2xy - y^2 + 2 - 12\varepsilon y)e^{-x/\varepsilon} + O(\varepsilon^2).$$

(45) $u^c(x,y) = x^2 - 4xy + 4y^2 + 4x - 8y + 2 + 34\varepsilon(1 - y)$
$$- (x^2 + 4xy + 3x + 6y + 1 + 34\varepsilon)e^{-y/(4\varepsilon)}$$
$$+ [-4xy - 4y^2 + 3x + 6y - 1 + 34\varepsilon(y - 1)]e^{-2x/\varepsilon} + O(\varepsilon^2).$$

CHAPTER 14

(1) $u(x,y) = x^2 - 4y^2 + 2.$

(3) $u(x,y) = -2x^2 + 3xy + y^2 + y - 2.$

(5) $u(x,y) = 3x^2 - 4xy - y^2 + 2x - 3y.$

(7) $u(x,y) = 2x^2 - xy + y^2.$

(9) $u(x,y) = x^2 + 4xy - 2y^2 - y.$

(11) $u(x,y) = x^2 - 3xy + 3y^2 + 2x - 4.$

(13) $u(x,y) = x^2 - 2xy + 3y^2 + x - 2y + 1.$

(15) $u_1(x,y) = -x^2 + y^2 + 2x, \quad u_2(x,y) = 2xy + x.$

(17) $u_1(x,y) = 4xy - x + 2, \quad u_2(x,y) = (1/5)(-14x^2 - 4y^2 + 14).$

(19) $u_1(x,y) = 2x^2 - 2xy - 2y^2 + 1, \quad u_2(x,y) = (1/7)(x^2 - 28xy + y^2 + 7x - 1).$

(21) $u_1(x,y) = (1/7)(3x^2 + 14xy + 3y^2 + 7y + 11),$
$\quad u_2(x,y) = (1/7)(27x^2 - 14xy - 15y^2 - 6).$

(23) $u_1(x,y) = (1/3)(x^2 - 6xy - 5y^2 + 2), \quad u_2(x,y) = (1/3)(-8x^2 + 4y^2 + 3y + 5).$

Appendix

A1. Useful Integrals

For all $m, n = 1, 2, \ldots,$

$$\int_0^L \cos\frac{n\pi x}{L}\,dx = 0; \quad \int_{-L}^L \sin\frac{n\pi x}{L}\,dx = 0;$$

$$\int_0^L \sin\frac{n\pi x}{L}\sin\frac{m\pi x}{L}\,dx = \begin{cases} 0, & n \neq m, \\ L/2, & n = m; \end{cases}$$

$$\int_0^L \cos\frac{n\pi x}{L}\cos\frac{m\pi x}{L}\,dx = \begin{cases} 0, & n \neq m, \\ L/2, & n = m; \end{cases}$$

$$\int_{-L}^L \sin\frac{n\pi x}{L}\cos\frac{m\pi x}{L}\,dx = 0;$$

$$\int_0^L \sin\frac{(2n-1)\pi x}{2L}\sin\frac{(2m-1)\pi x}{2L}\,dx = \begin{cases} 0, & n \neq m, \\ L/2, & n = m; \end{cases}$$

$$\int_0^L \cos\frac{(2n-1)\pi x}{2L}\cos\frac{(2m-1)\pi x}{2L}\,dx = \begin{cases} 0, & n \neq m, \\ L/2, & n = m. \end{cases}$$

For all real numbers a, b, c, and $p \neq 0$,

$$\int (ax^2 + bx + c)\cos(px)\,dx$$
$$= \frac{1}{p^2}(2ax + b)\cos(px) + \frac{1}{p^3}\left[p^2(ax^2 + bx + c) - 2a\right]\sin(px) + \text{const};$$

$$\int (ax^2 + bx + c)\sin(px)\,dx$$
$$= -\frac{1}{p^3}\left[p^2(ax^2 + bx + c) - 2a\right]\cos(px) + \frac{1}{p^2}(2ax + b)\sin(px) + \text{const};$$

$$\int e^{ax}\cos(px)\,dx = \frac{e^{ax}}{a^2 + p^2}\left[a\cos(px) + p\sin(px)\right] + \text{const};$$

$$\int e^{ax}\sin(px)\,dx = \frac{e^{ax}}{a^2 + p^2}\left[-p\cos(px) + a\sin(px)\right] + \text{const}.$$

A2. Table of Fourier Transforms

$f(x) = \mathcal{F}^{-1}[F](x)$	$F(\omega) = \mathcal{F}[f](\omega)$
1 $f'(x)$	$-i\omega F(\omega)$
2 $f''(x)$	$-\omega^2 F(\omega)$
3 $f(ax+b)$ $(a>0)$	$\dfrac{1}{a}e^{-i(b/a)\omega}F(\omega/a)$
4 $(f*g)(x)$	$F(\omega)G(\omega)$
5 $\delta(x)$	$\dfrac{1}{\sqrt{2\pi}}$
6 $e^{iax}f(x)$	$F(\omega+a)$
7 $e^{-a^2x^2}$	$\dfrac{1}{\sqrt{2}a}e^{-\omega^2/(4a^2)}$
8 $xe^{-a^2x^2}$ $(a>0)$	$\dfrac{i}{2\sqrt{2}a^3}\omega e^{-\omega^2/(4a^2)}$
9 $x^2e^{-a^2x^2}$ $(a>0)$	$\dfrac{1}{4\sqrt{2}a^5}(2a^2-\omega^2)e^{-\omega^2/(4a^2)}$
10 $\dfrac{1}{x^2+a^2}$ $(a>0)$	$\sqrt{\dfrac{\pi}{2}}\dfrac{1}{a}e^{-a|\omega|}$
11 $\dfrac{x}{x^2+a^2}$ $(a>0)$	$-i\sqrt{\dfrac{\pi}{2}}\dfrac{1}{2a}\omega e^{-a|\omega|}$
12 $H(a-|x|)=\begin{cases}1, & |x|\le a \\ 0, & |x|>a\end{cases}$	$\sqrt{\dfrac{2}{\pi}}\dfrac{\sin(a\omega)}{\omega}$
13 $xH(a-|x|)=\begin{cases}x, & |x|\le a \\ 0, & |x|>a\end{cases}$	$i\sqrt{\dfrac{2}{\pi}}\dfrac{1}{\omega^2}\left[\sin(a\omega)-a\omega\cos(a\omega)\right]$
14 $e^{-a|x|}$	$\sqrt{\dfrac{2}{\pi}}\dfrac{a}{a^2+\omega^2}$
15 $e^{-(x+b)^2/(4a)}+e^{-(x-b)^2/(4a)}$	$2\sqrt{2a}\,e^{-a\omega^2}\cos(b\omega)$
16 $\operatorname{erf}(ax)$	$i\sqrt{\dfrac{2}{\pi}}\dfrac{1}{\omega}e^{-\omega^2/(4a^2)}$

A3. Table of Fourier Sine Transforms

$f(x) = \mathcal{F}_S^{-1}[F](x)$	$F(\omega) = \mathcal{F}_S[f](\omega)$
1 $f'(x)$	$-\omega \mathcal{F}_C[f](\omega)$
2 $f''(x)$	$\sqrt{\dfrac{2}{\pi}}\,\omega f(0) - \omega^2 F(\omega)$
3 $f(ax)$ $(a > 0)$	$\dfrac{1}{a} F(\omega/a)$
4 $f(ax)\cos(bx)$ $(a, b > 0)$	$\dfrac{1}{2a}\left[F\left(\dfrac{\omega+b}{a}\right) + F\left(\dfrac{\omega-b}{a}\right) \right]$
5 1	$\sqrt{\dfrac{2}{\pi}}\dfrac{1}{\omega}$
6 e^{-ax} $(a > 0)$	$\sqrt{\dfrac{2}{\pi}}\dfrac{\omega}{a^2 + \omega^2}$
7 xe^{-ax} $(a > 0)$	$\sqrt{\dfrac{2}{\pi}}\dfrac{2a\omega}{(a^2 + \omega^2)^2}$
8 $x^2 e^{-ax}$ $(a > 0)$	$2\sqrt{\dfrac{2}{\pi}}\dfrac{3a^2\omega - \omega^3}{(a^2 + \omega^2)^3}$
9 $\dfrac{x}{x^2 + a^2}$	$\sqrt{\dfrac{\pi}{2}}\,e^{-a\omega}$
10 $H(a - x) = \begin{cases} 1, & 0 \le x \le a, \\ 0, & x > a \end{cases}$	$\sqrt{\dfrac{2}{\pi}}\dfrac{1}{\omega}\left[1 - \cos(a\omega)\right]$
11 $xH(a - |x|) = \begin{cases} x, & |x| \le a \\ 0, & |x| > a \end{cases}$	$\sqrt{\dfrac{2}{\pi}}\dfrac{1}{\omega^2}\left[\sin(a\omega) - a\omega\cos(a\omega)\right]$
12 $\operatorname{erfc}(ax)$ $(a > 0)$	$\sqrt{\dfrac{2}{\pi}}\dfrac{1}{\omega}\left[1 - e^{-\omega^2/(4a^2)}\right]$
13 $xe^{-a^2 x^2}$	$\dfrac{1}{2\sqrt{2}}\dfrac{1}{a^3}\omega e^{-\omega^2/(4a^2)}$
14 $\tan^{-1}(x/a)$ $(a > 0)$	$\sqrt{\dfrac{\pi}{2}}\dfrac{1}{\omega}e^{-a\omega}$

A4. Table of Fourier Cosine Transforms

$f(x) = \mathcal{F}_C^{-1}[F](x)$	$F(\omega) = \mathcal{F}_C[f](\omega)$
1 $f'(x)$	$-\sqrt{\dfrac{2}{\pi}} f(0) + \omega \mathcal{F}_S[f](\omega)$
2 $f''(x)$	$-\sqrt{\dfrac{2}{\pi}} f'(0) - \omega^2 F(\omega)$
3 $f(ax) \quad (a > 0)$	$\dfrac{1}{a} F(\omega/a)$
4 $f(ax)\cos(bx) \quad (a, b > 0)$	$\dfrac{1}{2a}\left[F\left(\dfrac{\omega + b}{a}\right) + F\left(\dfrac{\omega - b}{a}\right) \right]$
5 $e^{-ax} \quad (a > 0)$	$\sqrt{\dfrac{2}{\pi}} \dfrac{a}{a^2 + \omega^2}$
6 $xe^{-ax} \quad (a > 0)$	$\sqrt{\dfrac{2}{\pi}} \dfrac{a^2 - \omega^2}{(a^2 + \omega^2)^2}$
7 $x^2 e^{-ax} \quad (a > 0)$	$2\sqrt{\dfrac{2}{\pi}} \dfrac{a^3 - 3a\omega^2}{(a^2 + \omega^2)^3}$
8 $e^{-a^2 x^2}$	$\dfrac{1}{\sqrt{2}} \dfrac{1}{|a|} e^{-\omega^2/(4a^2)}$
9 $\dfrac{1}{x^2 + a^2} \quad (a > 0)$	$\sqrt{\dfrac{\pi}{2}} \dfrac{1}{a} e^{-a\omega}$
10 $\dfrac{1}{(a^2 + x^2)^3} \quad (a > 0)$	$\sqrt{\dfrac{\pi}{2}} \dfrac{1}{8a^5} (a^2\omega^2 + 3a\omega + 3)e^{-a\omega}$
11 $\dfrac{x^2}{(a^2 + x^2)^3} \quad (a > 0)$	$\sqrt{\dfrac{\pi}{2}} \dfrac{1}{8a^3} (-a^2\omega^2 + a\omega + 1)e^{-a\omega}$
12 $\dfrac{x^4}{(a^2 + x^2)^3} \quad (a > 0)$	$\sqrt{\dfrac{\pi}{2}} \dfrac{1}{8a} (a^2\omega^2 - 5a\omega + 3)e^{-a\omega}$
13 $H(a - x) = \begin{cases} 1, & 0 \le x \le a, \\ 0, & x > a \end{cases}$	$\sqrt{\dfrac{2}{\pi}} \dfrac{1}{\omega} \sin(a\omega)$
14 $\begin{cases} (1/b)e^{-bx}\cosh(ab), & x \ge a, \\ (1/b)e^{-ab}\cosh(bx), & x < a \\ \qquad\qquad (a, b > 0) \end{cases}$	$\sqrt{\dfrac{2}{\pi}} \dfrac{\cos(a\omega)}{b^2 + \omega^2}$

A5. Table of Laplace Transforms

$f(t) = \mathcal{L}^{-1}[F](t)$	$F(s) = \mathcal{L}[f](s)$
1 $f^{(n)}(t)$ (nth derivative)	$s^n F(s) - s^{n-1} f(0) - \cdots$ $- f^{(n-1)}(0)$
2 $H(t-a)f(t-a)$	$e^{-as} F(s)$
3 $e^{at} f(t)$	$F(s-a)$
4 $(f * g)(t)$	$F(s)G(s)$
5 1	$\dfrac{1}{s}$ $(s > 0)$
6 t^n (n positive integer)	$\dfrac{n!}{s^{n+1}}$ $(s > 0)$
7 e^{at}	$\dfrac{1}{s-a}$ $(s > a)$
8 $\sin(at)$	$\dfrac{a}{s^2 + a^2}$ $(s > 0)$
9 $\cos(at)$	$\dfrac{s}{s^2 + a^2}$ $(s > 0)$
10 $\sinh(at)$	$\dfrac{a}{s^2 - a^2}$ $(s > |a|)$
11 $\cosh(at)$	$\dfrac{s}{s^2 - a^2}$ $(s > |a|)$
12 $\delta(t-a)$ $(a \geq 0)$	e^{-as}
13 $e^{a^2 t} \operatorname{erfc}(a\sqrt{t})$ $(a > 0)$	$\dfrac{1}{s + a\sqrt{s}}$
14 $\dfrac{a}{2\sqrt{\pi}} t^{-3/2} e^{-a^2/(4t)}$ $(a > 0)$	$e^{-a\sqrt{s}}$
15 $\operatorname{erfc}\left(\dfrac{a}{2\sqrt{t}}\right)$ $(a > 0)$	$\dfrac{1}{s} e^{-a\sqrt{s}}$
16 $-a\sqrt{\dfrac{t}{\pi}} e^{-a^2/(4t)} + \left(\tfrac{1}{2}a^2 + t\right) \operatorname{erfc} \dfrac{a}{2\sqrt{t}}$ $(a > 0)$	$\dfrac{1}{s^2} e^{-a\sqrt{s}}$

A6. Second-Order Linear Equations

If

$$Au_{xx} + Bu_{xy} + Cu_{yy} + Du_x + Eu_y + Fu = G,$$

if new variables

$$r = r(x,y), \quad s = s(x,y)$$

are defined by the characteristic equations

$$\frac{dy}{dx} = \frac{B - \sqrt{B^2 - 4AC}}{2A}, \qquad \frac{dy}{dx} = \frac{B + \sqrt{B^2 - 4AC}}{2A},$$

and if

$$u(x,y) = u\big(x(r,s), y(r,s)\big) = v(r,s),$$

then

$$\bar{A}v_{rr} + \bar{B}v_{rs} + \bar{C}v_{ss} + \bar{D}v_r + \bar{E}v_s + \bar{F}v = \bar{G},$$

where

$$\bar{A} = A(r_x)^2 + Br_xr_y + C(r_y)^2,$$

$$\bar{B} = 2Ar_xs_x + B(r_xs_y + r_ys_x) + 2Cr_ys_y,$$

$$\bar{C} = A(s_x)^2 + Bs_xs_y + C(s_y)^2,$$

$$\bar{D} = Ar_{xx} + Br_{xy} + Cr_{yy} + Dr_x + Er_y,$$

$$\bar{E} = As_{xx} + Bs_{xy} + Cs_{yy} + Ds_x + Es_y,$$

$$\bar{F} = F,$$

$$\bar{G} = G.$$

Bibliography

L.C. Andrews, *Elementary Partial Differential Equations with Boundary Value Problems,* Academic Press, Orlando, FL, 1986.

J.W. Brown and R.V. Churchill, *Fourier Series and Boundary Value Problems,* 6th ed., McGraw–Hill, New York, NY, 2000.

P. DuChateau and D. Zachmann, *Applied Partial Differential Equations,* Harper & Row, New York, NY, 1989.

D.G. Duffy, *Solutions of Partial Differential Equations,* TAB Books, Blue Ridge Summit, PA, 1986.

S.J. Farlow, *Partial Differential Equations for Scientists and Engineers,* Dover, New York, NY, 1993.

P.R. Garabedian, *Partial Differential Equations,* 2nd ed., American Mathematical Society, Providence, RI, 1999.

K.E. Gustafson, *Introduction to Partial Differential Equations and Hilbert Space Methods,* 3rd ed., Dover, Mineola, NY, 1999.

R. Haberman, *Elementary Applied Partial Differential Equations with Fourier Series and Boundary Value Problems,* 3rd ed., Prentice Hall, Upper Saddle River, NJ, 1998.

M.K. Keane, *A Very Applied First Course in Partial Differential Equations,* Prentice Hall, Upper Saddle River, NJ, 2002.

J. Kevorkian and J.D. Cole, *Perturbation Methods in Applied Mathematics,* Springer, New York, NY, 1985.

L.D. Kovach, *Boundary–Value Problems,* Addison–Wesley, Reading, MA, 1984.

J.A. Leach and D.J. Needham, *Matched Asymptotic Expansions in Reaction–Diffusion Theory,* Springer, London, 2004.

T. Myint-U and L. Debnath, *Linear Partial Differential Equations for Scientists and Engineers,* 4th ed., Birkhäuser, Boston, MA, 2007.

A.H. Nayfeh, *Introduction to Perturbation Techniques,* Wiley, New York, NY, 1981.

I. Stakgold, *Green's Functions and Boundary Value Problems,* Wiley–Interscience, New York, NY, 1979.

D. Vvedensky, *Partial Differential Equations with Mathematica,* Addison–Wesley, Reading, MA, 1993.

F.Y.M. Wan, *Mathematical Models and Their Analysis,* Harper & Row, New York, NY, 1989.

E. Zauderer, *Partial Differential Equations of Applied Mathematics,* 3rd ed., Wiley–Interscience, New York, NY, 2006.

Index

asymptotic series 263

Bessel
  equation 43, 120, 268
  function 43, 119, 230, 268
  integrals 43
body force 73
boundary layer 275
Brownian motion 78

canonical form 227
  reduction to 229
    for elliptic equations 238
    for hyperbolic equations 231
    for parabolic equations 235
Cauchy–Riemann relations 286
characteristics 231
condition
  boundary 29, 62, 69, 75
    Dirichlet 69
    mixed 134
    Neumann 69
    Robin 69
    time-independent 136
  initial 62, 69, 74
conservation of heat energy 60, 67
convolution 167, 190

Dirac delta 188
displacement
  complex 292
  vector 291
  vertical 73
distribution 188

divergence theorem 214

eigenfunction 29
  expansion 143
    for the heat equation 143
    for the Laplace equation 152
    for the wave equation 149
  linearly independent 37, 39
eigenvalue 29
  problem 29
    two-dimensional 115
elastic
  attachment 75
  body 291
  string 73
equation
  Bessel 43, 120, 268
  biharmonic 80, 287
  Black–Scholes 79
  Cauchy–Euler 6, 106, 123
  characteristic 3, 231
  convection 199
  diffusion 62
  elliptic 70, 238
  Euler–Tricomi 80
  first-order 1
  heat 62
  Helmholtz 80, 266
  hyperbolic 75, 231
  Klein–Gordon 79
  Laplace 67, 287
    in a circular disk 105
    in a rectangle 101

linear 2, 8
    homogeneous 3
    nonhomogeneous 5
    normal form 2
    modified Helmholtz 80
    nonlinear 8
    ordinary differential 1
    parabolic 63, 235
    partial differential 59
    Poisson 70
    quasilinear 248
    Schrödinger 79
    second-order 3
    telegraph 79
    transcendental 34
    wave
        one-dimensional 74, 249
        signal problem for 192
        with more than two variables 113
Euler's formula 51, 122, 165
evolution of a quantum state 79
extension
    even 22
    odd 18
    periodic 12

flux prescribed at the
        endpoints 132
Fourier's law of
        heat conduction 61, 68
Fourier series
    coefficients 13
    convergence 23
    cosine 21
    differentiation 23

double 117, 209
    full 11
    generalized 37, 45
    pointwise convergence 14
    sine 17
Fourier transform
    cosine 172
    full 166
    sine 172
Fourier transformation
    cosine 173
        inverse 173
    full 165
        inverse 167
    parameter 167
    sine 173
        inverse 173
function
    absolutely integrable 167
    Airy stress 80
    analytic 286
    Bessel 43, 119, 230, 268
    complementary error 171
    error 171
    even 17
    generalized 188
    Heaviside 187
    holomorphic 286
    influence 170
    magnitude of 263
    odd 17
    of a complex variable 286
    periodic 11
    piecewise continuous 14
    piecewise smooth 14

square-integrable 54
unit step 187

Gauss–Weierstrass kernel 170
Green's formula 206
  for functions of two
    space variables 214
Green's function
  for the heat equation 207, 211
  for the Laplace equation 215
  for the wave equation 219

heat
  energy density 60
  equation 62
    higher-dimensional 67
    two-dimensional 68
  flux 60, 67
  sources 60
    time-independent 213
  specific 61
  transfer coefficient 63
heat conduction
  in a circular annulus 135
  in a circular ring 93
  in a one-dimensional rod 59
    with insulated endpoints 86
    with an endpoint in a
      zero-temperature medium 91
    with mixed homogeneous
      boundary conditions 88
    with temperature prescribed
      on the boundary 196
    with zero temperature at
      the endpoints 83
  in a semi-infinite rod 174, 194

integrating factor 2

jump discontinuity 14

Laplace transform 189
  inverse 189
Laplace transformation 189
  inverse 189
  parameter 189
Laplacian 68, 291
Legendre
  associated functions 50
  functions of the second kind 47
  generating function for 48
  polynomials 47
loss transmission line 79, 201

magnitude
  of functions 263
  order of 263
mass density 61, 73
mathematical model 41, 59, 63
mean value property 108
membrane
  vibrating circular 117
  vibrating rectangular 113
method
  asymptotic 263
  of characteristics 241
  complex variable 285
    for first-order linear
      equations 241
    for first-order quasilinear
      equations 248
  of eigenfunction expansion 143
    for the heat equation 143

for the Laplace equation 152
for the wave equation 149
of Green's functions 205
for the heat equation 205
for the Laplace equation 213
for the wave equation 217
of multiple scales 272
of separation of variables
for the heat equation 83
for the Laplace equation 101
for the wave equation 95
perturbation 263
motion of a quantum scalar field 79

Newton's
law of cooling 63, 69
second law 74
normal modes of vibration 97

operator
linear 7
nonlinear 7
symmetric 27
orthogonal set 28
orthogonality
formulas 50
of eigenfunctions 33

parameter
large 264
small 264
transformation 167, 189
particular integral 5
period 11
plane
deformation 291

transonic flow 80
principle
causality 212
maximum 71
of superposition 9
problem
boundary value 63
Cauchy 169, 179
diffusion–convection 78, 198
eigenvalue 28
two-dimensional 115
initial boundary value 63
initial value 63
linear nonhomogeneous 131
perturbation
regular 265, 266
singular 265, 274
reduced 265
signal 192
Sturm–Liouville 27
periodic 39
regular 29
singular 40
unperturbed 265

Rayleigh quotient 32
Rodrigues formula 47

secular term 271
separation
constant 84
of variables 83
for other equations 109
for the heat equation 83
for the Laplace equation 101
for the wave equation 95

small perturbation 72
solution
  classical 64, 71, 76
  composite 276
  d'Alembert 222, 249, 251
  dependence on the data 72
  inner 275
  outer 275
  stable 72
  time-independent 69
  unique 66
spherical harmonics 50
  orthogonality formula for 51
stock market prices 78
string
  elastic 73
  vibrating 73
    finite 255
    infinite 171
    semi-infinite 175
    with fixed endpoints 95
    with free endpoints 98
systems of equations 291

table
  of Fourier cosine transforms 316

of Fourier sine transforms 315
of Fourier transforms 314
of Laplace transforms 317
Taylor series 11
temperature 61
  equilibrium 69
    in a semi-infinite strip 176
  prescribed at the endpoints 131
  steady-state 69
tension 73
thermal
  conductivity 61
  diffusivity 62
transform domain 169
transverse vibrations of a rod 80

wave
  dissipative 79, 110
  equation 73
  fixed-shape 250
  front 200
  one-dimensional 256
  reflected 254
  scattered 80
  spherical 259
  standing 97, 252